HANDBOOK ON
BIOLOGICAL NETWORKS

World Scientific Lecture Notes in Complex Systems – Vol. 10

WORLD SCIENTIFIC LECTURE NOTES IN COMPLEX SYSTEMS

Editor-in-Chief: A.S. Mikhailov, *Fritz Haber Institute, Berlin, Germany*

B. Huberman, *Hewlett-Packard, Palo Alto, USA*
K. Kaneko, *University of Tokyo, Japan*
Ph. Maini, *Oxford University, UK*
Q. Ouyang, *Peking University, China*

AIMS AND SCOPE

The aim of this new interdisciplinary series is to promote the exchange of information between scientists working in different fields, who are involved in the study of complex systems, and to foster education and training of young scientists entering this rapidly developing research area.

The scope of the series is broad and will include: Statistical physics of large nonequilibrium systems; problems of nonlinear pattern formation in chemistry; complex organization of intracellular processes and biochemical networks of a living cell; various aspects of cell-to-cell communication; behaviour of bacterial colonies; neural networks; functioning and organization of animal populations and large ecological systems; modeling complex social phenomena; applications of statistical mechanics to studies of economics and financial markets; multi-agent robotics and collective intelligence; the emergence and evolution of large-scale communication networks; general mathematical studies of complex cooperative behaviour in large systems.

Preface

Networks are all around us. Nevertheless, only very recently have scientists started to reconsider the traditional reductionism viewpoint that has always driven science. The accumulated evidence that systems as complex as a group of social animals, or as the cells of a living system, cannot be fully understood by simply reducing them to a sum of their fundamental parts, has produced an increasingly large interest in the study of complex systems where interaction networks are crucial, giving rise to the birth of a new science, the *science of complex networks*.

The massive and comparative analysis of real networks has produced, in the last ten years, a series of unexpected and striking results in various fields, and the identification of basic principles common to all the networks are considered. The lifesciences community has certainly been one of the principal actors in the complex networks revolution, and we can say that the network approach has finally affected modern biology. Networks allow in fact a systemic approach to biological problems able to overcome the evident limitations of the strict reductionism of the past twenty years. Biologists have found that several systems can be represented as networks, and they have understood that there is a lot to be learned by studying those networks. Taking into consideration not only the single elements but also the whole complicated web of links connecting the different parts of biological entities is a radical change, a change that makes a substantial difference. The outcomes have been rewarding. New features, hardly detected by standard "mean field" approaches, have emerged. The applications are wide, and they all demonstrate how the network approach has become today an invaluable method to comprehend the functions of biological systems.

Biological networks in a broad sense, the subject of this book, have been among the most studied complex networks, and the field has benefited from many important contributions from physicists, mathematicians and computer scientists. Here, we have collected, in a single book, most of the relevant results and novel insights provided by network theory in the biological sciences, putting together topics at the forefront of current research. The subjects covered include: cortical and neural networks, cultured neural networks, functional connectivity in brain networks, Boolean dynamics, gene circuits, metabolic networks, protein folding, evolutionary dynamics, motion coordination, and ecosystems. The subjects have been organized in three main sections: (i) *networks at the cellular level*, (ii) *brain networks*, and

(iii) *networks at the individual and population levels.* The recurrent theme through-out the three sections is to explain how the structure of a biological system influences its function, and the other way around.

When this project was conceived, we considered that the plethora of available results deserved to be summarized and synthesized in a clear and concise source. With this book, we hope to have succeeded in producing a useful tool not only for researchers working on specific subjects of networks biology, but also for the whole community of physicists, biologists, engineers and computer scientists interested in the multidisciplinary applications of complex networks. We also expect that the book can be used as a reference text for academic purposes.

Finally, our heartfelt thanks go to the authors of the Chapters for their generous contributions and willingness to join this project.

S. Boccaletti, V. Latora and Y. Moreno

Contents

Chapter 1

Introduction

The book is organized in three main parts. *Part I* deals with *networks at the level of the cell*, the basic unit of any living organism. In the cell, all kinds of structural and functional processes are ruled by the intricate interaction of genes, proteins and other molecules. It is then easy to understand why the complex network approach has become a common and useful paradigm to investigate, model and understand cellular processes. The first part of the book covers various kinds of cellular networks such as: gene networks, protein networks, metabolic networks, protein-folding networks, as well as signaling networks.

The first Chapter, by *Brilli and Lió*, is a general overview on the structure and function of regulatory networks, gene coexpression networks and protein-protein interaction networks. Here, the reader will find a review of the main properties of a network, and the basic measures to characterize its topology, such as degree distributions, average shortest path lengths, clustering coefficient, assortativity and network motifs.

The first step to understand the behavior of gene regulatory systems, is to study the dynamics of small gene networks. Historically, one of the first examples of artificial gene circuits to be investigated was the repressilator, a synthetic network of three genes whose products inhibit the transcription of each other cyclically. In the second Chapter, *Ullner et al.* show how to describe the dynamics of coupled small synthetic gene networks through a set of ordinary differential equations. In particular, two different types of genetic oscillators, the repressilator and a relaxator oscillator, with two different types of coupling (namely a phase-attractive and a phase-repulsive coupling) are considered. This is enough to produce rich dynamical scenarios that include multistability, oscillation death, and quantized cycling.

The drawback of the differential equation-based approach is the necessity to know the kinetic details of molecular and cellular interactions, an information that is rarely available. There is increasing evidence, however, that the input-output curves of many regulatory relationships are strongly sigmoidal and can be approximated by step functions. In addition to this, regulatory networks often maintain their function even when faced with fluctuations in components and reaction rates. This explains the diffusion of coarse-grained methods, such as Boolean models. In a Boolean network the state of each node can assume one of two possible values, and the interaction between nodes is ruled by a set of Boolean functions. The Chapter by *Thakar and Albert* introduces the reader to Boolean modeling methods, showing how these can be used to infer causal relationships from expression data, and also to analyze the dynamics of systems whose network of interactions is known.

The global dynamics of gene regulatory networks must be robust in order to guarantee their stability under a broad range of external conditions. The effect of noise on the dynamics of Boolean networks is investigated in the Chapter by *Diaz-Guilera and Alvarez-Buylla*. In this case, the noise simulates errors and external stimuli affecting the transmission of signals in real living processes. The authors focus on an experimentally grounded gene regulatory network that describes the interactions required for primordial cell fate determination during early stages of flower development. This is a very instructive example, since it shows how the noise maximizes the capacity of the system to explore the state space, while at the same time the network is able to retain the observed steady states under different noisy regimes.

The Chapter by *Bongini and Casetti* is about protein folding, one of the most fundamental and challenging open questions in molecular biology. The core of the protein folding is to understand how the information contained in a sequence of aminoacids is translated into the three-dimensional native structure of a protein. And to clarify why all the natural selected sequences of aminoacids fold to a uniquely determined native state, while a generic polypeptide does not. With these questions in mind, the authors describe two different strategies to analyze the high-dimensional energy landscape of model proteins. The first approach is essentially topological in character, and amounts to define a network whose nodes are the minima of the potential energy and whose edges are the saddles connecting them. The second approach is based on the definition of global geometric quantities that characterize the folding landscape as a whole. The reader will lear that both methods can give interesting information on the differences between the landscapes of protein-like systems and those of generic polymers.

A simplified representation of the potential energy function of a protein near equilibrium can be obtained, at low computational cost, by using the so-called elastic network models (ENMs). In ENMs, the aminoacids are the nodes of the network, represented as point particles in three dimensions (3D), while the edges of the network are springs joining the nodes, representing harmonic restraints on displacements from the equilibrium structure. The attractive feature of ENMs is that they provide an intuitive and quantitative description of the behavior near equilibrium. Furthermore, the few parameters used in ENMs can be easily adjusted, giving uncommon adaptability to the method. There are, however, some limitations. Although ENMs robustly predict collective global motions, they do not provide reliable descriptions of local motions. Also, the harmonic approximation requires a potential minimum, limiting the utility of ENMs for modeling non-equilibrium dynamics. In the Chapter by *Lezon et al.* the theory behind the ENM and some of its extensions are reviewed. Finally, some recent applications, mainly focusing on two groups of proteins, membrane proteins and viruses, are presented. Those are among the most ubiquitous classes of proteins, and are difficult to examine by all-atom simulations due to their large sizes.

In the final Chapter of Part I, *Palumbo et al.* focus on metabolic networks. The Chapter try to answer the question: "how far can we go by knowing only the wiring diagram of metabolic networks?". For such a reason the authors compare purely topological information with physiological and kinetic information, highlighting the particular relations holding between the static and the dynamic approach in the biochemical regulation of cells.

Neuroscience is another field of biology and medicine where the complex networks approach has found wide applications. *Brain networks* ar the subject of *Part II* of the book. On one hand, scientists are indeed interested in disclosing the main structure at the basis of the functioning of the human brain network, in both modeling and analysis of high resolution data. On the other hand, laboratory experiments on cultured neural networks make today possible to closely investigate the basic response in the network's dynamics and organization to controlled external stimulations. This part of the book offers a review of the current state of the art on both these approaches.

In the first Chapter of Part II, *Sporns* reviews what is currently known on the human brain connectivity structure, and on how the latter plays an essential role in some functional brain dynamics. The Chapter reports a series of recent results, based on which scientists are now able to furnish a first description of the architecture of the human brain in terms of a network of small-world type, including the presence of modules and hub regions that shape the brain endogenous dynamics as well as its responses to external stimuli.

The brain network analysis from high resolution data is the subject, instead, of the Chapter by *De Vico Fallani and Babiloni,* in which it is pointed out that brain networks have no precise anatomical support, but they can be represented as functional networks, which could change in topology and properties according to a specific subject's behavior or task. The Chapter highlights how some of the peculiar features of these functional networks can be estimated from high-resolution EEG signals.

The Chapter by *Arenas et al.* reviews a recent approach to unravel the wiring connectivity in the nematode Caenorhabditis Elegans, which in biology is nowadays considered the benchmark to understand the mechanisms underlying a whole animal's behavior, at the molecular and cellular levels. The Chapter discusses the validity of an optimization approach with the actual neuronal layout data, and remarks how the current approach to optimization of neuronal layouts is still far from being conclusive.

The Chapter by *Raichman et al.* reviews extensively how complex patterns of activity and morphological memory are expressed in cultured neural networks cultivated on micro-electrode arrays. These cultures, indeed, are widely used as a tool for laboratory controlled non invasive investigations of network neuronal systems' organization and dynamics, allowing for chemical and electrical manipulations. The series of experiments described in this Chapter allows to look at cultured neuronal networks as complex dynamical biophysical systems that have some forms of intrinsic memory, information coding and self-regulation, and that further show repeating activity motifs and long term adaptation processes to changing environment, in response to different external stimuli, such as morphology constrains and thermal stimulations.

The second part of the book ends with the Chapter by *Memmesheimer and Timme*, presenting the state of the art on how patterns of precisely timed and synchronized spikes emerge in neural circuits. In this Chapter the reader will find an overview of a series of recent results on synchrony and spatio-temporal patterns in recurrent networks. Two classes of hypothesis that might be at the basis of the emergence of such precisely timed spikes are discussed: the possibility that some feed-forward anatomical structures are embedded in the cortical circuit supporting the propagation of synchronous spiking activity, and the alternative possibility that recurrent networks may collectively organize synchronous spikes without the need of a specific feed-forward anatomy.

Part III of the book is devoted to *networks at the individual and population levels*. The first Chapter of this part, by *Stouffer et al.* addresses ecological systems as made up of highly interconnected and complex networks of interactions between species, and not as independent patches. Although food webs, mutualistic networks and spatial ones are well described by their structural properties, nowadays little is known about these networks' dynamics. Given the importance of understanding as many aspects as possible of the complex interplay between the networks of interacting species with the spatial context in which they live, further developments along the lines exposed in this Chapter will prove crucial in the future.

The Chapter by *Bagnoli* discusses evolutionary models in simple biosystems. The Chapter deals with theoretical approaches to self-organization in evolutionary population dynamics, including evolution on a fitness landscape, dynamic ecosystems, and game theoretical models. Noticeable, the Chapter put into evidence the role and the emergence of network structure in systems that range several orders of magnitude, from the elementary constituents to whole ecosystems, with the aim of showing how macro-evolutionary patterns may arise from a simplified individual-based dynamics.

The Chapter by *Pacheco et al* focuses on the evolution of cooperation in adaptive social networks. The tools used are those of evolutionary game dynamics, coupled to recent advances in network modeling. Specifically the authors discuss the problem of how cooperation arises in networks that are adaptive and dynamic in nature. They start from the observation that the structure of many modern networks of interactions are not stable in the long term and develop a two-player dilemma-like model, to account for the influence of a changing topology on the level of cooperation achieved by the system. The conclusions point out that such a modeling approach might be an important ingredient towards more realistic models of cultural dynamics.

The Chapter by *Frasca et al.* deals with models of collective behaviors in animal groups. The social complex organization of an animal group undergoes many shape and structural changes over time and space, and has been the subject of study from many years as a way to learn from natural systems how motion of different units can be coordinated. The Chapter address recent findings about both the structure and the dynamics of coordination models. The Chapter concludes with a discussion about the applications of the models treated and their usefulness in many contexts such as distributed sensing, search and rescue, environmental modeling and surveillance and what should be the minimum ingredients needed to design decentralized coordination and control strategies in engineering systems.

The last Chapter of the book, contributed by *T. Gross*, addresses one of the first problems in which the influence of the network topology was evidenced — the spreading of a disease. The author discusses the role of state-topology interplay in epidemics dynamics by working out a model in which not only the structure affects the way the epidemics spread but also the dynamics induce topological changes. In particular, the conceptual model introduced incorporates a mechanism of rewiring that depends on the dynamical states of the network nodes and produces an increase of the invasion threshold concurrent to a persistence threshold which is below the former. The author concludes the Chapter by discussing several aspects of the model analyzed and many open questions on the epidemiology of adaptive networks.

PART 1
Networks at the Cellular Level

Chapter 2

The Structural Network Properties of Biological Systems

Matteo Brilli[1] and Pietro Lió[2]

[1] *Laboratoire Biométrie et Biologie Evolutive UMR CNRS 5558, Université Lyon*
[2] *Computer Laboratory, University of Cambridge, Cambridge, UK*

2.1. Introduction

Many significant bio-medical innovations of the last years have resulted from a more accurate understanding of the fundamental properties of complex systems that reside inside the cell, such as the topological properties of biological networks, that is crucial for exploiting an integrated knowledge of living beings.

The NCBI Genomes database records data for 49 archaeal, 940 bacterial and 162 eukaryotic genomes in September 2008. While giving an enormous amount of information on the biological properties of many organisms, genome sequences can't tell us directly which metabolites, molecular structures or organelles are present, the genes that are expressed in a given conditions, the possible splicing variants and the post-translational modifications of proteins. Moreover, cell physiology depends on thousands of genes and proteins that interact on several levels giving rise to a plethora of networks. Genetic, biochemical and molecular biology techniques have been used for decades to identify these interactions but only newly developed high-throughput methodologies allow inferring genome-wide interaction maps. These, paired with computational approaches can be used to infer networks of interactions and causal relationships within the cell. The challenge of systems biology is to integrate the different information to understand dynamical properties of cellular systems [1] and be able to design semi-synthetic organisms or parts thereof.

Biological networks have evolved in billions of years under the pressure of natural selection which promoted the emergence of several important features. These features are now inspiring the design of technological systems with high efficiency and robustness.

For sake of simplicity biological networks can be grouped in several categories some of which will be discussed in more detail, by focusing on the experimental, the computational and the integrated approach for their construction and subsequent analysis.

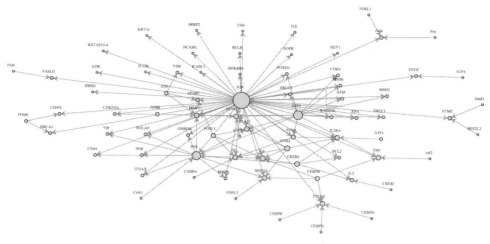

Fig. 2.1.   The transcription regulatory network of the JUN TF in human. Only direct JUN targets and TFs coregulating genes with JUN are shown.

2.2. Regulatory Networks

Living cells are endowed with specific sensory and signalling systems to obtain and transmit information from their environment in order to adjust cellular metabolism, growth, and development to environmental alterations. Such molecular communications can be triggered by small molecules (i.e. nutrients, ions, drugs), and by physical parameters (i.e. temperature and pressure). Moreover, there are several systems that continuously monitor the intracellular milieu, e.g. energetic charge and redox state, and accordingly modulate cell physiology. Similarly, the extremely different cell types in higher eukaryotes are the results of expression pattern differences, as well as cellular proliferation and differentiation, that are controlled by regulatory circuits giving origin to space- and time-dependent activation patterns.

Several systems directly modulate the response to extra- and intra-cellular signals; one of the most important acts on the rate of transcription of a gene [2–9]. The genetic basis for the dynamic expression profile of each gene resides in part within its promoter and in part within the many other segments of the genome that encode transcription factors (TF) which are able to recognize specific promoter sequences. TFs interact with promoters and cis-regulatory modules, and, within such elements TFs are able to bind specific elements (TF binding sites). The genome of the unicellular yeast *Saccharomyces cerevisiae* encodes ∼ 200 predicted TFs [10], which is ∼ 3% of all protein-coding genes; the relatively simple metazoan nematode *Caenorhabditis elegans* contains 934 predicted TFs (∼ 5% of all protein-coding genes [11]); humans may devote up to 10% of their coding potential to regulatory TFs, in agreement with a positive correlation between the fraction of TF-encoding genes in a genome and the complexity of an organism. Determining the binding affinities of TFs to different DNA sequences has been a significant technical challenge in molecular biology for many years. The global relationships between TFs and their targets on the genome can be described in the framework of network theory: *transcription regulatory networks* (TRN) are formed by two types of nodes: TFs and their target genes; edges are directed and always start from a TF. Many different strategies have been employed to reconstruct TRNs, including computational and experimental methods.

2.2.1. *Experimental approach*

Protein-DNA interactions are required for accessing and protecting genetic information within the cell and they have been studied in the past using genetic, biochemical, and structural methods producing qualitative or semiquantitative data. Today, and even more in the future, the focus will be on high throughput high quality data, a goal that would allow the reliable modeling of a huge number of interactions. The last ten years have witnessed the development and refinement of methods to crosslink proteins to DNA *in vivo* through which we can measure the interaction of sequence-specific DNA-binding proteins with genomic DNA. The methodology was then improved by combining DNA microarray technology with *in vivo* crosslinking [12]. TRN identification today is mainly accomplished with the combination of microarray techniques with chromatin immunoprecipitation (ChIP) and, whereas initial uses of ChIP required prior knowledge of a potential binding target DNA sequence to identify binding of TFs to DNA regions of interest, today it allows the reconstruction of a TRN almost without prior knowledge. In the standard ChIP-on-chip protocol a DNA binding protein is covalently attached to its target DNA using formaldehyde. After shearing of the chromatin, the protein-DNA adduct is immunoprecipitated from cell extracts using an antibody specific for the protein in question or for an epitope sequence appended to the end of the protein (generally the hemagglutinin or myc tags). After immunoprecipitation, the enriched DNA sample can be amplified using a variety of methods. DNA samples (enriched and reference) are fluorescently labeled and then they are hybridized to various microarrays. By using the ChIP-chip technique, Lee and collaborators [13] recovered promoters bound to each TF encoded by the *Saccharomyces cerevisiae* genome. In addition to providing an outline of global transcriptional regulation in yeast, these results represent a quantitative estimate of the amount of combinatorial regulation in yeast and they allowed to obtain a comprehensive set of DNA regulatory motifs that are specific for each TF. Of course, there are still gaps in our understanding of this TRN (presumably to be filled by more traditional, small scale genetic studies).

2.2.2. *Computational approach*

Identifying the repertoire of regulatory elements in a genome is one of the major challenges in modern biology and it is today approached by using several high-throughput methodologies followed by a detailed computational analysis to (i) reduce the noise, (ii) to normalize and finally (iii) to extract statistically meaningful biological information from such data. One such type of information is the TRN. The DNA sequence elements representing binding sites for TFs are responsible for the coordinated expression of genes possessing a given TF binding site i.e. TRNs can be deciphered by finding all true TFs binding sites in a given genome. These regulatory sites are however short, degenerated and embedded into large regions of non–coding DNA. The shortness of TF binding sites translates in a high number of false-positives which is moreover increased by the well known degeneracy of regulatory motifs. A relatively novel approach to improve the computational identification of TF binding sites takes advantage of the identification of clusters of co-expressed genes over a large number of different conditions. The upstream regions of co-expressed genes can then be analyzed for the presence of shared sequence motifs which might explain the observed co-regulation.

Identifying Candidate Motifs: Several computational methods for the discovery of transcription binding sites (TFBSs) have been described [14]. They can be schematically divided into two main approaches. The first one evaluates the frequencies of all possible sequences of length n (n-mers) iteratively and at each step it updates a position weighted probability matrix (PWM) corresponding to the candidate motif. A background distribution of n-mers is calculated for a set of non-coregulated genes so that n-mers that are more abundant than expected can be identified [15–18]. The second approach specifies the n-mer as a PWM and utilizes an iterative algorithm, typically represented as an expectation maximization [19] or Gibbs sampling procedure [20–22]. These iterative methods can extend the window size to lengths $n > 8$ nucleotides which is an upper limit for enumerative techniques; on the other hand, these methods might be trapped at local optima. Improved models of the background sequence distribution using high order Markov models [23] to represent the null distribution increase the accuracy. A score reflecting how well a DNA string matches each candidate motif can be calculated by taking into account its PWM, the background model and the number of motif occurrences for each promoter sequence. One of the most commonly used scoring functions is the following: let be μ a motif of length w, and occurrences μ_i in upstream sequence g; the corresponding PWM is $\mathbf{M}_\mu$, of size $w \times n$, where $n = 4$, one column for each possible nucleotide. We calculate the motif probability on $\mathbf{M}_\mu$, $P(\mu_i|M_\mu)$ and on the background model, $P(\mu_i|M_{BKG})$. Then, the scoring function is:

$$S_{\mu,g} = log_2 \left[\sum_{i=1}^{i} \frac{P(\mu_i|M_\mu)}{P(\mu_i|M_{BKG})} \right].$$

This formula is widely used e.g. MotifRegressor [24] and MotifScorer [25]. Another possible scoring function is the one proposed by [26], which does not require a background model:

$$S_i = \frac{1}{L} \sum_j [2 + log_2(F_{ij})] \,,$$

where F_{ij} is the frequency of the ith base at the jth position. S_i is in this case an information-based measure of potential binding sites.

2.2.3. *Integrated approach*

A computationally efficient way to quantify the extent to which regulatory sequence elements can explain changes in genome-wide expression data is to fit the logarithm of the expression ratio of a gene to the score of its promoter for a given motif's PWM. The simplest case involve a single candidate motif which is fitted on a single expression condition e.g. using simple linear regression. If the motif has a statistically significant regression coefficient, than it is likely to be involved in determining at least some of the observed expression changes. By increasing the number of motifs and the number of expression conditions it would be possible to decipher whole-genome regulatory networks. In this case, regression procedures involve a great number of predictors (scores for all the motif considered) and response variables (all the expression conditions) and linear models are not appropriated. For such a reason we might identify two procedures with good performance using high dimensional data, such as Partial Least Squares [27] or Bayesian variable selection methods [28].

2.3. Gene Coexpression Networks

Gene coexpression networks are intimately dependent on the regulatory network, but they can reflect subtle relationships that cannot be easily described starting from the regulatory network. In Fig. 2.2 we show the simplest case of a direct correspondence between regulatory and coexpression networks, common in Bacteria. However, especially in eukaryotes, gene regulation can be *combinatorial*, so that two genes controlled by a common TF can have several additional regulators, and divergent expression patterns in different conditions, reducing the correspondence between regulatory and coexpression networks.

2.3.1. *Experimental approach*

Microarray gene expression profiling is a high throughput system providing large amounts of quantitative data. The analysis of the data mainly looks at common patterns within a defined group of biological samples, so that more focused experiments are possible. Molecular data obtained can be absolute or relative gene expression changes in a genome-wide fashion so that the *transcriptome*, the globality of mRNAs present in the cell at a given condition/moment, can be explored.

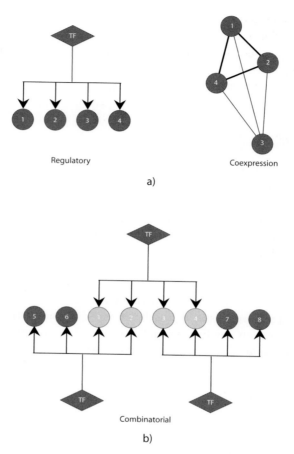

Fig. 2.2.   a) Schematic view of the direct correspondence existing between regulatory and co-expression network in a simple case of regulation: the TF is the only regulator of genes 1 to 4 and moreover it has a positive effect on the expression of each of them. In this case genes 1 to 4 have highly correlated gene expression profiles and thus form a tight cluster in the coexpression network. This is true in Bacteria, where combinatorial promoters are less common. b) In most eukaryotes combinatorial promoters (i.e. promoters controlled by more transcription factors) can be common, making this relationship less clear and often dynamic (the expression of different TF in time could mean that different coexpression groups can be formed from the same set of genes).

A scheme of the basic approach is measuring mRNA levels in a cell culture in two different conditions. One of the two is intended as a *control*, the other as a *treated* sample. The treatment is here intended in the general sense of a perturbation of the original system, i.e. presence of a given compound in the culture medium or a genetic alteration, so that it is possible to monitor the genes significantly changing their expression levels with respect to the control situation. Genes that are more expressed in the treated sample can be further analyzed with the idea in mind that they have a function related to the treatment. Schematically total mRNA is extracted from control and treated cells and the two mRNAs are labeled with two fluorescent dyes. The mRNA is then hybridized onto a microarray and scanning allows recording the fluorescent signal associated to each gene. The procedure is intrinsically noisy so that it is common practice to do the entire process in parallel for at least two-three replicates, even more if it is possible. By repeating the protocol it is possible to obtain time dependent changes in gene expression.

When promoters respond to a single specific TF there is a direct correspondence between the regulatory and the coexpression networks; the first example has been reported by Maas and Clark in the sixties to describe the genetic and regulatory properties of the genes of Arginine biosynthesis (for a review see [29]). In the bacterium *E. coli arg* genes are scattered over the chromosome, but they are expressed together when required thanks to the action of the ArgR TF. The dependency of the expression level from ArgR is illustrated by the presence, upstream of these genes, of a conserved motif called *ARG box* (e.g. see [30]).

However, this simple arrangement of a promoter characterizes a small fraction of genes, e.g. see Fig. 2.3, and, especially in eucaryotes, TFs very often cooperate/ compete for a promoter giving rise to combinatorial regulation.

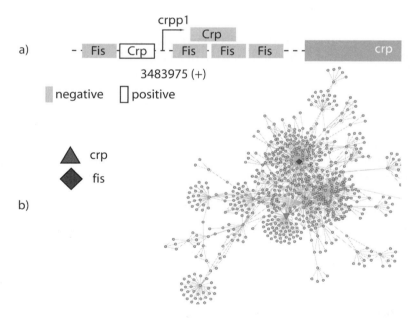

Fig. 2.3. a) The crpp1 promoter, upstream of the *E. coli* gene encoding CRP (catabolite repressor protein), a global regulator, showing the presence of two types of binding sites: 4 are specific for the protein Fis, another global regulator which exerts a negative control on crp expression, and 2 are binding targets of CRP itself, which moreover exerts a dual control on its own transcription rate. b) The known interaction network of *E. coli* where the two master regulators CRP and Fis are indicated.

2.3.2. *Computational approach*

Coexpression networks are built on the basis of the similarity between all pairs of expression profiles; nodes correspond to genes but more precisely to their expression profile, e.g. obtained by recording genome-wide gene expression levels in many different conditions, or at different times; edges represent a measure of similarity between expression profiles. Each edge has a weight corresponding to the similarity coefficient between neighbor nodes. Weights can be filtered with a thresholding procedure: the coexpression similarities (s_{ij}) are used to derive the a_{ij} components of a matrix describing the adjacencies in the coexpression network. The simplest strategy originates a binary graph: the adjacency between gene expression profiles x_i and x_j of genes i and j can be defined on the basis of a minimum threshold (τ): $a_{ij} = 1$ if $s_{ij} \geq \tau$. Following this scheme, two genes are linked ($a_{ij} = 1$) if the correlation between their expression profiles exceeds the threshold. This type of thresholding is called *hard* and leads to simple network concepts (e.g., node connectivity simply amounts to the number of direct neighbors). On the converse, it also lead to a loss of information, because the similarity between expression profiles is encoded as a continuous variable. To preserve the continuous nature of the correlation between pairs of expression profiles, one could simply define a weighted adjacency matrix where $a_{ij} = |s_{ij}|^\beta$ with $\beta \geq 1$, eventually after elimination of negative s_{ij}s. This *soft* thresholding approach leads to a weighted gene coexpression network. A novel approach giving a geometric interpretation of co-expression network analysis takes advantage of the interpretation of the correlation coefficient between expression profiles as the angle between the two vectors of expression values [31].

Regulatory Networks from Expression Compendium: During recent times several algorithms devoted to the identification of TRNs from microarray data have been proposed. An algorithm with good performances has been recently proposed by Faith and collaborators [32], called the CLR algorithm (context likelihood relatedness), an extension of relevance network algorithms. In this case the coexpression network has been constructed calculating pairwise relationships between gene expression profiles using a measure taken from information theory, the Mutual Information (MI).

MI indicates how much one random variable tells us about another random variable, so that it can be seen as a reduction in the uncertainty of a variable if another is known.

The CLR algorithm then takes as input the matrix of MI scores calculated for all pairs of genes and it estimates the likelihood of each edge. To do so, the algorithm evaluates the MI score for a particular pair of genes by comparing it to a background distribution. By taking advantage of the sparsity of biological regulatory networks, it is possible to approximate the background distribution as a joint normal distribution with MI_i and MI_j as independent variables. By doing so, the final form of this likelihood estimate can be expressed in the following way $f(Z_i, Z_j) = \sqrt{Z_i^2 + Z_j^2}$, where Z_i and Z_j are z-scores from the marginal distributions, and $f(Z_i, Z_j)$ is the joint likelihood measure.

Pathway Analysis: Genes encode proteins that in most cases take part to well defined metabolic, regulatory or signalling pathways. The correct functioning of a pathway is guaranteed by the overlap of the expression profiles of all genes involved in that process. In other words, when the pathway is required, all the proteins must be present. It follows that finding groups of coexpressed genes facilitates the identification of genes belonging to the same cellular process i.e. the same pathway. In the network framework, this corresponds to the identification of densely inter-connected regions in the coexpression network. There are many different algorithms to cluster high dimensional data, and most of them are implemented in most common commercial and freeware mathematical/statistical software. Some of the most known algorithms are K-means, fuzzy c-means and quality threshold (QT) clustering [33]. The QT is an alternative method of partitioning data, invented for gene clustering, which does not require specifying the number of clusters. One recent algorithm having the same good property is the Markov Clustering (MCL) [34] that has been used with success to cluster networks based on both sequence homologies, e.g. [35–37] or, as in this case, expression profiles [38, 39].

MCL Clustering: A natural property of *clusters* is that most edges are intra-cluster and only a few are inter-cluster. This implies that random walks on the graph will rarely go from one cluster to another. This feature is exploited by the MCL algorithm, that computes the probabilities of random walks through the graph performing iterations of two operators, expansion and inflation, on a stochastic matrix. Expansion of a stochastic matrix corresponds to computing long random walks. It associates new probabilities with all pairs of connected nodes, where one node is the point of departure and the other is the destination. Since higher length paths are more common within than between clusters, intra-cluster probabilities will be often relatively large. Inflation will then have the effect of boosting the probabilities of intra-cluster walks and demoting inter-cluster walks. After several iterations, each one followed by normalization of the matrix to keep it stochastic, the graph will be eventually separated into different clusters. The only parameter of the MCL is the *granularity* which has the effect of changing the strength of the inflation step, increasing the tightness of clusters. Additionally, also the parameter controlling expansion can be changed.

2.4. Protein-Protein Interaction Networks

Protein interactions are crucial for all levels of cellular function, including architecture, regulation, metabolism, and signaling. Therefore, protein interaction maps represent essential components of post-genomic toolkits needed for understanding biological processes at a systems level. A protein-protein interaction network (PIN) consists of all reported protein-protein interactions in an organism detected using specific assays, such as yeast two-hybrid (now implemented in bacterial systems too [40]), immuno-precipitation and tandem-affinity purification [41–43]. In general, the data are only of a qualitative nature; that is, interactions are either present or not but their strength is not quantified. The collection, verification and validation of such data pose considerable statistical challenges, and together form an active field of bioinformatics research.

2.4.1. *Experimental approach*

The advent of the yeast two-hybrid (Y2H) system in 1989 marked a milestone in proteomics. Exploiting the modular nature of TFs, Y2H allows measurements of the activation of reporter genes based on interactions between two chimeric proteins of interest. The Y2H system is today increasingly used in high-throughput applications intended to map genome-scale PPIs from viruses to humans. Although some significant technical limitations apply, Y2H has made a great contribution to our general understanding of the topology of interaction networks [44]. In addition to Y2H, especially in the past decade a variety of methods have been developed to detect and quantify PPIs, including surface plasmon resonance spectroscopy, NMR, peptide tagging combined with mass spectrometry and fluorescence-based technologies (see Table 2.1).

2.4.2. *Computational approach*

The prediction of protein-protein interactions is certainly amongst the most ambitious targets of modern bioinformatics. The increase of available genomic sequences has boosted the progress for detecting functions of genes and proteins and also allowed the development of complementary approaches to the high throughput experimental methods. Several approaches attempt to reconstruct the PIN and the most recent methods can be defined as based on the genomic-context [63], with four major families: methods based on phylogenetic profiles, methods based on gene clustering, gene fusion methods, and gene neighbor methods. Each of them mainly takes advantage of one specific genomic feature but in recent times there have been proposals of methods exploiting an integration of different genomic features two of which are the joint observation method and the algorithm implemented in the web server STRING. The joint observation method selects the PPIs that are predicted or identified by more than one method. Its rationale is based on the understanding that the confidence of PPI prediction relies on the amount of supporting evidence, and that the confidence increases with more evidence (i.e., methods). STRING calculates a combined score for each pair of proteins assuming that the features from various sources are independent [64].

Table 2.1. Experimental methods for studying PPI networks (modified from [62]). Abbreviations: FPR, false positive rate, BKG background.

Methods	Pros	Cons	Refs.
In vivo			
Yeast two-hybrid	Mature; robust; can be automated	High FPR	[45, 46]
Split-ubiquitin system	Interactions in different cellular compartments	High FPR; high BKG	[47–50]
Split-luciferase system	Very low BKG	No information on subcellular localization	[51–53]
In vitro			
Native chromatography or electrophoretic purification	Simple	Potentially poor resolution of complexes; high sample complexity	[54]
Single affinity purification-tagging	Efficient; flexible, low cost	Non-specific protein contamination	[54]
Tandem affinity purification tagging	High sample purity; reduces non-specific proteins	Weak interactions can be disrupted	[55, 56]
Stable-isotope labeling of amino acids in cell culture	Uniform direct isotopic labeling without chemical differences	Incomplete (80%) isotopic labeling; expensive	[57]
^{15}N-labeling	Inexpensive, complete labeling		[58, 59]
Chemical crosslinking-MS with PI reporter	Simple; large-scale	Problematic with complex protein mixtures	[60, 61]

2.5. Evolutionary Networks

The analysis of proteins encoded in different genomes has revealed that several of them group in families of sequences with common ancestry. There are two types of evolutionary relationships within these families: *orthology* and *paralogy*. Two genes (but a corresponding definition exists for proteins) are defined to be orthologous if they derive from a common ancestral gene in consequence of speciation events. On the contrary, two genes are paralogous when they derive from a common ancestor sequence by means of a DNA duplication event. Orthologs groups of proteins are generally considered to have the same molecular function in different organisms; however, recent analysis show that exceptions are quite common. On the other hand, paralogous genes generally have different even if related functions. However, there are also cases (especially if the duplication is recent) of paralogous genes encoding proteins with the same function as the ancestral gene. This relatedness is quantifiable in a simple way in term of identity percentage shared by two protein sequences, that while being a very intuitive measure lacks a lot of information concerning the evolutionary processes that shaped a gene (protein) family.

2.5.1. *Computational approach*

In the last years great improvements in phylogenetic methods has made it possible to run very complex algorithms to understand the relationships existing within a protein family. However, these methods could not use the enormous number of proteins today deposited in the databases mainly because they need multiple alignments, whose phylogenetic signal becomes deteriorated when very distant or partial sequences are added to the analysis. On the contrary methods based on relatively simpler algorithms, such as the local alignment tools Blast and Fasta, are able to manage such large numbers of sequences. These, combined with network theory and clustering algorithms, allow describing the structure of the evolutionary network without the need to make multiple alignments and to manually refine them. Moreover, when comparing very large numbers of sequences in this way it is also possible to have both related and unrelated sequences within the same dataset and these will be classified at the clustering stage. In the simplest example, the output of a similarity search may be recoded as an adjacency matrix where each element a_{ij} is a measure of the similarity shared by proteins i and j and can be used as input for a clustering algorithm to infer communities of related sequences. This strategy has been used to construct the Clusters of Orthologous Groups (COG, [65]) and OrthoMCL databases [35]. Advanced clustering techniques have been implemented to better identify clusters of related proteins in large scale similarity searches; tribeMCL [36] and orthoMCL [35] both implement a relatively recent procedure, the Markov Clustering algorithm (Section 2.3.2) to partition proteins into families.

The a_{ij} element of the matrix used for clustering can be the normalized score for the local alignment between sequence i and j; raw blast scores are not suitable because they are the sum of substitution and gap scores taken from a given substitution matrix (e.g. Blosum, Dayhoff) and they are consequently dependent on the alignment length; this causes the risk of over-scoring long alignments; for this reason it is recommended to normalize the matrix row by row on the basis of the diagonal element $s_{ii} = s_{i\,max}$ (the score of a self aligned protein is the maximum attainable for each row); in this way each element becomes $s_{ij}^{norm} = \frac{s_{ij}}{s_{i\,max}}$. Alternatively, the score can be transformed in bit form, which is normalized with respect to the scoring system and thus can be used to compare alignment scores from different searches; for Blast, $S_{bit}^{norm} = \frac{\lambda S - ln(K)}{ln(2)}$, where λ and K are parameters of the Blast run; bit scores subsume the statistical essence of the scoring system employed, so that to calculate significance one needs to know in addition only the size of the search space.

After this transformation, the matrix encodes a network of normalized global sequence relationships: nodes are proteins with edges between homologous proteins,

Table 2.2.   The γ parameter inferred for several types of networks. PI: Protein Interaction.

Network Type	γ	Reference
Yeast paralogs PI divergence (n=274)	1.38	[68]
Yeast PI (data from [69])	2.80	[70]
Yeast PI (core data from [71])	2.2	[72]
C. elegans PI (data from [73])	1.8	[74]
Metabolic network *E. nidulans*	2.2	[72]
Protein domains family size (*T. maritima*)	3	[75]
Protein domains family size (*C. elegans*)	1.9	[75]
Homology of Plasmid encoded proteins (γ–proteobacteria; n = 3393)	1.32	[37]
Global Homologies of Nitrogen fixation proteins (n = 4299)	2.3	[37]

and edge values reflecting the degree of similarity between two sequences. The following step is the clustering of the network, that can be done using one of the algorithms cited above. Following the clustering procedure it is possible to analyze the global topological properties of the network or to use data mining procedures to characterize the function(s) of proteins of each cluster.

2.6. Statistical Properties of Biological Networks

Here we discuss the biological significance of complex network concepts.

Degree: The degree or connectivity (k) of a node in its simplest form is the number of links starting or reaching it. In directed graphs we can define an *incoming* degree (k_{in}), i.e. the links received by a node, and an *outgoing* degree (k_{out}) summarizing the number of links sent by the node. In a regulatory network, the indegree concerns regulated genes, while the outdegree concerns only TFs. The degree distribution ($P(k)$) gives the probability that a node has k links. It is obtained by counting the number of nodes with a certain number of links and dividing by the total number of connected nodes. The degree distribution of most biological networks fits a power-law (i.e. $P(k) \simeq k^{-\gamma}$, with $\gamma \simeq 2$) indicating the presence of rare very connected nodes (*hubs*) [66] in a vast majority of nodes with only a few connections. The value of γ determines many properties of the network; if it is very small, the network's topology is highly dependent on the role of the hubs, whereas for $\gamma > 3$ highly connected nodes are not relevant; in this case the network's behavior approaches the random type. For $2 < \gamma < 3$ there is a hierarchy of hubs, with the most connected being in contact with a small fraction of all nodes; for $\gamma = 2$ a *hub-and-spoke* network emerges, with the largest hub contacting a large fraction of nodes (the so-called *giant component*).

In general, the unusual properties of scale-free networks are valid only for $\gamma < 3$ when the dispersion of the degree distribution ($\sigma^2 = \langle k^2 \rangle - \langle k \rangle^2$) diverges as the number of nodes increases; this results in a series of unexpected features, such as a high robustness against random node removal [67]. In biological networks this might correspond to the random deletion (or loss of function) of a gene, so that all edges passing through it are interrupted. In the case of regulatory networks, the disappearance of a TF determines the loss of its control over the target genes. On the contrary, scale-free networks are highly sensitive to directed attack, i.e. removal of *hubs*, a strategy that can be exploited in drug design.

In a regulatory network the outdegree corresponds to the number of genes a given transcription factor is able to regulate, so that TFs with high degree can be thought as the *master regulators*; on the converse, the indegree of a gene corresponds to the complexity of its regulation.

The properties of hub proteins are of particular interest, not least because, being involved in a number of cellular processes, they are attracting candidates for antimicrobial agents. Studying protein-protein interaction networks, it has been claimed that hubs might be physiologically more important (i.e., less dispensable) [76–79] and slow evolving [80–83]. Not all analysis, however, support this idea probably because of biased and less-than reliable global PPI data; for this reason Batada et al. [84] studied a comprehensive literature-curated dataset of well-substantiated protein interactions in *Saccharomyces cerevisiae* finding a relatively robust correlation between degree and dispensability (i.e. highly connected nodes are more often essential). In contrast, no correlation with evolutionary rates has been found.

Average Shortest Path Length, ASPL: Path length tells us how many links we need to pass through to travel between two nodes and it is used as a measure of distance in graphs. The paths connecting two nodes can be many, so that it is common practice to use the shortest path. ASPL is the average over all shortest paths in the network giving a measure of networks' navigability.

In a recent work the dynamics of a biological network on a genomic scale have been addressed [85] by integrating transcriptional regulatory information and gene-expression data for multiple conditions in *Saccharomyces cerevisiae*. We will illustrate the results of that work and the interpretation given by the authors. The networks reconstructed from different conditions have large topological differences: regulatory interactions conserved in different conditions mostly regulate house-keeping functions and, while a few TFs are always highly connected, most of them becomes active only in a few cases. Luscombe and collaborators [85] identify five condition-specific sub-networks which they further classify into *endogenous*, e.g. cell cycle and sporulation responding to an internal programme, and *exogenous*, e.g. diauxic shift, DNA damage and stress response, controlled by extracellular signals. The differences between the two classes of sub-networks can be summarized in four main points:

(1) the average in-degree decreases by 20% from endogenous to exogenous conditions indicating that TFs are going to regulate genes in simpler combinations;

(2) the average out-degree double from endogenous to exogenous conditions: each TF has greater regulatory influence by targeting more genes simultaneously;

(3) the ASPL (here equal to the number of intermediate regulators between a TF and a target gene) halves from endogenous to exogenous conditions, implying a faster propagation of the regulatory signal;

(4) the average clustering coefficient (indicating the level of transcription factor inter-regulation) nearly halves from endogenous to exogenous conditions.

Clustering Coefficient: The clustering coefficient has been introduced by Watts and Strogatz in 1988 to determine whether or not a graph has small-world properties [86]. Roughly speaking, the clustering coefficient of a node depends on the number of a node's neighbors that share a connection. More rigorously, $C_I = 2n_I/k(k-1)$, where n_I is the number of links connecting the k_I neighbors of node I to each other. C_I gives the number of "triangles" going through node I, whereas $k_I(k_I-1)/2$ is the total number of triangles passing through node I if all of I's neighbors are interconnected. The average clustering coefficient ($\langle C \rangle$) indicates the tendency of the nodes of a network to form well-defined clusters. The average clustering coefficient can also be calculated by considering nodes with the same degree ($C(k)$), and can be used as an indication of hierarchical behavior of a network [87, 88]. Network topology may be said to correspond to a *small world* if the network's clustering coefficient is much greater than that of equivalent random controls $C \gg C_{random}$, while their path lengths are comparable $ASPL \sim ASPL_{random}$. The small-world index $\sigma_{small\ world}$, introduced by [89], is defined as: $\sigma_{small\ world} = \frac{C}{C_{random}} \times \frac{ASPL}{ASPL_{random}}$.

Protein-protein interaction and gene regulatory networks often have small-world topology; one explanation for an evolutionary advantage of this topology is that it is very robust to perturbations. If this is the case, it would provide an advantage to biological systems that are subject to damage e.g. by mutation or viral infection. Moreover, in power law small-world networks the deletion of a random node rarely causes a dramatic increase in ASPL (or a significant decrease of C). This is because (i) most paths pass through the hubs, and (ii) hubs are very rare, which means that most of the times random deletions will involve loosely connected nodes, maintaining unaltered the paths in the large majority of the network.

Assortativity Coefficient: Assortativity is a measure of how much edges in a network tend to connect *similar* nodes with respect to a certain feature [90, 91]. The so-called *assortative mixing* of complex networks has recently gained a lot of attraction because it has a profound impact on the topological properties of a network. In the case of perfect assortative mixing there are parts of the network that are completely isolated, but within those parts there are dense interconnections. The most basic form of assortativity concerns degree-degree correlation; in this case the *assortativity coefficient* ($r \in [-1, 1]$) is the *correlation coefficient* between the degrees of connected nodes, see Eq. 2.1 where L is the total number of edges, j_i and

k_i are the degrees of vertices at the ends of the ith edge. Positive values of r indicate frequent connections between nodes of similar degree, while negative values indicate relationships between nodes of different degree. When $r = 1$, the network is said to have perfect assortative mixing, implying a high modularity of the network and the presence of isolated clusters composed by nodes with the same degree i.e. layers. On the converse, for $r = -1$ the network is said to be completely disassortative; there are no more layers, and connections maintain compact the network.

$$r = \frac{L^{-1}\sum_i j_i k_i - [L^{-1}\sum_i \frac{1}{2}(j_i + k_i)]^2}{L^{-1}\sum_i \frac{1}{2}(j_i^2 + k_i^2) - [L^{-1}\sum_i \frac{1}{2}(j_i + k_i)]^2}. \qquad (2.1)$$

The PIN in yeast turned out to be *disassortative* i.e. hubs tend to be connected to more peripheral nodes. The same has been observed for a network of the synaptic connections in the neural network of the nematode *C. elegans* and for most of the biological networks studied so far [90, 91]. Newman [90, 91] has observed that several key properties of networks change when the model used to simulate network growth includes different levels of assortative mixing and may thus be important in biological networks evolution. One of the properties varying much is the size of the *giant component* that is smaller if the network is assortative, reflecting a better *percolation* of assortative networks. Assortativity with respect to degree can tell us several information on network topology; however, the most important feature of nodes in biological networks is that they represent real biological entities participating in the physiology of the cell. Thus it is interesting to study the correlations existing between properties of the nodes when a certain topology of the network is specified, i.e. the PIN or the TRN. The categories can be the molecular process in which proteins are involved. The work of Newman [91] improves the previous coefficient developing a way to quantify the assortativity when discrete nodes characteristics are accounted for.

With a similar purpose Park and Barabási introduced in the biological sciences the concept of dyadicity [92]. The simplest case is of un undirected graph where each node can take only two values, say 1 or 0 (e.g. is this gene essential? Is this gene involved in 'cell-cycle'?). Let us call n_1 (n_0) the number of nodes of each group ($N = n_1 + n_0$). This allows for three kinds of *dyads*: (1-1), (0-1), and (0-0) that are present in the network in m_{11}, m_{10}, m_{00} fractions, respectively ($M = m_{11} + m_{10} + m_{00}$). The expected values of m_{11} and m_{10} [92] under the null hypothesis that the property is distributed randomly on nodes can be obtained with the hypergeometric distribution:

$$\bar{m}_{11} = \binom{n_1}{2} \times p = \frac{n_1(n_1 - 1)}{2}p \qquad (2.2)$$

and:

$$\bar{m}_{10} = \binom{n_1}{1}\binom{n_0}{1} \times p = n_1(N - n_1)p \qquad (2.3)$$

where $p = \frac{2M}{N(N-1)}$ is the *connectance*, i.e. the average probability that two nodes are connected. Statistically significant deviations from these expected values imply that property is not randomly distributed. The authors introduce *dyadicity* (D) and *heterophilicity* (H) to classify a property's behavior:

$$D \equiv \frac{m_{11}}{\bar{m}_{11}} \quad and \quad H \equiv \frac{m_{10}}{\bar{m}_{10}} \tag{2.4}$$

a property is called *dyadic* if $D > 1$ (*antidyadic* if $H < 1$), indicating that the nodes with property 1 tend to connect more (less) densely among themselves than expected for random configuration. Similarly the property is *heterophilic* if $H > 1$ (*heterophobic* if $H < 1$), meaning that the nodes with property 1 have more (fewer) connections to nodes with property 0 than expected randomly. By studying the *S. cerevisiae* PPI network for proteins involved in cellular communications and signal transduction, Park and Barabási [92] calculated that this class of proteins is more dyadic and less heterophilic than expected; this describes the tendency of proteins belonging to this functional class to have interacting partners of the same class, and it moreover highlights the modular structure of the class.

2.7. Functional Networks Motifs

Motifs can be defined as a set of nodes in a biological network with specific molecular functions which are arranged together and perform some 'useful' process. The behaviors of motifs are generally not separable from the rest of the system and they constitute only part of a recognizable systems level function. There are several known motifs in biological networks (e.g. see [93, 94] e.g. in regulatory network we might identify switches, amplitude filters, oscillators, frequency filters, noise filters and many others. Shen Orr and collaborators [94] identified three highly frequent motifs in a regulatory network of the bacterium *Escherichia coli*. The first motif, the *feedforward loop*, is defined by X, a TF that regulates a second TF Y, such that both TFs jointly regulate gene or operon Z. Moreover, if X and Y both positively regulate Z, and X positively (negatively) regulates Y, the feedforward loop is coherent (incoherent). The second motif, termed *single-input module* (SIM), is defined by a set of genes controlled by a single TF. Moreover, the TF controlling a SIM has the same effect (activation or inhibition) on all the component genes, which lack any other transcriptional regulation. The third motif, the *dense overlapping regulons*, is a layer of overlapping interactions between operons and a group of input TFs.

By developing an algorithm to evaluate the statistical significance of all motifs composed by three and four nodes, Milo *et al.* [93] analyzed different types of networks. Their datasets were from biology (PPI and nervous system) as well as from technology (electronic circuits and the world wide web). Different motif sets were found which differ in different networks, suggesting that motifs can define broad classes of networks, each with specific types of elementary structures. The motifs

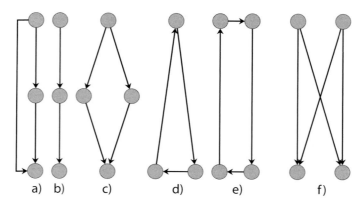

Fig. 2.4. Network motifs, examples [93]: a – the feedforward loop, b – three nodes chain, c - bi-parallel, d – three nodes feedback loop, e – four nodes feedback loop, f – bi-fan. The different motifs are differentially enriched in different networks, suggesting the use of different building blocks for networks with different *purposes*, see text for details.

reflect the underlying processes that generated each type of network; for example, food webs evolve to allow a flow of energy from the bottom to the top of food chains, whereas gene regulation and neuron networks evolve to process information. Information processing seems to give rise to significantly different structures than does energy flow.

2.8. Conclusion

Biological research in the past focused on single gene/protein - analysis so that for a small fraction of all known genes/proteins there is a lot of information available. However, it is today clear that many biological processes cannot be studied with such a reductionistic approach. Protein and DNA interact at the level of promoters to finely tune gene expression; this is mainly accomplished in response to specific environmental conditions that are integrated in cell behavior through interaction networks composed by signal transduction cascades. In some sense the environmental information flows through different biological networks to elicit a specific response. One meaningful way to study the properties of these intricate relationships is in the framework of network theory which allows the qualitative and quantitative characterization of the most important features of a network by means of well-defined statistics. In the last years, the availability of techniques for high throughput analysis in biology enormously boosted such approach which also took advantage of the development of computational methods that today represent essential components of biological research; the challenge of *systems biology* is to integrate the different -*omics* to fully elucidate the functioning of living cells and progressing to effective medicine therapies.

References

[1] Barabási, A.L., Oltvai, Z.N.: Network biology: understanding the cell's functional organization. *Nat. Rev. Genet.* **5** (2004) 101–113.

[2] Latchman, D.: *Eukaryotic Transcription Factors* (Academic Press, San Diego, California, 1998).

[3] Weinzierl, R.O.J.: *Mechanisms of Gene Expression* (Imperial College Press, London, 1999).

[4] Carey, M., Smale, S.T.: *Transcriptional Regulation in Eukaryotes: Concepts, Strategies, and Techniques* (Cold Spring Harbor Laboratory Press, Cold Spring Harbor, New York, 2000).

[5] Lee, T.I., Young, R.A.: Transcription of eukaryotic protein-coding genes. *Annual Review of Genetics* **34** (2000) 77–137.

[6] Lewin, B.: *Genes VII* (Oxford University Press, Oxford, 2000).

[7] Davidson, E.H.: *Genomic Regulatory Systems: Development and Evolution* (Academic Press, San Diego, CA, 2001).

[8] Locker, J.: *Transcription Factors* (Academic Press, San Diego, California, 2001).

[9] White, R.J.: *Gene Transcription: Mechanisms and Control* (Blackwell Science, Malden, MA, 2001).

[10] Harbison, C.T., Gordon, D.B., Lee, T.I., Rinaldi, N.J., Macisaac, K.D., Danford, T.W., Hannett, N.M., Tagne, J.B., Reynolds, D.B., Yoo, J., Jennings, E.G., Zeitlinger, J., Pokholok, D.K., Kellis, M., Rolfe, P.A., Takusagawa, K.T., Lander, E.S., Gifford, D.K., Fraenkel, E., Young, R.A.: Transcriptional regulatory code of a eukaryotic genome. *Nature* **431** (2004) 99–104.

[11] Reece-Hoyes, J.S., Deplancke, B., Shingles, J., Grove, C.A., Hope, I.A., Walhout, A.J.: A compendium of *Caenorhabditis elegans* regulatory transcription factors: a resource for mapping transcription regulatory networks. *Genome Biol.* **6** (2005) R110.

[12] Ren, B., Robert, F., Wyrick, J.J., Aparicio, O., Jennings, E.G., Simon, I., Zeitlinger, J., Schreiber, J., Hannett, N., Kanin, E., Volkert, T.L., Wilson, C.J., Bell, S.P., Young, R.A.: Genome-wide location and function of DNA binding proteins. *Science* **290** (2000) 2306–2309.

[13] Lee, T.I., Rinaldi, N.J., Robert, F., Odom, D.T., Bar-Joseph, Z., Gerber, G.K., Hannett, N.M., Harbison, C.T., Thompson, C.M., Simon, I.: Transcriptional regulatory networks in *S. cerevisiae*. *Science* **298** (2002) 799–804.

[14] Ohler, U., Niemann, H.: Identification and analysis of eukaryotic promoters: recent computational approaches. *Trends Genet.* **17** (2001) 56–60.

[15] Atteson, K.: Calculating the exact probability of language-like patterns in biomolecular sequences. *Proc. Int. Conf. Intell. Syst. Mol. Biol.* **6** (1998) 17–24.

[16] Tompa, M.: An exact method for finding short motifs in sequences, with application to the ribosome binding site problem. *Proc. Int. Conf. Intell. Syst. Mol. Biol.* **7** (1999) 262–271.

[17] Brazma, A., Jonassen, I., Vilo, J., Ukkonen, E.: Predicting gene regulatory elements in silico on a genomic scale. *Genome Res.* **8** (1998) 1202–1215.

[18] Jensen, L.J., Knudsen, S.: Automatic discovery of regulatory patterns in promoter regions based on whole cell expression data and functional annotation. *Bioinformatics* **16** (2000) 326–333.

[19] Bailey, T., Elkan, C.: Unsupervised learning of multiple motifs in biopolymers using expectation maximization. *Machine Learning* **21** (1995) 51–80.

[20] Hughes, J.D., Estep, P.W., Tavazoie, S., Church, G.M.: Computational identification of cis-regulatory elements associated with groups of functionally related genes in *Saccharomyces cerevisiae*. *J. Mol. Biol.* **296** (2000) 1205–1214.

[21] Workman, C.T., Stormo, G.D.: Ann-spec: a method for discovering transcription binding sites with improved specificity. *Pac. Symp. Biocomput.* **5** (2000) 467–478.

[22] Lawrence, C.E., Altschul, S.F., Boguski, M.S., Liu, J.S., Neuwald, A.F., Wootton, J.C.: Detecting subtle sequence signals: a gibbs sampling strategy for multiple alignment. *Science* **262** (1993) 208–214.

[23] Thijs, G., Lescot, M., Marchal, K., Rombauts, S., Moor, B.D., Rouzé, P., Moreau, Y.: A higher-order background model improves the detection of promoter regulatory elements by gibbs sampling. *Bioinformatics* **17** (2001) 1113–1122.

[24] Conlon, E.M., Liu, X.S., Lieb, J.D., Liu, J.S.: Integrating regulatory motif discovery and genome-wide expression analysis. *Proceedings of the National Academy of Science (USA)* **100** 44 (2003) 3339–3333.

[25] Brilli, M., Fani, R., Lió, P.: Motifscorer: using a compendium of microarrays to identify regulatory motifs. *Bioinformatics* **23** (2007) 493–495.

[26] Schneider, T.D., Stormo, G.D., Gold, L., Ehrenfeucht, A.: Information content of binding sites on nucleotide sequences. *Journal of Molecular Biology* **188** (1986) 415–431.

[27] Abdi, H.: Partial least squares regression (pls-regression). In: *Encyclopedia for Research Methods for the Social Sciences* (Thousand Oaks (CA): Sage, 2003).

[28] Tadesse, M.G., Vannucci, M., Lió, P.: Identification of DNA regulatory motifs using bayesian variable selection. *Bioinformatics* **20** (2004) 2553–2561.

[29] Maas, W.K.: The arginine repressor of *Escherichia coli*. *Microbiol. Rev.* **58** (1994) 631–640.

[30] Caldara, M., Charlier, D., Cunin, R.: The arginine regulon of *Escherichia coli*: whole-system transcriptome analysis discovers new genes and provides an integrated view of arginine regulation. *Microbiology* **152** (2006) 3343–3354.

[31] Horvath, S., Dong, J.: Geometric interpretation of gene coexpression network analysis. *PLoS Comput. Biol.* **4** (2008) e1000117.

[32] Faith, J.J., Hayete, B., Thaden, J.T., Mogno, I., Wierzbowski, J., Cottarel, G., Kasif, S., Collins, J.J., Gardner, T.S.: Large-scale mapping and validation of *Escherichia coli* transcriptional regulation from a compendium of expression profiles. *PLoS Biol.* **5** (2007) e8.

[33] Heyer, L.J., Kruglyak, S., Yooseph, S.: Exploring expression data: identification and analysis of coexpressed genes. *Genome Res.* **9** (1999) 1106–1115.

[34] van Dongen, S.: Graph clustering by flow simulation. PhD thesis available at http://igitur-archive. library. uu. nl/dissertations/1895620/inhoud. htm (2000).

[35] Li, L., Stoeckert, C.J.J., Roos, S.D.: Orthomcl: Identification of ortholog groups for eukaryotic genomes. *Genome Res.* **13** (2003) 2178–2189.

[36] Enright, A.J., Dongen, S.V., Ouzounis, C.A.: An efficient algorithm for large-scale detection of protein families. *Nucleic Acids Res.* **30** (2002) 1575–1584.

[37] Lió, P., Brilli, M., Fani, R.: Topological metrics in blast data mining: plasmid and nitrogen-fixing proteins case studies. In: *BIRD 2008, CCIS 13* (Springer-Verlag, Berlin, Heidelberg, 2008).

[38] Caretta-Cartozo, C., Rios, P.D.L., Piazza, F., Lió, P.: Bottleneck genes and community structure in the cell cycle network of *S. pombe*. *PLoS Comput. Biol.* **3** (2007) e103.

[39] Freeman, T.C., Goldovsky, L., Brosch, M., van Dongen, S., Maziére, P., Grocock, R.J., Freilich, S., Thornton, J., Enright, A.J.: Construction, visualisation, and clustering of transcription networks from microarray expression data. *PLoS Comput. Biol.* **3** (2007) 2032–2042.

[40] Lallo, G.D., Ghelardini, P., Paolozzi, L.: Two-hybrid assay: construction of an *Escherichia coli* system to quantify homodimerization ability in vivo. Microbiology **145** (1999) 1485–1490

[41] Parrish, J.R., Gulyas, K.D., Finley, R.L.J.: Yeast two-hybrid contributions to interactome mapping. *Curr. Opin. Biotechnol.* **17** (2006) 387–393.

[42] Causier, B.: Studying the interactome with the yeast two-hybrid system and mass spectrometry. *Mass Spectrom Rev.* **23** (2004) 350–367.

[43] Miller, J., Stagljar, I.: Using the yeast two-hybrid system to identify interacting proteins. *Methods Mol. Biol.* **261** (2004) 247–262.

[44] Ratushny, V., Golemis, E.: Resolving the network of cell signaling pathways using the evolving yeast two-hybrid system. *Biotechniques* **44** (2008) 655–662.

[45] Causier, B.: Analysing proteinprotein interactions with the yeast two-hybrid system. *Plant Mol. Biol.* **50** (2002) 855–870.

[46] Causier, B.: Studying the interactome with the yeast two-hybrid system and mass spectrometry. *Mass Spectrom. Rev.* **23** (2004) 350–367.

[47] Michnick, S.W.: Protein fragment complementation strategies for biochemical network mapping. *Curr. Opin. Biotechnol.* **14** (2003) 610–617.

[48] Reinders, A., Schulze, W., Kühn, C., Barker, L., Schulz, A., Ward, J.M., Frommer, W.B.: Proteinprotein interactions between sucrose transporters of different affinities colocalized in the same enucleate sieve element. *Plant Cell* **14** (2002) 1567–1577.

[49] Schulze, W.X., Reinders, A., Ward, J., Lalonde, S., Frommer, W.B.: Interactions between co-expressed arabidopsis sucrose transporters in the split-ubiquitin system. *BMC Biochem.* **4** (2003) 3.

[50] Obrdlik, P., El-Bakkoury, M., Hamacher, T., Cappellaro, C., Vilarino, C., Fleischer, C., Ellerbrok, H., Kamuzinzi, R., Ledent, V., Blaudez, D., Sanders, D., Revuelta, J.L., Boles, E., André, B., Frommer, W.B.: K+ channel interactions detected by a genetic system optimized for systematic studies of membrane protein interactions. *Proc. Natl. Acad. Sci. USA* **101** (2004) 12242–12247.

[51] Paulmurugan, R., Gambhir, S.: Firefly luciferase enzyme fragment complementation for imaging in cells and living animals. *Anal. Chem.* **77** (2005) 1295–1302.

[52] Paulmurugan, R., Gambhir, S.: Combinatorial library screening for developing an improved split-firefly luciferase fragment-assisted complementation system for studying protein–protein interactions. *Anal. Chem.* **79** (2007) 2346–2353.

[53] Fujikawa, Y., Kato, N.: Split luciferase complementation assay to study protein–protein interactions in arabidopsis protoplasts. *Plant J.* **52** (2007) 185–195.

[54] Berggard, T., Linse, S., James, P.: Methods for the detection and analysis of protein–protein interactions. *Proteomics* **7** (2007) 2833–2842.

[55] Brown, A.P., Affleck, V., Fawcett, T., Slabas, A.R.: Tandem affinity purification tagging of fatty acid biosynthetic enzymes in *Synechocystis* sp. pcc6803 and *Arabidopsis thaliana*. *J. Exp. Bot.* **57** (2006) 1563–1571.

[56] Rohila, J.S., Chen, M., Chen, S., Chen, J., Cerny, R., Dardick, C., Canlas, P., Xu, X., Gribskov, M., Kanrar, S., Zhu, J.K., Ronald, P., Fromm, M.E.: Protein–protein interactions of tandem affinity purification-tagged protein kinases in rice. *Plant J.* **46** (2006) 1–13.

[57] Mann, M.: Functional and quantitative proteomics using silac. *Nat. Rev. Mol. Cell Biol.* **7** (2006) 952–958.

[58] Huttlin, E.L., Hegeman, A.D., Harms, A.C., Sussman, M.R.: Comparison of full versus partial metabolic labeling for quantitative proteomics analysis in arabidopsis thaliana. *Mol. Cell. Proteomics* **6** (2007) 860–881.

[59] Nelson, C.J., Huttlin, E.L., Hegeman, A.D., Harms, A.C., Sussman, M.R.: Implica-

tions of 15n-metabolic labeling for automated peptide identification in *Arabidopsis thaliana. Proteomics* **7** (2007) 1279–1292.

[60] Tang, X., Munske, G.R., Siems, W.F., Bruce, J.E.: Mass spectrometry identifiable cross-linking strategy for studying protein–protein interactions. *Anal Chem.* **77** (2005) 311–318.

[61] Anderson, G.A., Tolic, N., Tang, X., Zheng, C., Bruce, J.E.: Informatics strategies for large-scale novel cross-linking analysis. *J. Proteome Res.* **6** (2007) 3412–3421.

[62] Morsy, M., Gouthu, S., Orchard, S., Thorneycroft, D., Harper, J.F., Mittler, R., Cushman, J.C.: Charting plant interactomes: possibilities and challenges. *Trends in Plant Science* **13** (2008) 1360–1385.

[63] Shi, T.L., Li, Y.X., Cai, Y.D., Chou, K.C.: Computational methods for protein-protein interaction and their application. *Curr. Protein Pept. Sci.* **6** (2005) 443–449.

[64] Sun, J., Sun, Y., Ding, G., Liu, Q., Wang, C., He, Y., Shi, T., Li, Y., Zhao, Z.: Inpreppi: an integrated evaluation method based on genomic context for predicting protein-protein interactions in prokaryotic genomes. *BMC Bioinformatics* **26** (2007) 414.

[65] Tatusov, R.L., Fedorova, N.D., Jackson, J.D., Jacobs, A.R., Kiryutin, B., Koonin, E.V., Krylov, D.M., Mazumder, R., Mekhedov, S.L., Nikolskaya, A.N., Rao, B.S., Smirnov, S., Sverdlov, A.V., Vasudevan, S., Wolf, Y.I., Yin, J.J., Natale, D.A.: The cog database: an updated version includes eukaryotes. *BMC Bioinformatics* **4** (2003) 41.

[66] Albert, R., Barabási, A.L.: Statistical mechanics of complex networks. *Rev. Mod. Phys.* **74** (2002) 47.

[67] Albert, R., Jeong, H., Barabási, A.L.: Error and attack tolerance of complex networks. *Nature* **406** (2000) 378–382.

[68] Zhang, Y., Wolf-Yadlin, A., Ross, P.L., Pappin, D.J., Rush, J., Lauffenburger, D.A., White, F.M.: Time-resolved mass spectrometry of tyrosine phosphorylation sites in the epidermal growth factor receptor signaling network reveals dynamic modules. *Mol. Cell Proteomics* **4** (2005) 1240–1250.

[69] Uetz, P., Giot, L., Cagney, G., Mansfield, T.A., Judson, R.S., Knight, J.R., Lockshon, D., Narayan, V., Srinivasan, M., Pochart, P., Qureshi-Emili, A., Li, Y., Godwin, B., Conover, D., Kalbfleisch, T., Vijayadamodar, G., Yang, M., Johnston, M., Fields, S., Rothberg, J.M.: A comprehensive analysis of protein-protein interactions in *Saccharomyces cerevisiae. Nature* **403** (2000) 623–627.

[70] Wagner, A.: The yeast protein interaction network evolves rapidly and contains few duplicate genes. *Molecular Biology and Evolution* **18** (2001) 1283–1292.

[71] Ito, T., Chiba, T., Ozawa, R., Yoshida, M., Hattori, M., Sakaki, Y.: A comprehensive two-hybrid analysis to explore the yeast protein interactome. *Proc. Natl. Acad. Sci. USA* **98** (2001) 4569–4574.

[72] Goh, K.I., Oh, E., Jeong, H., Kahng, B., Kim, D.: Classification of scale-free networks. *Proc. Natl. Acad. Sci. USA* **99** (2002) 12583–12588.

[73] Li, S., Armstrong, C.M., Bertin, N., Ge, H., Milstein, S., Boxem, M., Vidalain, P.O., Han, J.D.J., Chesneau, A., Hao, T., Goldberg, D.S., Li, N., Martinez, M., Rual, J.F., Lamesch, P., Xu, L., Tewari, M., Wong, S.L., Zhang, L.V., Berritz, G.F., Jacotot, L., Vaglio, P., Reboul, J., Hirozane-Kishikawa, T., Li, Q., Gabel, H.W., Elewa, A., Baumgartner, B., Rose, D.J., Yu, H., Bosak, S., Sequerra, R., Fraser, A., Mange, S.E., Saxton, W.M., Strome, S., van den Heuvel, S., Piano, F., Vandenhaute, J., Sardet, C., Gerstein, M., Doucette-Stamm, L., Gunsalus, K.C., Harper, J.W., Cusick, M.E., Roth, F.P., Hill, D.E., Vidal, M.: A map of the interactome network of the metazoan *C. elegans. Science* **303** (2004) 540–543.

[74] Hughes, A.L., Friedman, R.: Gene duplication and the properties of biological networks. *J. Mol. Evol.* **61** (2005) 758–764.

[75] Koonin, E.V., Wolf, Y.I., Karev, G.P.: The structure of the protein universe and genome evolution. *Nature* **420** (2002) 218–223.

[76] Jeong, H., Mason, S.P., Barabási, A.L., Oltvai, Z.N.: Lethality and centrality in protein networks. *Nature* **411** (2001) 41–42.

[77] Chen, Y., Xu, D.: Understanding protein dispensability through machine-learning analysis of high-throughput data. *Bioinformatics* **21** (2005) 575–581.

[78] Hahn, M.W., Kern, A.D.: Comparative genomics of centrality and essentiality in three eukaryotic protein-interaction networks. *Mol. Biol. Evol.* **22** (2005) 803–806.

[79] Pereira-Leal, J.B., Audit, B., Peregrin-Alvarez, J.M., Ouzounis, C.A.: An exponential core in the heart of the yeast protein-interaction network. *Mol. Biol. Evol.* **22** (2005) 421–425.

[80] Fraser, H.B., Wall, D.P., Hirsh, A.E.: A simple dependence between protein evolution rate and the number of protein–protein interactions. *BMC Evol. Biol.* **3** (2003) 11.

[81] Fraser, H.B., Hirsh, A.E., Steinmetz, L.M., Scharfe, C., Feldman, M.W.: Evolutionary rate in the protein-interaction network. *Science* **296** (2002) 750–752.

[82] Saeed, R., Deane, C.M.: Proteinprotein interactions, evolutionary rate, abundance, and age. *BMC Bioinformatics* **7** (2006) 21.

[83] Makino, T., Gojobori, T.: The evolutionary rate of a protein is influenced by features of the interacting partners. *Mol. Biol. Evol.* **23** (2006) 784–789.

[84] Batada, N.N., Hurst, L.D., Tyers, M.: Evolutionary and physiological importance of hub proteins. *PLoS Comput. Biol.* **2** (2006) e88.

[85] Luscombe, N.M., Babu, M.M., Yu, H., Snyder, M., Teichmann, S.A., Gerstein, M.: Abstract genomic analysis of regulatory network dynamics reveals large topological changes. *Nature* **431** (2004) 308–312.

[86] Watts, D.J., Strogatz, S.H.: Collective dynamics of 'small-world' networks. *Nature* **393** (1988) 440–442.

[87] Ravasz, E., Barabási, A.L.: Hierarchical organization in complex networks. *Phys. Rev. E. Stat. Nonlin. Soft Matter Phys.* **67** (2003) 026112.

[88] Ravasz, E., Somera, A.L., Mongru, D.A., Oltvai, Z.N., Barabási, A.L.: Hierarchical organization of modularity in metabolic networks. *Science* **297** (2002) 1551–1555.

[89] Humphries, M.D., Gurney, K., Prescott, T.J.: The brainstem reticular formation is a small-world, not scale-free, network. *Proc. Roy Soc. B Biol. Sci.* **273** (2006) 503–511.

[90] Newman, M.E.: Assortative mixing in networks. *Phys. Rev. Lett.* **89** (2002) 208701.

[91] Newman, M.E.: Mixing patterns in networks. *Phys. Rev. E Stat. Nonlin. Soft Matter Phys.* **67** (2003) 026126.

[92] Park, J., Barábasi, A.L.: Distribution of node characteristics in complex networks. *Proc. Natl. Acad. Sci. USA* **104(46):17916-20** (2007) 17916–17920.

[93] Milo, R., Shen-Orr, S., Itzkovitz, S., Kashtan, N., Chklovskii, D., Alon, U.: Network motifs: simple building blocks of complex networks. *Science* **298** (2002) 824–827.

[94] Shen-Orr, S., Milo, R., Mangan, S., Alon, U.: Network motifs in the transcriptional regulation network of *Escherichia coli*. *Nat. Genet.* **31** (2002) 64–68.

Chapter 3

Dynamics of Multicellular Synthetic Gene Networks

Ekkehard Ullner[1,2,9], Aneta Koseska[3], Alexey Zaikin[4,5], Evgenii Volkov[6],
Jürgen Kurths[7,8,9], Jordi García-Ojalvo[1]

[1] *Departament de Física i Enginyeria Nuclear, Universitat Politècnica de Catalunya, Colom 11, E–08222 Terrassa, Spain*
[2] *Institute for Complex Systems and Mathematical Biology, Kings College, University of Aberdeen, Aberdeen AB24 3UE, UK*
[3] *Center for Dynamics of Complex Systems, University of Potsdam, D–14469 Potsdam, Germany*
[4] *Department of Mathematics, University of Essex, Wivenhoe Park, Colchester CO4 3SQ, UK*
[5] *Department of Mathematics & Institute for Women's Health, University College London, Gower Street, London WC1E 6BT, UK*
[6] *Department of Theoretical Physics, Lebedev Physical Inst., Leninskii 53, Moscow, Russia*
[7] *Institute of Physics, Humboldt University Berlin, D-10099 Berlin, Germany*
[8] *Potsdam Institute for Climate Impact Research, D-14412 Potsdam, Germany*
[9] *Institute of Medical Sciences, Foresterhill, University of Aberdeen, Aberdeen AB25 2ZD, UK*

3.1. Introduction

Living systems are driven by intricate networks of genes and proteins, whose dynamical behavior underlies all kinds of structural and functional processes in cells. Understanding the dynamics that emerges from such complex networks has benefited greatly in recent years by synthetic approaches, through which simpler network modules have been built that perform natural-like dynamical processes without interfering with, nor being perturbed by, natural cellular processes [for a review, see e.g. Sprinzak and Elowitz (2005)].

One of the first examples of an artificial gene circuit was the *repressilator*, a synthetic biological oscillator developed in *E. coli* from a network of three transcriptional repressors that inhibit one another in a cyclic way [Elowitz and Leibler (2000)]. Spontaneous oscillations were initially observed in individual cells within a growing culture, although substantial variability and noise was present among the different cells. After its conception, the repressilator immediately has become a milestone example of how natural dynamical processes can be mimicked within cells through the design of artificial circuits built from standard genetic parts. Other examples of such genetic gene circuits included a toggle switch [Gardner et al. (2000)], a metabolic relaxator [Fung et al. (2005)], or a relaxation oscillator [Atkinson et al. (2003)].

Natural genetic networks, however, do not usually operate in isolation. Not only in multicellular higher organisms, but even in bacterial populations, cells conspicuously communicate among each other by different means, e.g. electrically or chemically. A particularly useful (and common) means of communication between bacteria is *quorum sensing*, which relies on the relatively free diffusion of small molecules, known as autoinducers, through the bacterial membrane. When such an autoinducer is part of a feedback loop that regulates the expression of certain genes, bacteria are able to determine the local density of similar cells around them by monitoring the level of expression of these autoinducer-controlled genes [Miller and Bassler (2001)]. An example of this mechanism is provided by the Lux system, used by the bacterium *Vibrio fischeri* to provide bioluminiscence only when the bacterial density is high (which happens within specialized light organs of certain marine organisms with whom the bacteria live in symbiosis).

Cell-cell coupling often leads to exceptional examples of cooperative behavior. In order to understand how such collective phenomena emerge from passive intercellular communication, it seems natural to make use of the synthetic approaches described above. The Lux system described above has been used, for instance, as a communication module to build a synthetic mechanism for programmed population control in a bacterial population [You et al. (2004)]. In this Chapter, we review recent developments that are helping us to understand the rich dynamical behavior that can be produced in coupled synthetic gene networks. We concentrate on two different types of genetic oscillators, the repressilator and a relaxator oscillator, and consider two different types of coupling, namely a phase-attractive and a phase-repulsive coupling, both resulting from the autoinducer diffusion. As we will see, many different dynamical scenarios arise from these types of coupling, including multistability, oscillation death, and quantized cycling, among others.

3.2. Coupled Repressilators

As mentioned above, the repressilator is a synthetic network of three genes whose products inhibit the transcription of each other cyclically [Elowitz and Leibler (2000)] (see left module of Fig. 3.1). A readout module using fluorescent proteins provides access to the time-resolved dynamics of the repressilator proteins. Experiments reveal oscillations with a period of the order of an hour, i.e. slower than the cell-division cycle. The limited number of interacting genes and proteins and the well-understood interactions between them enable a precise theoretical description of this oscillator by means of coupled differential equations.

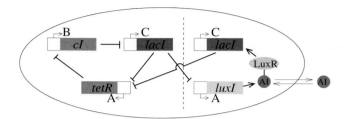

Fig. 3.1. Scheme of the repressilator network coupled to a quorum-sensing mechanism. The original repressilator module is located at the left of the vertical dashed line, while the coupling module appears at the right.

3.2.1. *Phase-attractive coupling*

Quorum sensing has been theoretically shown to lead to synchronization in ensembles of identical genetic oscillators [McMillen *et al.* (2002)]. The oscillators considered in that work were relaxational, analogous to neural oscillators. The repressilator, on the other hand, is sinusoidal rather than relaxational. Furthermore, in the experimental implementation of the repressilator [Elowitz and Leibler (2000)], individual cells were found to oscillate in a "noisy" fashion, exhibiting cell-cell variation in period length, as well as variation from period to period within a single cell.

Accordingly, it seems natural to consider the effect of inter-cell signaling on a population of non-identical and noisy repressilators coupled by reinforcing quorum sensing. Using computational modeling, García-Ojalvo *et al.* (2004) showed that a diverse population of such oscillators is able to self-synchronize, even if the periods of the individual cells are broadly distributed. The onset of synchronization is sudden, not gradual, as a function of varying cell density. In other words, the system exhibits a phase transition to mutual synchrony. This behavior has been experimentally reported in the zebrafish somitogenesis clock [Riedel-Kruse *et al.* (2007)].

The coupling also has a second beneficial effect: it reduces the system's noisiness, effectively transforming an ensemble of "sloppy" clocks into a very reliable collective oscillator [Enright (1980); Somers and Kopell (1995); Needleman *et al.* (2001)]. The results of García-Ojalvo *et al.* (2004) suggest that the constraints that local cell oscillators have to face in order to be noise resistant, could be relaxed in the presence of intercell coupling, since coupling itself provides a powerful mechanism of noise resistance.

3.2.1.1. *Model*

The repressilator consists of three genes, *lacI*, *tetR*, and *cI*, whose protein products repress transcription of the genes cyclically [Elowitz and Leibler (2000)]. García-Ojalvo *et al.* (2004) proposed to incorporate the quorum-sensing system of the bacterium *Vibrio fischeri* as an inter-cell signaling module, by placing the gene that encodes LuxI under the control of the repressilator protein LacI, as shown in Fig. 3.1. LuxI synthesizes a small molecule, the autoinducer (AI), that diffuses freely among the cells and thus couples them to one another. A second copy of another of the repressilator's genes (such as *lacI*) is inserted into the genetic machinery of the *E. coli* cell in such a way that its expression is induced by the complex LuxR-AI. The result is the appearance of a feedback loop in the repressilator, which is reinforced the more similar among neighboring cells the levels of LacI are.

The mRNA dynamics is governed by degradation and repressible transcription for the repressilator genes, plus transcriptional activation of the additional copy of the *lacI* gene:

$$\frac{da_i}{dt} = -a_i + \frac{\alpha}{1 + C_i^n} \tag{3.1}$$

$$\frac{db_i}{dt} = -b_i + \frac{\alpha}{1 + A_i^n}, \tag{3.2}$$

$$\frac{dc_i}{dt} = -c_i + \frac{\alpha}{1 + B_i^n} + \frac{\kappa S_i}{1 + S_i}. \tag{3.3}$$

Here a_i, b_i, and c_i are the concentrations in cell i of mRNA transcribed from *tetR*, *cI*, and *lacI*, respectively, and the concentration of the corresponding proteins are represented by A_i, B_i, and C_i (note that the two *lacI* transcripts are assumed to be identical). The concentration of AI inside each cell is denoted by S_i. A certain amount of cooperativity is assumed in the repression mechanisms via the Hill coefficient n, whereas the AI activation is chosen to follow a standard Michaelis-Menten kinetics. The protein and AI concentrations are scaled by their Michaelis constants. α is the dimensionless transcription rate in the absence of a repressor, and κ is the maximal contribution to *lacI* transcription in the presence of saturating amounts of AI. The protein dynamics is given by:

$$\frac{dA_i}{dt} = \beta_a(a_i - A_i), \tag{3.4}$$

and similarly for B_i (with b_i) and C_i (with c_i). The parameter β_a is the ratio between the mRNA and protein lifetimes of A (resp. β_b and β_c, all three are considered equal in this Section). The mRNA concentrations have been rescaled by their translation efficiency (proteins produced per mRNA, assumed equal for the three genes).

Finally, the dynamical evolution of the intracellular AI concentration is affected by degradation, synthesis and diffusion toward/from the intercellular medium. The dynamics of TetR and LuxI can be assumed identical if their lifetimes are considered to be the same, and hence we will use the same variable to describe both variables. Consequently, the synthesis term of the AI rate equation will be proportional to A_i:

$$\frac{dS_i}{dt} = -k_{s0}S_i + k_{s1}A_i - \eta(S_i - S_e), \tag{3.5}$$

where $\eta = \sigma \mathcal{A}/V_c \equiv \delta/V_c$ measures the diffusion rate of AI across the cell membrane, with σ representing the membrane permeability, $\mathcal{A}$ its surface area, and V_c the cell volume. The parameters k_{s0}, k_{s1}, and η have been made dimensionless by time rescaling. S_e represents the extracellular concentration of AI, whose dynamics is given by

$$\frac{dS_e}{dt} = -k_{se}S_e + \eta_{\text{ext}} \sum_{j=1}^{N} (S_j - S_e) \equiv -k_{se}S_e + k_{\text{diff}}(\overline{S} - S_e), \qquad (3.6)$$

where $\eta_{\text{ext}} = \delta/V_{\text{ext}}$, with V_{ext} being the total extracellular volume, and $\overline{\cdots}$ indicates average over all cells. The diffusion rate is given by $k_{\text{diff}} = \eta_{\text{ext}}N$ and the degradation rate by k_{se}. This approach assumes a uniform AI concentration throughout the cell culture, which describes reasonably well the situation encountered in a well-controlled chemostat.

In the quasi-steady-state approximation [McMillen *et al.* (2002); Dockery and Keener (2001)], the extracellular AI concentration can be approximated by

$$S_e = \frac{k_{\text{diff}}}{k_{se} + k_{\text{diff}}} \overline{S} \equiv Q\,\overline{S}. \qquad (3.7)$$

From the definition of k_{diff} given above, we note that Q depends on the cell density $N/(V_{\text{ext}} + V_c) \approx N/V_{\text{ext}}$ according to

$$Q = \frac{\delta N/V_{\text{ext}}}{k_{se} + \delta N/V_{\text{ext}}}. \qquad (3.8)$$

In other words, Q is linearly proportional to the cell density provided $\delta N/V_{\text{ext}}$ is sufficiently smaller than the extracellular AI degradation rate k_{se}. In the following the effect of reinforcing quorum-sensing coupling on the collective behavior of model (3.1)-(3.5) is analyzed, with S_e defined by (3.7)-(3.8), considering Q (and hence the cell density) as a control parameter.

3.2.1.2. *Transition to synchronization*

In the hypothetical case of infinite cell dilution ($Q \to 0$), the system consists of a population of uncoupled limit-cycle oscillators. Each individual cell clock is an extension of the original repressilator [Elowitz and Leibler (2000)], where a new degree of freedom has been added to the original six-dimensional phase space to represent the intracellular AI dynamics (3.5). The resulting dynamical system exhibits limit-cycle oscillations in a wide region of parameter space. The characteristic oscillations of the repressilator [Elowitz and Leibler (2000)] do not change qualitatively in the presence of the AI dynamics.

The oscillator population will likely contain substantial differences from cell to cell (e.g. extrinsic noise [Elowitz *et al.* (2002)]), giving rise to a relatively broad distribution in the frequencies of the individual clocks at any given time. The variability in the cell population is modeled by considering that β ($\equiv \beta_a = \beta_b = \beta_c$) is non-uniformly distributed among the repressilators following a Gaussian law with

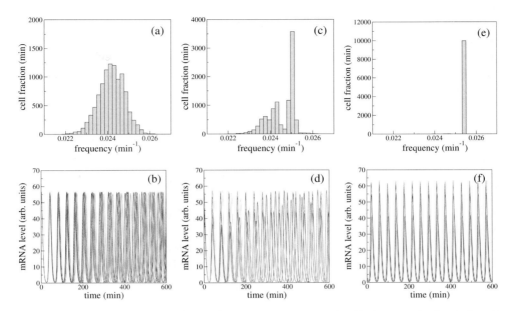

Fig. 3.2.   Frequency histogram (a, c, e) and time evolution of $b_i(t)$ for 10 cells (b, d, f) and
increasing cell density: (a, b) $Q = 0.4$, (c, d) $Q = 0.63$, (e, f) $Q = 0.8$. Other parameters are
$N = 10^4$, $\alpha = 216$, $\kappa = 20$, $n = 2.0$, $k_{s0} = 1$, $\eta = 2.0$ and $k_{s1} = 0.01$. The lifetime ratio β in
the different cells is chosen from a random Gaussian distribution of mean $\bar{\beta} = 1.0$ and standard
deviation $\Delta\beta = 0.05$.

standard deviation $\Delta\beta$. The corresponding frequency distribution of a group of 10^4
uncoupled cells for $\Delta\beta/\beta = 0.05$ is shown in Fig. 3.2(a). The temporal evolution of
the cI mRNA concentration in 10 of those cells is plotted in Fig. 3.2(b), showing how
the global operation of the system is completely disorganized, so that no collective
rhythm can exist under these conditions.

As the cell density increases, diffusion of extracellular AI molecules into the
cells provides a mechanism of intercell coupling, which leads to partial frequency
locking of the cells [Figs. 3.2(c,d)]. Finally, when the cell density is large enough
[Figs. 3.2(e,f)] perfect locking and synchronized oscillations are observed. In that
case the system behaves as a macroscopic clock with a well-defined period, even
though it is composed of a widely varied collection of oscillators. This results
indicate that a transition from an unsynchronized to a synchronized regime exists
as the strength of coupling increases (due to an increase in cell density). This
behavior is robust in the presence of noise. In fact, noise can be seen to enhance the
collective coherence of the system, leading to a better clock [Ullner $et\ al.$ (2009)].

3.2.2.   *Phase-repulsive coupling*

We now show how significantly can cell-cell coupling influence the dynamics of
synthetic gene network. Only one rewiring in the connectivity between the ba-

sic repressilator and the quorum sensing module, with respect to the case of the previous Section, alters the coupling from its original reinforcing character to a phase-repulsive one [Ullner *et al.* (2007)]. As a consequence, the previously favored in-phase regime becomes now unstable, and many new dynamical regimes appear.

To create a phase-repulsive coupling, one can modify the initial scheme (Fig. 3.1) by placing the gene *luxI* under inhibitory control of the repressilator protein TetR. The proposed 'rewiring' between the repressilator and the quorum sensing module introduces a feedback loop that competes with the overall negative feedback loop along the repressilator ring, resulting in a phase-repulsive intercellular coupling.

The mRNA and protein dynamics are described by Eqs. (3.1)–(3.4) above. In contrast to Section 3.2.1, we assume here different lifetime ratios for the protein/mRNA pairs, which results in a weak relaxator-like dynamics of the repressilator. The rewiring affects the equation of the AI concentration. Now the AI concentration S_i in cell i is generated at a rhythm proportional to B_i:

$$\dot{S}_i = -k_{s0}S_i + k_{s1}B_i - \eta(S_i - S_e).\tag{3.9}$$

A moderate increase of the Hill coefficient to $n = 2.6$, a value in agreement with recent experimental measures [Rosenfeld *et al.* (2005)], together with different lifetime ratios $\beta_a = 0.85$, $\beta_b = 0.1$, and $\beta_c = 0.1$, increase the nonlinear character of the repressilator dynamics, leading to the appearance of two time scales in the time series, with a fast concentration increase and a relative slow decay. The slower protein decay increases the period of the repressilator by a factor of approximately three.

3.2.2.1. *Bifurcation analysis for two coupled repressilators*

A first glimpse into the effect of coupling on the dynamics of inter-cell genetic networks can be obtained by investigating a minimal system of only two oscillators. Figure 3.3 shows representative time traces, obtained by direct numerical calculations of a population of $N = 2$ coupled repressilators for increasing coupling strength. The different dynamical regimes found are self-sustained oscillatory solutions [Fig. 3.3(a)], inhomogeneous limit cycles (IHLC) [Fig. 3.3(b)], inhomogeneous steady states (IHSS) [Fig. 3.3(c)] and homogeneous steady states (HSS) [Fig. 3.3(d)], all of which exist for biologically realistic parameter ranges.

A detailed bifurcation analysis allows to determine the origin of these different solutions and the transition scenarios between them, thus providing deeper qualitative and quantitative conclusions about the structure and dynamical behavior of the system. This analysis can be performed with public software such as the XPPAUT package [Ermentrout (2002)]. In the bifurcation analysis below we use the coupling strength Q [Eq. (3.7)] as a biologically relevant parameter to obtain one-parameter continuation diagrams. Starting from the homogeneous unstable steady state of isolated oscillators ($Q = 0$), Fig. 3.4 shows the basic continuation curve containing the homogeneous and inhomogeneous stable steady states.

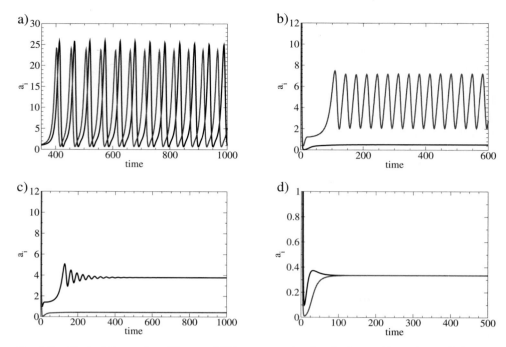

Fig. 3.3. Typical time series of the a_i mRNA concentration for the four stable regimes: a) $Q = 0.1$ – oscillatory, b) $Q = 0.3$ – inhomogeneous limit cycle, c) $Q = 0.4$ – inhomogeneous steady state, and d) $Q = 0.4$ – homogeneous steady state. The common parameters are: $N = 2$, $n = 2.6$, $\alpha = 216$, $\beta_a = 0.85$, $\beta_b = 0.1$, $\beta_c = 0.1$, $\kappa = 25$, $k_{s0} = 1.0$, $k_{s1} = 0.01$, $\eta = 2.0$.

The basic continuation curve is characterized by two important properties: (1) the presence of broken symmetry bifurcations (BP_1 and BP_2 in Fig. 3.4) where inhomogeneous solutions arise, and (2) the stabilization of the homogeneous state for large coupling values ($Q > 0.129$). The HSS solution is characterized by a constant protein level concentration, stabilized through a saddle node bifurcation (LP_1 in Fig. 3.4). A typical time series of this regime can be seen in Fig. 3.3(d). Additionally, another HSS branch is found between LP_4 and HB_4 (Fig. 3.4), but it is located outside the biologically relevant range (since $Q > 1$).

As a result of the symmetry breaking of the system through a pitchfork bifurcation (BP_1 in Fig. 3.4), the unstable steady state splits in two additional branches, giving rise to an inhomogeneous steady state (IHSS). This particular phenomenon is model-independent, persisting for large parametric regions in several models of diffusively coupled chemical [Bar-Eli (1985); Dolnik and Marek (1988); Crowley and Epstein (1989)] or biological oscillators [Kuznetsov *et al.* (2004); Tsaneva-Atanasova *et al.* (2006)]. The IHSS in the present model is manifested through two distinct steady protein concentration levels [Fig. 3.3(c)], gaining stability through a Hopf bifurcation, denoted as HB_1 in Fig. 3.4, and thus leading to the so-called "oscillation death" (*OD*) regime. This regime arises at a critical coupling $Q_{crit} = 0.3588$ for

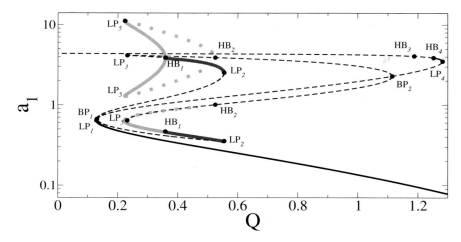

Fig. 3.4. Bifurcation diagram obtained by variation of Q, illustrating the stable steady state regimes (HSS and IHSS) and the inhomogenous limit cycle (IHLC). For parameters values see Fig. 3.3. Here, thin solid lines denote the HSS, thick blue solid lines the IHSS, thick solid orange line the stable IHLC, and dashed lines denote the unstable steady states especially the dashed orange line the unstable IHLC. The same bifurcation diagram is valid for the second repressilator.

the set of parameters used here, and is stable until LP_2 at $Q = 0.5548$. The IHSS solution coexists in the Q parameter space with the HSS (Fig. 3.4). For example, for $Q = 0.37$ there is a coexistence of 9 steady state solutions, 3 of them stable and 6 unstable.

The next step of the bifurcation analysis is to study the limit cycles that arise from the Hopf bifurcations found on the basic continuation curve. In particular, the Hopf bifurcation HB_1 gives rise to a branch of stable inhomogeneous periodic solutions, known in the literature as inhomogeneous limit cycle (IHLC) [Tyson and Kauffman (1975)]. The manifestation of this regime is however different in different systems: for two identical diffusively coupled Brusselators, e.g., it is defined to be a periodic solution of the system of oscillators rotating around two spatially non uniform centers [Tyson and Kauffman (1975); Volkov and Romanov (1995)]. For the model investigated here, the manifestation of the IHLC is somewhat different: the IHLC is characterized by a complex behavior, where one of the oscillators produces very small oscillations of the protein level, whereas the other one oscillates in the vicinity of the steady state with an amplitude 4 times smaller than that of an isolated oscillator [see Fig. 3.3(b)]. The IHLC is stable for values of Q between HB_1 and LP_5 (Fig. 3.4). In the case of the two-oscillator system considered here, each oscillator has the same probability to occupy and stay in the upper or lower state, due to the symmetry of the system. The initial conditions are the only factor determining the separation of the oscillators.

For coupling values smaller than a given critical value $Q_{crit} < 0.129$, the system is characterized by a self-oscillatory solution. For two coupled oscillators, this regime

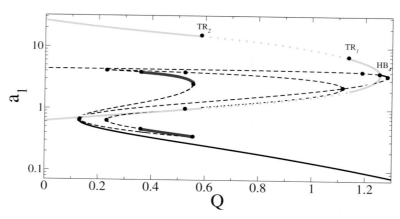

Fig. 3.5. Bifurcation diagram versus coupling Q, focusing on the stable anti-phase oscillations (thick yellow line). Parameters are those of Fig. 3.3.

corresponds to anti-phase oscillations. As shown on Fig. 3.5, this state belongs to a branch of periodic orbits originating at the Hopf bifurcation HB_4. Fig. 3.5 illustrates in detail the bifurcation structure of the antiphase dynamics when Q is being varied. Stable anti-phase oscillations are observed between HB_4 ($Q = 1.253$) and TR_1 (torus bifurcation for $Q = 1.137$), and from $Q = 0$ until TR_2 ($Q = 0.5848$). As demonstrated, this solution loses its stability for $0.5848 < Q < 1.137$. Direct numerical simulations revealed the existence of complex behavior in the latter range of Q values, which we discuss briefly in Sec. 3.2.2.3.

In contrast to the case of positively coupled repressilators [García-Ojalvo et al. (2004)], where coupling was seen to provide coherence enhancement, investigations of the dynamical structure of the system with phase-repulsive coupling by means of direct calculations [Ullner et al. (2007)] did not reveal the presence of a stable in-phase regime (synchronous oscillations over the entire cell population). The present bifurcation analysis confirms this result: a branch of synchronous periodic oscillations is in fact seen to emanate from HB_3, but it is unstable (data not shown, see Ullner et al. (2008)). The bifurcation analysis confirmed that the in-phase regime is unstable for all values of α and Q studied, in contrast to the anti-phase limit cycle oscillations, which arise even for small α values. The existence of this anti-phase (or phase-shifted) solution is a clear manifestation of the phase repulsive character of the AI-mediated coupling, which enhances the phase difference between the oscillators in the model, until the maximal phase difference of $\frac{\pi}{2}$ is reached.

3.2.2.2. Comparison between bifurcation analysis and direct calculations

Bifurcation analyses reveal all solutions, their stability, and the connecting bifurcation points. Special interest evokes the ranges of multi-stability, i.e. the coexistence of dynamical regimes, because it offers opportunities of the biological system to

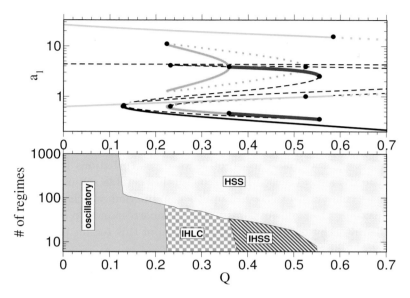

Fig. 3.6. Comparison between the bifurcation analysis (top) and the direct calculation with random initial conditions (bottom). Note the logarithmic scale of both ordinates in the two plots. The oscillatory regime is represented by a yellow solid line (top) and a yellow area (bottom); the IHLC by solid orange lines (top) and a orange-white chess board pattern (bottom); the IHSS by solid blue lines (top) and a small blue striped area (bottom); and finally the HSS is illustrated by a solid black line (top) and a grey area (bottom). Parameters are those of Fig. 3.3.

adapt or to store information. On the other hand, only stable regimes with a sufficient basin of attractions play a role in biological systems, an information that is not in the scope of the bifurcation analysis. The basins of attraction can be quantified in direct numerical simulations from the probability of occurrence of the different dynamical regimes for a set of randomly and appropriately drawn initial conditions. In what follows, we show results for 1000 time series with random initial conditions. Figure 3.6 shows a histogram of the resulting regimes as the bifurcation parameter Q is varied (bottom), compared with the bifurcation plot resulting from the continuation analysis described in the previous Section. Both methods indicate that for small coupling, $Q < 0.129$, anti-phase self-oscillations are the only stable regime. At $Q = 0.129$ the homogenous steady state stabilizes through a limit point bifurcation (LP_1 in Fig. 3.4), coexisting with an oscillatory solution. The direct calculations reveal the dominance of the single-fixed-point solution, which has a larger basin of attraction: at $Q = 0.2$, for instance, only about 70 of the total 1000 random initial conditions result in the oscillatory state, while the other remaining 930 result in HSS. For $Q \in [0.2236, 0.3588]$, direct calculations show the existence of an inhomogenous limit cycle (orange white chessboard pattern in Fig. 3.6,bottom) that coincides with the region where a stable IHLC solution was found by the bifurcation analysis (solid orange line in Fig. 3.6,top). One can see a very good coincidence of the stability ranges of the IHLC and the IHSS predicted by the bifurcation analysis and shown by the direct calculation. Both regimes have a small basin of attraction.

3.2.2.3. *Chaos provoked by repressive cell-to-cell communication*

The bifurcation analysis (Fig. 3.5) predicts unstable anti-phase oscillations between the torus bifurcation points TR_2 and TR_1. To find the stable solutions emerging from those bifurcations, one can perform direct simulations starting with small coupling Q, and trace the self-oscillatory regime up to strong coupling. The resulting self-oscillations are stable and resistant to small perturbations in the initial conditions and to dynamical noise. Interestingly, these stable self-oscillations display very different dynamics with erratic amplitude and period, which is associated with a positive maximal Lyapunov exponent, and thus corresponds to chaotic dynamics. For a detailed description of the chaotic features of this regime and its validation see Ullner *et al.* (2008).

3.2.2.4. *Large system sizes*

Typically, bacterial colonies consist of many cells and hence the results of the minimal system with $N = 2$ repressilators have to be validated in large ensembles. Here we show results for an ensemble of $N = 100$ coupled identical cells obtained from direct calculations with random initial conditions. Figure 3.7 plots the resulting frequency of stable regimes for increasing Q. The four main regimes HSS, IHSS, IHLC and self-oscillations already observed in the minimal system can be detected in the large systems too.

The results shown in Fig. 3.7 reveal a transition from self-oscillations to a single stable fixed point as the coupling Q increases. This transition is gradual, and

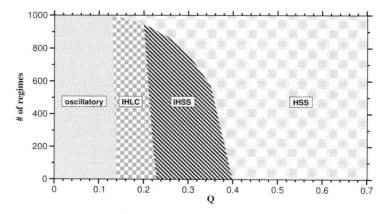

Fig. 3.7. Distribution of stable regimes for increasing coupling strength Q. The parameters are: $N = 100$, $n = 2.6$, $\alpha = 216$, $\beta_a = 0.85$, $\beta_b = 0.1$, $\beta_c = 0.1$, $\kappa = 25$, $k_{s0} = 1.0$, $k_{s1} = 0.01$, and $\eta = 2.0$.

exhibits a multiplicity of regimes. For $Q \lesssim 0.13$ only self-oscillations are found. As in the case $N = 2$, this regime is characterized by large oscillations with the same amplitude and period for all repressilators. The repressive character of the coupling destabilizes the in-phase dynamics, and leads to a spreading of the phases among all oscillators. After a certain transient time, oscillatory clusters appear [Golomb *et al.* (1992); Kaneko and Yomo (1994); Wang *et al.* (2000)]. The population self-organizes into three clusters of cells that oscillate with a phase difference close to $2\pi/3$. The separation into three clusters could provide the population of cells with high reliability and stress resistance, because at any given time the cells in the different clusters are in different states of the limit cycle, and hence each cluster will be affected differently by sudden environmental stresses such as chemicals or lack of nutrients.

At $Q \approx 0.13$ the basin of the self-oscillatory regime disappears abruptly, and a new dynamical regime arises in which some of the cells become trapped in a quasi-steady state with a negligible amplitude, while the rest undergo small amplitude oscillations in protein concentration. This dynamical regime corresponds to an inhomogeneous limit cycle (IHLC), in which cells do not switch from one regime to the other, i.e. there is no mixing of the two populations. As in the minimal case of $N = 2$, the basin of the IHLC coexists with the basin of the HSS, as shown in Fig. 3.7. This single fixed point attractor becomes more likely for larger coupling strengths Q.

At $Q \approx 0.2$, a second abrupt transition takes place, through which the IHLC disappears and the IHSS regime, corresponding to the fully developed oscillation death, arises. In this regime all cells stop oscillating, but they do so differentiating into two different clusters. Since each cluster is specialized in the production of a different protein, this regime could be interpreted as a mechanism of artificial differentiation in an isogenic population of cells. As in the case of the IHLC, cells may distribute into the two clusters at high and low CI levels in many different ratios which differ slightly in the constant protein levels. Hence in fact many different attractors exist, and a fine tuning of protein production can be achieved. Again, the basin of the IHSS regime described coexists with the basin of the single fixed point attractor, which becomes increasingly more likely for increasing Q, until it turns into the dominant attractor of the system for $Q \gtrsim 0.4$.

The dynamical regimes described above and their multistability persist even in a noisy environment. For instance, protein fluctuations larger than 25% of their mean level do not alter the clustering attractor in the multistable parameter range. Interestingly, a comparison of Figs. 3.6 and 3.7 show that the IHLC and the IHSS regimes become much more likely in large systems, at the expense of the HSS. Furthermore, those two regimes appear for smaller coupling in large systems. Together, these results show that the IHLC and the IHSS regimes become more likely in a large ensemble of identical cells than in a small one.

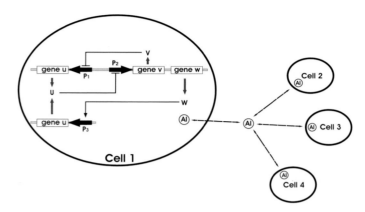

Fig. 3.8. Schematic diagram of the network of genetic relaxation oscillators. u, v and w denote the genes, and P_1, P_2 and P_3 the corresponding promoters.

3.3. Genetic Relaxation Oscillators

Different types of genetic circuit architectures, besides the repressilator, can give rise to oscillations and dynamical behavior. We now consider a different kind of network, consisting of coupled hysteresis-based genetic relaxation oscillators [Kuznetsov *et al.* (2004)]. Studying this system allows the identification of the intercellular mechanisms responsible for multirhythmicity in coupled genetic circuits. Additionally, this system exhibits a dynamical behavior closely related to a known biological problem, namely the existence of quantized cycles in cellular processes.

3.3.1. *Dynamical regimes of coupled relaxators*

Recently, Kuznetsov *et al.* (2004) proposed a model of hysteresis-based relaxation genetic oscillators coupled via quorum-sensing. This oscillator can be constructed, as shown in Fig. 3.8, by combining two engineered gene networks, the toggle switch [Gardner *et al.* (2000)] and an intercell communication system, which have been previously implemented experimentally in *E. coli* by Kobayashi *et al.* (2004), and in *V. fischeri* by Fuqua and Greenberg (2002), respectively. The synthesis of the two repressor proteins, which constitute the toggle switch, are regulated such that the expression of the two genes is mutually exclusive, which leads to bistability. The second network is based on the dynamics of an AI, which on the one hand drives the toggle switch through the hysteresis loop, and on the other hand provides an intercell communication by diffusion through the cell membrane. The time evolution of the system is governed by the dimensionless equations [Kuznetsov *et al.* (2004)]:

$$\frac{du_i}{dt} = \alpha_1 f(v_i) - u_i + \alpha_3 h(\omega_i) \tag{3.10}$$

$$\frac{dv_i}{dt} = \alpha_2 g(u_i) - v_i \tag{3.11}$$

$$\frac{d\omega_i}{dt} = \varepsilon(\alpha_4 g(u_i) - \omega_i) + 2d(\omega_e - \omega_i) \tag{3.12}$$

$$\frac{d\omega_e}{dt} = \frac{d_e}{N} \sum_{i=1}^{N} (\omega_i - \omega_e) \tag{3.13}$$

where N is the total number of cells, u_i and v_i represent the proteins from which the toggle switch is constructed in the i-th cell, ω_i represents the intracellular, and ω_e the extracellular AI concentration. The mutual influence of the genes is defined by the functions:

$$f(v) = \frac{1}{1 + v^\beta}, \qquad g(u) = \frac{1}{1 + u^\gamma}, \qquad h(w) = \frac{w^\eta}{1 + w^\eta}.$$

Here β, η and γ are the parameters of the corresponding activatory or inhibitory Hill functions.

In the Eqs. (3.10)-(3.13), the dimensionless parameters α_1 and α_2 regulate the repressor operation in the toggle switch, α_3 denotes the activation due to the AI, and α_4 the repression of the AI. The coupling coefficients in the system are given by d and d_e (intracellular and extracellular) and depend mainly on the diffusion properties of the membrane, as well as on the ratio between the volume of the cells and the extracellular volume [Kuznetsov *et al.* (2004)]. If the parameter ε is small ($\varepsilon \ll 1$), as in our case, the evolution of the system splits into two well-separated time-scales, a fast dynamics of u_i, v_i and ω_e, and a slow dynamics of ω_i. Due to the presence of multiple time scales, the system can produce relaxation oscillations.

The particular organization of the intercellular signaling mechanism in this case allows coupling to be organized through the slow recovery variable in the genetic network. As is known from oscillation theory, such coupling has the phase-repulsive property and can be referred to as inhibitory. On the other hand, local coupling of limit cycles via inhibitory variables has been reported to yield a coexistence of different stable attractors [Volkov and Stolyarov (1991, 1994)], thus leading typically to multirhythmicity.

The main manifestation of multistability in systems of globally coupled oscillators is clustering, defined as a dynamical state characterized by the coexistence of several subgroups, where the oscillators exhibit identical behavior. Oscillator clustering has been proved theoretically for identical phase oscillators [Okuda (1993)], observed experimentally for salt-water oscillators [Miyakawa and Yamada (2001)] and electrochemical oscillators [Wang *et al.* (2001); Kiss and Hudson (2003)]. For a detailed recent review of synchronization in oscillatory networks see [Osipov *et al.* (2007)]. As already mentioned in the repressilator case, the effects of multirhythmicity and multistability can be very important in understanding of evolutionary mechanisms behind cell differentiation and genetic clocks.

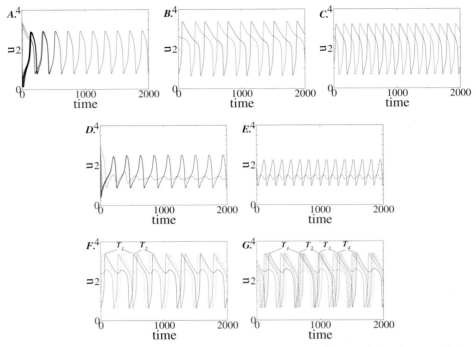

Fig. 3.9. Different oscillatory clusters for a system of $N = 8$ oscillators. A: In-phase oscillations for $\alpha_1 = 3, d = 0.005, d_e = 1$. B, C: Anti-phase oscillations with different distributions of the oscillators between the clusters, for $\alpha_1 = 3.3$, $d = 0.001$. D, E: Asymmetric solution with different distribution of the oscillators, for $\alpha_1 = 2.868$, $d = 0.001$. F: Three oscillatory clusters for $\alpha_1 = 3.3$, $d = 0.00105$. G: Five oscillatory clusters for $\alpha_1 = 3.3$, $d = 0.001$.

We discuss here two main phenomena. First, we show the existence of different possible modes of organized collective behavior in the system of globally coupled relaxation genetic oscillators. We distinguish between two different types of clusters: (i) steady-state clusters, and (ii) oscillatory clusters. Second, for each separate cluster formation, we demonstrate how the dependence on initial conditions can lead to different distributions of the oscillators between the clusters. In general, a system consisting of N oscillators can exhibit $N - 1$ different distributions of the oscillators among the clusters.

When the cells are identical, the coupled system is symmetric and identical behavior of the cells is a solution (Fig. 3.9A), though not necessarily stable. The inhibitory coupling and the presence of multiple time scales, as previously discussed, create the possibility for multistability and multirhythmicity, resulting in the generation of various dynamical regimes, among which oscillatory clusters are formed. For $d < 0.01$, the system can exhibit anti-phase oscillations, with oscillators distributed between the two oscillatory clusters (Fig. 3.9B,C). An important feature to be mentioned is the characterization of different distributions with different periods of the limit cycle, providing more complex dynamics with different rhythms: com-

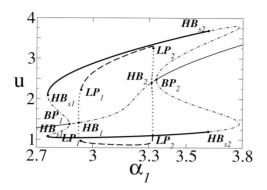

Fig. 3.10. Coexistence of five different states for increased coupling strength $d = 0.3$. Other parameters are: $\alpha_2 = 5$, $\alpha_3 = 1$, $\alpha_4 = 4$, $\beta = \eta = \gamma = 2$, $d_e = 1$ and $\varepsilon = 0.05$. Coexistence of the OD and the in-phase oscillatory regime is also shown.

pare for instance Fig. 3.9B (5:3 distribution) with period $T = 364.15$ and Fig. 3.9C (4:4 distribution) with period $T = 256.27$.

Another possible collective behavior of this system consists in asymmetric oscillations (for $d < 0.003$), when some of the oscillators in the system perform large excursions, while the rest oscillate in the vicinity of a stable steady state with small amplitude. This results in the presence of two oscillatory clusters, (Fig. 3.9D,E). Again, the number of possible different distributions for a system of N oscillators is $N - 1$, and each has different oscillation period: compare Fig. 3.9D (1:7) with period $T = 216.95$ and Fig. 3.9E (4:4) with $T = 141.01$.

The oscillators in the system can be also ordered in multiple cluster regimes; we present only two examples here: three (Fig. 3.9F) and five (Fig. 3.9G) oscillatory clusters. Again, different distributions of the oscillators between the clusters are possible in this case. To illustrate this, we present here a 3:3:2 distribution when three oscillatory clusters are formed (Fig. 3.9F), and a 1:2:2:2:1 distribution when five oscillatory clusters are created (Eig. 3.9G).

3.3.2. *Bifurcation analysis*

Bifurcation analysis can be used to identify and characterize the different dynamical solutions described above. When applied to the case $N = 2$, it shows that already two oscillators provide a large variety of possible regimes, as shown in Fig. 3.10. The OD regime, similarly to the IHSS one, is a result of the symmetry breaking in the system through a pitchfork bifurcation (labeled BP_1 in Fig. 3.10). The unstable steady-state splits into two branches that gain stability through Hopf bifurcations, denoted as HB_{s1} and HB_{s2} in Fig. 3.10. The solution coexists in the α_1-parameter space with different oscillatory solutions, e.g. in-phase oscillations (marked with dashed lines), as shown in Fig. 3.10. The true IHLC that emerges from HB_{s1} is unstable in this model and not shown in Fig. 3.10.

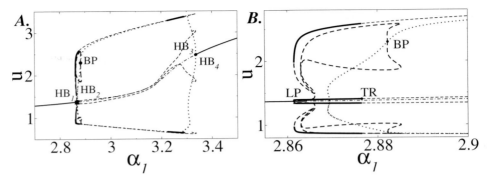

Fig. 3.11.   **A**, bifurcation diagram obtained by variation in α_1. Other parameters: $\alpha_2 = 5$, $\alpha_3 = 1$, $\alpha_4 = 4$, $\beta = \eta = \gamma = 2$, $d = 0.001$, $d_e = 1$ and $\epsilon = 0.01$. **B**, detailed view of the region where stable asymmetric solution exists. Between LP and TR, one oscillator has a large amplitude and the other oscillates with small amplitude.

The Hopf bifurcations labeled HB_1 and HB_2 in Fig. 3.10 give rise to a branch of periodic orbits, corresponding to a synchronous in-phase solution (see Fig. 3.9A). The stability of this region is determined with two saddle-node bifurcations LP_1 and LP_2. It is important to note that the in-phase oscillations present in the system are stable for all values of d, in contrast with the case of coupled repressilators discussed above.

For small coupling ($d < 0.01$) anti-phase oscillations arise (Fig. 3.9B,C). The periodic branch giving rise to the anti-phase solution is limited again by two Hopf bifurcations: HB_2 at $\alpha_1 = 2.869$, and HB_3 at $\alpha_1 = 3.336$. However, their stability region is significantly smaller than the corresponding stability region in the repressilator model discussed in Sec. 3.2.2.

Another mode of collective behavior is the asymmetric regime, characterized by the presence of large and small amplitude oscillations (see Fig. 3.9D,E). Although this solution resembles the IHLC shown in 3.2.2, its bifurcation structure here is completely different and very complex. In particular, for $\alpha_1 = 2.882$ a pitchfork bifurcation (labeled BP in Fig. 3.11A,B) is found on the bifurcation branch that gives rise to the anti-phase oscillations. Starting from this bifurcation point, a secondary bifurcation branch with a complex structure is observed (Fig. 3.11A). The stable asymmetric solution lies within this branch; the stability region is depicted with thick lines in Fig. 3.11B (zoomed region where a stable asymmetric solution exists), the unstable asymmetric solution is shown with the dashed line. The asymmetric regime is stabilized through a torus bifurcation at $\alpha_1 = 2.877$ (labeled as TR in Fig. 3.11B). This bifurcation leads to two incommensurate frequencies. For isolated oscillators ($d = 0$) and for $\alpha_1 > \alpha_{HB_1}$, the first frequency is that of a large cycle, and the second one is determined by the eigenvalues of the unstable focus. Slight diversity in the ensemble of relaxators does not alter the behavior shown above (results not shown) and confirms the relevance of these findings for biological networks.

3.3.3. *Response to external noise: quantized cycling time*

The presence of multistability influences the response of the system to external stimuli, in particular noise. This response can be modeled by substituting Eq. (3.12) above by:

$$\frac{d\omega_i}{dt} = \varepsilon(\alpha_4 g(u_i) - \omega_i) + 2d(\omega_e - \omega_i) + \xi_i(t). \tag{3.14}$$

Let us consider the case when all oscillators are confined to the oscillatory region. In order to establish the effect of noise in a population of such genetic units, we quantify the histogram of cycling times, analogous to the inter-spike interval (ISI) histograms used in studies of neural dynamics. We find that noise contributes to the establishment of variability and leads to multiple frequencies [Fig. 3.12(a,b)], even when the oscillators are initially synchronized. The cycling is now quantized, having either a bimodal [Fig. 3.12(a)] or a polymodal [Fig. 3.12(b)] distribution of periods. Thus, choosing slightly different α_1 values, one can effectively switch between different multipeak distributions. The ISI peaks observed are determined by the probability density to find phase points near the jumping threshold between the stochastic version of the attractors revealed by the bifurcation analysis above [Koseska *et al.* (2007a)]. The modes in the polymodal histogram might be separated by almost equal intervals if one of the stochastic attractors dominates over the others, or by different intervals in the opposite case. The same interplay between attractors disrupts the exponential decay of the peak amplitudes that is typical for a noisy attractor under the influence of a periodic signal [Longtin (1995)].

These results indicate that the interplay between intercell signaling and stochasticity might explain the emergence of quantized cycles, a concept that is central in the research of time-dependent biological processes, such as the cell cycle [Lloyd and Volkov (1990)]. Clear experimental evidence for quantized cycles has been obtained

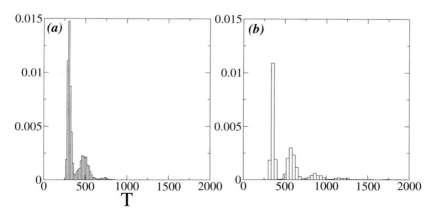

Fig. 3.12. (a) Bimodal ISI distribution for 8 identical oscillators ($\alpha_1 = 3.3$), and (b) polymodal ISI distribution ($\alpha_1 = 3.328$). The noise intensity is $5 \cdot 10^{-7}$.

for *Chinese hamster V*79 cells [Klevecz (1976)] and *wee1⁻ cdc25Δ* fission yeast cells [Sveiczer *et al.* (1996)], among others.

The variability in the system behavior can be significantly enhanced when the network becomes slightly inhomogeneous (due for instance to different α_1 values in different cells) in the presence of noise. Another important effect that arises in this system is the possibility to observe maximal variability for an optimal noise intensity. This is in contrast to the well-known effect of coherence resonance [Pikovsky and Kurths (1997)], where for intermediate noise intensities, maximal order can be achieved in systems with underlying nonlinear dynamics [Koseska *et al.* (2007b)]. The results also show that, although organized in a population, different oscillators are characterized by different ISI distributions, as a consequence of the specific, repulsive coupling considered.

3.4. Conclusions and Discussion

The concept of synthetic genetic networks is becoming increasingly exploited as a basic step to understand how cellular processes arise from the connectivity of genes and proteins. The ability of these circuits to produce different rhythms, as has been shown in this Chapter, could have important applications in functional genomics, gene and cell therapy, etc., since the multistability and multirhythmicity of synthetic genetic networks leads to an extended functionality, improved adaptation and ability to store information. On the other hand, one could more easily relate different biological phenomena and extract functional conclusions by observing a highly-adaptive synthetic genetic network, instead of a network producing a unified rhythm.

Here we have reviewed the possibility to use a modular coupling mechanism via quorum sensing, which leads to synchronization under realistic conditions in an ensemble of existing synthetic repressilators. By its design, the communication module can be added directly to existing repressilator strains and mimic natural multicellular clocks that operate on mean periods resulting from averaging multiple cells [Liu *et al.* (1997); Herzog *et al.* (1998); Honma *et al.* (1998); Nakamura *et al.* (2001); Herzog *et al.* (2004)]. Besides its efficiency, the synchronization reported here has been seen to lead to the generation of a global rhythm in a highly heterogeneous ensemble of genetic oscillators. The resulting clock behavior is seen to be highly robust to random phase drifts of the individual oscillators due to noise. In the light of these results, one might speculate whether natural biological clocks have evolved in this same way, i.e. by using inter-cell communication to couple an assembly of originally independent sloppy clocks. The cell-to-cell communication module can also be coupled with the individual genetic circuit in such a way that coupling is phase-repulsive [Ullner *et al.* (2007)].

Beside its biological consequences and extended functionality, the coupling mechanism discussed here leads to new phenomena from a general nonlinear dynamics viewpoint. First, the oscillation death (OD) described above is stable far from any Hopf bifurcation in a wide range of parameter space. This contrasts with other situations [Herrero *et al.* (2000); Wang *et al.* (2000)], where OD occurred only in a small range close to a Hopf bifurcation. Second, the phase-repulsive character of the coupling leads to multistability between the regimes of OD, IHLC and the single fixed point. The simultaneous availability of these different dynamical regimes to the cellular population improves its adaptability and robustness. Such an improved efficiency induced by coupling can probably exist in natural genetic networks, and can be definitely exploited in synthetic devices. The theoretical predictions reported here are amenable to experimental observation at the single-cell level via time-lapse fluorescence microscopy [Rosenfeld *et al.* (2005)]. This technique is very useful to experimentally test theoretical predictions in genetic networks [Süel *et al.* (2007)].

The results discussed here lead to several open questions in the field of synthetic biology of genetic networks. One of them is the influence of stochasticity arising from the small number of reactant molecules involved in gene regulation (sometimes around 1 mRNA molecule per cell in average), which can lead to significant fluctuations in intracellular mRNA and protein concentrations [Ozbudak *et al.* (2002); Elowitz *et al.* (2002)]. Hence it is important to understand how the variety of dynamical regimes discussed here will change in the presence of noise. Here one should distinguish intrinsic and extrinsic noise acting upon the gene regulation process [Swain *et al.* (2002)]. For the simulations with intrinsic noise usually the Gillespie algorithm is used [Gillespie (1977)], whereas in some situations the chemical Langevin equation approach can be employed [Gillespie (2000)]. In the system presented here, the dynamics can be expected to be quite complicated and counter-intuitive, if extrinsic noise leads to noise-induced ordering. It has been reported that noise may induce a bistable behaviour qualitatively different from what is possible deterministically [Samoilov *et al.* (2005)], induce stochastic focusing [Paulsson *et al.* (2000)], or increase the robustness of oscillations. Especially interesting would be to identify mechanisms through which noise-resistance appears due to the phase-repulsive property of the coupling. Taking into account the fact that stochastic effects in biomolecular systems have been recognized as a major factor, functionally and evolutionarily important, and that only a small amount of the recently discovered noise-induced phenomena in general dynamical systems have been identified in gene expression systems, this opens very wide perspectives for further research.

Another interesting question regards the influence of time delay on the phenomena discussed above. This issue has been discussed in single genetic oscillators [Chen and Aihara (2002)], where it has been seen that time delay generally increases the stability region of the oscillations, thereby making them more robust. In coupled oscillators, such as the ones discussed above, the effect of delay could be much more complicated. In particular, it was reported that delay in coupling may suppress synchronization without suppression of the individual oscillations [Rosenblum and Pikovsky (2004)]. Interestingly, delay in the coupling can seemingly change the coupling from phase-attractive to phase-repulsive and *vice versa*. Since the multistability and multirhythmicity described here are the result of phase-repulsive interaction, time delay can probably induce such effects also in systems with phase-attractive coupling. Even more interesting would be to investigate the combined effect of delay, intrinsic noise, and cell-cell coupling. Recently it was shown that time delay in gene expression can induce oscillations even when system's deterministic counterpart exhibits no oscillations [Bratsun *et al.* (2005)].

An important aspect of synthetic biology is the design of smart biological devices or new intelligent drugs, through the development of *in vivo* digital circuits [Weiss *et al.* (2001)]. If living cells can be made to function as computers, one could envisage, for instance, the development of fully programmable microbial robots that are able to communicate with each other, with their environment and with human operators. These devices could then be used, e.g., for detection of hazardous substances or even to direct the growth of new tissue. In that direction, pioneering experimental studies have shown the feasibility of programmed pattern formation [Basu *et al.* (2005)], and the possibility of implementing logical gates and simple devices within cells [Hasty *et al.* (2002)]. We identify three perspective directions of this research. First is the construction of new biological devices capable to solve or compute certain problems [see e.g. Haynes *et al.* (2008)]. A second direction would be the identification of new dynamical regimes with extended functionality using standard genetic parts, as we have discussed here. Finally, it should be possible to add more levels of control, e.g. spatiotemporal control [Basu *et al.* (2004)] or temporal light-dependent control via encapsulation [Antipov and Sukhorukov (2004)] for precise regulation of synthetic genetic oscillators.

Finally it is worth noting that the investigation of synthetic genetic oscillators can profit greatly from techniques and methods transferred from other fields of science. Two areas are particularly relevant in this context: neural and electronic networks. Both neural and genetic networks make use of feedback and coupling mechanisms, and are significantly noisy [Swain and Longtin (2006)]. However, neural networks have attracted in recent years much more attention than genetic networks from scientists working in nonlinear dynamics. Neuroscientists have access to relatively long and clean time series of neural activity; such type of data are only now beginning to appear for genetic systems. This outlines a promising future to the combination of efforts in these two fields. On the other hand, direct analogies can be drawn between synthetic biology and established techniques in electrical engineering [Hasty *et al.* (2002)]. As a test bed of complicated experiments in the implementation of complex gene networks, electronic circuits provide much easier possibilities to investigate complex networks with similar topology and demonstrating complex dynamical phenomena [Buldú *et al.* (2005)].

Acknowledgments

E.U. acknowledges financial support from the Alexander von Humboldt Foundation and from SULSA. Financial support for E.U. and J.G.O. was provided by the European Commission (GABA project, contract FP6-2005-NEST-Path-043309). A.K. and J.K. acknowledge the GoFORSYS project funded by the Federal Ministry of Education and Research Grant Nr. 0313924 and the Network of Excellence BioSim (contract No. LSHB-CT-2004-005137), funded by the European Commission. A.Z. acknowledges financial support from Volkswagen-foundation and the UCLH/UCL NIHR Comprehensive Biomedical Research Centre. E.V. from the Program Radiofizika (Russian Academy) and from RFBR Grant No. RFBR 08-02-00682. J.G.O. also acknowledges support from MEC (Spain, project FIS2006-11452 and I3 program).

References

Antipov, A. and Sukhorukov, G. (2004). Polyelectrolyte multilayer capsules as vehicles with tunable permeability, *Adv. Coll. Interface Sci.* **111**, pp. 49–61.

Atkinson, M. R., Savageau, M. A., Myers, J. T. and Ninfa, A. J. (2003). Development of genetic circuitry exhibiting toggle switch or oscillatory behavior in *escherichia coli*, *Cell* **113**, pp. 597–607.

Bar-Eli, K. (1985). On the stability of coupled chemical oscillators, *Physica D* **14**, pp. 242–252.

Basu, S., Gerchman, Y., Collins, C., Arnold, F. and Weiss, R. (2005). A synthetic multicellular system for programmed pattern formation, *Nature* **434**, pp. 1130–1134.

Basu, S., Mehreja, R., Thiberge, S., Chen, M. and Weiss, R. (2004). Spatiotemporal control of gene expression with pulse-generating networks, *Proc. Natl. Acad. Sci. U.S.A.* **101**, pp. 6355–6360.

Bratsun, D., Volfson, D., Tsimring, L. and Hasty, J. (2005). Delay-induced stochastic oscillations in gene regulations, *Proc. Natl. Acad. Sci. U.S.A.* **102**, pp. 14593–14598.

Buldú, J., García-Ojalvo, J., Wagemakers, A. and Sanjuán, M. (2005). Electronic design of synthetic genetic networks, *Int. J. Bif. Chaos* **17**, pp. 3507–3511.

Chen, L. and Aihara, K. (2002). A model of periodic oscillations for genetic regulatory systems, *IEEE Trans. Circ. Syst.* **49**, pp. 1429–1436.

Crowley, M. F. and Epstein, I. R. (1989). Experimental and theoretical studies of a coupled chemical oscillator: phase death, multistability and in-phase and out-of-phase entrainment, *J. Phys. Chem.* **93**, 6, pp. 2496–2502.

Dockery, J. D. and Keener, J. P. (2001). A mathematical model for quorum sensing in pseudomonas aeruginosa, *Bull. Math. Biol.* **63**, pp. 95–116.

Dolnik, M. and Marek, M. (1988). Extinction of oscillations in forced and coupled reaction cells, *J. Phys. Chem.* **92**, 9, pp. 2452–2455.

Elowitz, M. and Leibler, S. (2000). A synthetic oscillatory network of transcriptional regulators, *Nature* **403**, pp. 335–338.

Elowitz, M., Levine, A., Siggia, E. and Swain, P. (2002). Stochastic gene expression in a single cell, *Science* **297**, pp. 1183–1186.

Enright, J. (1980). Temporal precision in circadian systems: a reliable neuronal clock from unreliable components? *Science* **209**, 4464, pp. 1542–1545.

Ermentrout, B. (2002). *Simulating, analyzing and animating dynamical systems: a guide to XPPAUT for researchers and Students* (SIAM).

Fung, E., Wong, W. W., Suen, J. K., Bulter, T., Lee, S.-G. and Liao, J. C. (2005). A synthetic gene-metabolic oscillator, *Nature* **435**, pp. 118–122.

Fuqua, C. and Greenberg, P. (2002). Listening in on bacteria: acyl-homoserine lactone signaling, *Nat. Rev. Mol. Cell Biol.* **3**, pp. 685–695.

García-Ojalvo, J., Elowitz, M. B. and Strogatz, S. H. (2004). Modeling a synthetic multicellular clock: Repressilators coupled by quorum sensing, *Proc. Natl. Acad. Sci. U.S.A.* **101**, 30, pp. 10955–10960.

Gardner, T. S., Cantor, C. R. and Collins, J. J. (2000). Construction of a genetic toggle switch in *escherichia coli, Nature* **403**, pp. 339–342.

Gillespie, D. (1977). Exact stochastic simulation of coupled chemical reactions, *J. Phys. Chem.* **81**, pp. 2340–2361.

Gillespie, D. (2000). The chemical langevin equation, *J. Chem. Phys.* **113**, pp. 297–306.

Golomb, D., Hansel, D., Shraiman, B. and Sompolinsky, H. (1992). Clustering in globally coupled phase oscillators, *Phys. Rev. A* **45**, 6, pp. 3516–3530.

Hasty, J., McMillen, D. and Collins, J. J. (2002). Engineered gene circuits, *Nature* **420**, pp. 224–230.

Haynes, K., Broderick, M., Brown, A., Butner, T., Dickson, J., Harden, W. L., Heard, L., Jessen, E., Malloy, K., Ogden, B., Rosemond, S., Simpson, S., Zwack, E., Campbell, A. M., Eckdahl, T., Heyer, L. and Poet, J. (2008). Engineering bacteria to solve the burnt pancake problem, *J. Biol. Eng.* **2**, 1.

Herrero, R., Figueras, M., Rius, J., Pi, F. and Orriols, G. (2000). Experimental observation of the amplitude death effect in two coupled nonlinear oscillators, *Phys. Rev. Lett* **84**, 23, pp. 5312–5315.

Herzog, E. D., Aton, S. J., Numano, R., Sakaki, Y. and Tei, H. (2004). Temporal precision in the mammalian circadian system: A reliable clock from less reliable neurons, *J. Biol. Rhythms* **19**, pp. 35–46.

Herzog, E. D., Takahashi, J. S. and Block, G. D. (1998). Clock controls circadian period in isolated suprachiasmatic nucleus neurons, *Nature Neurosci.* **1**, pp. 708–713.

Honma, S., Shirakawa, T., Katsuno, Y., Namihira, M. and ichi Honma, K. (1998). Circadian periods of single suprachiasmatic neurons in rats, *Neurosci. Lett.* **250**, pp. 157–160.

Kaneko, K. and Yomo, T. (1994). Cell division, differentiation and dynamic clustering, *Physica D* **75**, pp. 89–102.

Kiss, I. Z. and Hudson, J. L. (2003). Chaotic cluster itinerancy and hierarchical cluster trees in electrochemical experiments, *Chaos* **13**, pp. 999–1008.

Klevecz, R. R. (1976). Quantized generation time in mammalian cells as an expression of the cellular clock, *Proc. Natl. Acad. Sci. USA* **73**, pp. 4012–4016.

Kobayashi, H., Kaern, M., Araki, M., Chung, K., Gardner, T. S., Cantor, C. R. and Collins, J. J. (2004). Programmable cells: Interfacing natural and engineered gene networks, *Poc. Natl. Acad. Sci. U.S.A.* **101**, pp. 8414–8419.

Koseska, A., Volkov, E., Zaikin, A. and Kurths, J. (2007a). Inherent multistability in arrays of autoinducer coupled genetic oscillators, *Phys. Rev. E* **75**, 3, p. 031916(8).

Koseska, A., Zaikin, A., García-Ojalvo, J. and Kurths, J. (2007b). Stochastic suppression of gene expression oscillators under intercell coupling, *Phys. Rev. E* **75**, p. 031917.

Kuznetsov, A., Kærn, M. and Kopell, N. (2004). Synchrony in a population of hysteresis-based genetic oscillators, *SIAM J. Appl. Math.* **65**, 2, pp. 392–425.

Liu, C., Weaver, D. R., Strogatz, S. H. and Reppert, S. M. (1997). Cellular construction of a circadian clock: Period determination in the suprachiasmatic nuclei, *Cell* **91**, pp. 855–860.

Lloyd, D. and Volkov, E. I. (1990). Quantized cell cycle times: interaction between a relaxation oscillator and ultradian clock pulses, *BioSystems* **23**, pp. 305–310.

Longtin, A. (1995). Mechanisms of stochastic phase-locking, *Chaos* **5**, pp. 209–215.

McMillen, D., Kopell, N., Hasty, J. and Collins, J. J. (2002). Synchronizing genetic relaxation oscillators by intercell signaling, *Proc. Natl. Acad. Sci. U.S.A.* **99**, 2, pp. 679–684.

Miller, M. B. and Bassler, B. L. (2001). Quorum sensing in bacteria, *Annu. Rev. Microbiol.* **55**, pp. 165–199.

Miyakawa, K. and Yamada, K. (2001). Synchronization and clustering in globally coupled salt-water oscillators, *Physica D* **151**, pp. 217–227.

Nakamura, W., Honma, S., Shirakawa, T. and ichi Honma, K. (2001). Regional pacemakers composed of multiple oscillator neurons in the rat suprachiasmatic nucleus, *Eur. J. Neurosci.* **14**, pp. 666–674.

Needleman, D. J., Tiesinga, P. H. E. and Sejnowski, T. J. (2001). Collective enhancement of precision in networks of coupled oscillators, *Physica D* **155**, pp. 324–336.

Okuda, K. (1993). Variety and generality of clustering in globally coupled oscillators, *Physica D* **63**, pp. 424–436.

Osipov, G. V., Kurths, J. and Zhou, C. (2007). *Synchronization in Oscillatory Networks*, Springer Series in Synergetics (Springer, Berlin, Germany).

Ozbudak, E., Thattai, M., Kurtser, I., Grossman, A. and van Oudenaarden, A. (2002). Regulation of noise in the expression of a single gene, *Nature Genet.* **31**, pp. 69–73.

Paulsson, J., Berg, O. and Ehrenberg, M. (2000). Stochastic focusing: fluctuation-enhanced sensitivity of intracellular regulation, *Proc. Natl. Acad. Sci. U.S.A.* **97**, 13, pp. 7148–7153.

Pikovsky, A. S. and Kurths, J. (1997). Coherence resonance in a noise-driven excitable system, *Phys. Rev. Lett.* **78**, pp. 775–778.

Riedel-Kruse, I. H., Muller, C. and Oates, A. C. (2007). Synchrony dynamics during initiation, failure, and rescue of the segmentation clock, *Science* **317**, 5846, pp. 1911–1915.

Rosenblum, M. and Pikovsky, A. (2004). Controlling synchronization in an ensemble of globally coupled oscillators, *Phys. Rev. Lett.* **92**, p. 114102.

Rosenfeld, N., Young, J. W., Alon, U., Swain, P. S. and Elowitz, M. B. (2005). Gene regulation at the single-cell level, *Science* **307**, pp. 1962–1965.

Samoilov, M., Plyasunov, S. and Arkin, A. (2005). Stochastic amplification and signalling in enzymatic futile cycles through noise-induced bistability with oscillations, *Proc. Natl. Acad. Sci. U.S.A.* **102**, pp. 2310–2315.

Somers, D. and Kopell, N. (1995). Waves and synchrony in networks of oscillators of relaxation and non-relaxation type, *Physica D* **89**, pp. 169–183.

Sprinzak, D. and Elowitz, M. B. (2005). Reconstruction of genetic circuits, *Nature* **438**, pp. 443–448.

Süel, G. M., Kulkarni, R. P., Dworkin, J., García-Ojalvo, J. and Elowitz, M. B. (2007). Tunability and noise dependence in differentiation dynamics, *Science* **315**, 5819, pp. 1716–1719.

Sveiczer, A., Novak, B. and Mitchison, J. M. (1996). The size control of fission yeast revisited. *J. Cell Sci.* **109**, pp. 2947–2957.

Swain, P., Elowitz, M. and Siggia, E. (2002). Intrinsic and extrinsic contributions to stochasticity in gene expressions, *Proc. Natl. Acad. Sci. U.S.A.* **99**, pp. 12795–12800.

Swain, P. and Longtin, A. (2006). Noise in genetic and neural networks, *Chaos* **16**, p. 026101.

Tsaneva-Atanasova, K., Zimliki, C. L., Bertram, R. and Sherman, A. (2006). Diffusion of calcium and metabolites in pancreatic islets: Killing oscillations with a pitchfork, *Biophys J.* **90**, pp. 3434—3446.

Tyson, J. and Kauffman, S. (1975). Control of mitosis by a continuous biochemical oscillation: Synchronization; spatially inhomogeneous oscillations, *J. Math. Biol.* **1**, pp. 289–310.

Ullner, E., Buceta, J., Diez–Noguera, A. and Garcia–Ojalvo, J. (2009). Noise-induced coherence in multicellular circadian clocks, *Biophys. J.* **96**, pp. 3573–3581.

Ullner, E., Koseska, A., Kurths, J., Volkov, E., Kantz, H. and García-Ojalvo, J. (2008). Multistability of synthetic genetic networks with repressive cell-to-cell communication, *Phys. Rev. E* **78**, p. 031904.

Ullner, E., Zaikin, A., Volkov, E. I. and García-Ojalvo, J. (2007). Multistability and clustering in a population of cellular genetic oscillators via phase repulsive cell-to-cell communication, *Phys. Rev. Lett.* **99**, p. 148103.

Volkov, E. and Stolyarov, M. (1991). Periodic attractors in two coupled relaxation oscillators, *Phys. Lett. A* **59**, pp. 61–66.

Volkov, E. and Stolyarov, M. (1994). Temporal variability in a system of coupled mitotic timers, *Biol. Cybern.* **71**, pp. 451–459.

Volkov, E. I. and Romanov, V. A. (1995). Bifurcations in the system of two identical diffusively coupled brusselators, *Physica Scripta* **51**, pp. 19–28.

Wang, W., Kiss, I. Z. and Hudson, J. L. (2000). Experiments on arrays of globally coupled chaotic electrochemical oscillators: Synchronization and clustering, *Chaos* **10**, pp. 248–256.

Wang, W., Kiss, I. Z. and Hudson, J. L. (2001). Clustering of arrays of chaotic chemical oscillators by feedback and forcing, *Phys. Rev. Lett.* **86**, pp. 4954–4957.

Weiss, R., Homsy, G. and Knight Jr., T. (2001). Toward *in vivo* digital circuits, in L. Landweber and E. Winfree (eds.), *Evolution as Computation (DIMACS Workshop)* (Springer-Verlag), pp. 275–295.

You, L., Cox III, R. S., Weiss, R. and Arnold, F. H. (2004). Programmed population control by cell-cell communication and regulated killing, *Nature* **428**, pp. 868–871.

Chapter 4

Boolean Networks in Inference and Dynamic Modeling of Biological Systems at the Molecular and Physiological Level

Juilee Thakar and Réka Albert

Department of Physics, 104 Davey Laboratory, Pennsylvania State University, University Park, PA 16802, USA

4.1. Introduction

Mathematical modeling is a valuable tool for understanding biological systems, especially since vast amounts of data are getting produced with increasing technical efficiency. Modeling frameworks range from empirical methods driven by the analysis of the available data to mechanistic models encapsulating prior knowledge. The latter models are usually classified along two axes: from discrete to continuous based on the resolution of the variables, and from deterministic to stochastic based on the incorporation of fluctuations and noise (Fig. 4.1). We will focus in this chapter on discrete and continuous-discrete hybrid models, with particular focus on the Boolean formalism that assumes only two levels for each variable. Boolean modeling methods are used both to infer causal relationships from expression data and to analyze the dynamic behavior of systems whose network of interactions is known.

Theoretical and computational analysis of biochemical networks has a long history (Heinrich *et al.* 1996; Voit 2000; Bower 2001; Fall 2002). Most theoretical models focus on experimentally well-studied cellular pathways that have relatively few components, and they use differential equations based on mass-action (or more general) kinetics for the production and decay of components (Tyson, Novak *et al.* 1996; Barkai and Leibler 1997; Spiro, Parkinson *et al.* 1997; Ciliberto and Tyson 2000; Tyson, Chen *et al.* 2001; Hoffmann, Levchenko *et al.* 2002; Dillon, Gadgil *et al.* 2003; Lee, Salic et al. 2003). The drawback of the differential equation-based method is the necessity to know the kinetic parameters of the underlying reactions. Parameter estimation or systematic search in parameter space is usually employed to find general features

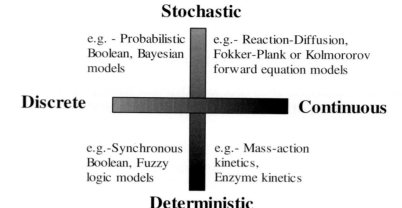

Fig. 4.1. Classification of mechanistic methods for modeling dynamics of biological systems. In each quadrant we have given some of the common examples of modeling methods used in that category. Most of the methods offer flexibility to incorporate resolution required for the specific questions under examination.

that are maintained in a range of possible values. The kinetic details of molecular and cellular interactions are rarely known. There is increasing evidence, however, that the input-output curves of many regulatory relationships are strongly sigmoidal and can be approximated by step functions (von Dassow *et al.* 2000; Bower 2001). Moreover, several models and experiments suggest that regulatory networks maintain their function even when faced with fluctuations in components and reaction rates (Alon, Surette *et al.* 1999; von Dassow, Meir *et al.* 2000; Eldar, Dorfman et al. 2002; Conant and Wagner 2004; Csete and Doyle 2004). These observations lend support to the applicability of Boolean and other qualitative models .

Boolean modeling is a top-down approach that describes the regulation between key players of the system and does not explicitly incorporate the underlying biochemical details. Such a method is a powerful way to convert a blackboard description of the biological system into a mathematical model. Though it is simple in the manner it describes each component, the interplay between the components of the system leads to rich emergent dynamic behaviors.

Boolean dynamic models have yielded significant insights into the behavior of complex biological systems and into understanding the evolutionary principles of biological networks. Most biological systems described by Boolean models are gene regulatory networks but one of the advantages of such coarse-grained method is that it can be easily extended to study systems at the physiological level. The system under consideration is described by its components such as proteins, transcription factors, cells, and by the relationships between those components, for example interaction between two proteins, activation of a transcription factor, or regulation of gene expression by a transcription factor. Such system can be represented by a network, where nodes of the network represent components and edges describe the processes connecting those nodes. The nodes are characterized by binary states, and the relationships among the nodes are incorporated into updating functions for the nodes' states. Boolean modeling has been pioneered by S. Kauffman and R. Thomas in the middle of the 20th century. Due to the lack of information about the architecture of biological networks most of the early research focused on the generic properties of networks governed by Boolean dynamics. The post-genomic revolution brought with it a resurgence of Boolean modeling and its successful application to a variety of biological systems.

4.2. Boolean Network Concepts and History

In a Boolean model, each node is assumed to have one of two states, denoted ON (1) and OFF (0). The ON state can correspond to a high gene expression, high concentration, open channel, or active transcription factor; the OFF state corresponds to low expression or concentration, closed channel or inactive transcription factor. The future state of each node is given by a Boolean function B_i, describing the regulation of node X_i by the other nodes, $X_i^{t+1} = B_i(X_1^t, X_2^t, ..., X_k^t)$. B_i is a statement whose inputs are the nodes that have edges directed toward node i (i.e. the regulators of i), and whose output is 1 (ON) or 0 (OFF). There are two main frameworks for Boolean models. In the first B_i is based on the Boolean (logical) operators AND, OR and NOT. In the second framework, called threshold Boolean networks (Derrida 1987; Kuerten 1988) B_i is a statement comparing the weighted sum of input signals to a node-specific threshold value. In both cases the function's output is 1(0) if the corresponding statement is true (false). In most Boolean models time is assumed to be quantized into discrete time-steps, thus the future state means the state at the next time-step. Starting from a given initial condition, the network then produces a dynamical sequence of network states, eventually reaching a periodic attractor (limit cycle) or a fixed point (Fig. 4.2). All initial conditions that evolve to a given attractor constitute its basin of attraction.

Most of the theory of Boolean networks was developed in the context of Random Boolean Networks (RBNs, also known as N-K models or Kauffman networks) introduced by S. Kauffman in 1969. RBNs are generic, because one

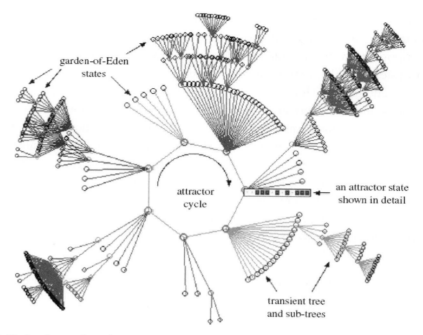

Fig. 4.2. Basin of attraction of a dynamical attractor of a random Boolean network. Network states (circles) and transitions between them are shown, which eventually reach a periodic attractor cycle. Some network states do not have any precursor state (garden-of-Eden states). Most states are transient states and form tree-like patterns of transient flows towards the attractor (adapted from Wuensche (1994)).

does not assume any particular functionality or connectivity for the nodes. In a RBN with N nodes and K inputs the state of each node in the network at time $t+1$ is determined by the states of its K inputs at time t through a randomly generated Boolean function. Typically the Boolean functions do not change throughout the lifetime of the network. The Boolean function for each node maps each of the 2^K possible input state combinations to an output state of 0 or 1 and can be represented with a look-up table. The expected attractor length of an RBN depends on the topology of the network. Early studies on random networks found that below a critical connectivity (average number of incoming links per node) $K_c = 2$ the network decouples into many disconnected regions, resulting in short transients and short attractors in its dynamics. The attractors are robust to small changes in initial conditions or to small perturbations in the state and hence in this regime the system can be classified as ordered. Above the connectivity K_c any local signal will initiate an avalanche of activity that may propagate throughout most of the system. This sensitivity to small perturbations led to the classification of this regime as chaotic. In this regime transients as well as attractor cycles tend to become quite long, their lengths increaseing exponentially with the number of nodes involved. A prominent feature of the dynamics of networks close to the critical connectivity is the relatively small number and intermediate length of attractors compared with the 2^N possible states of the network (Fig. 4.3). This feature motivated the hypothesis that a similar mechanism potentially could stabilize the macro-states of cellular regulation such as cell types (Kauffman 1993). While the notion of criticality is only well defined for random networks, it has long been argued that an intermediate range of activity is particularly suitable for efficient information processing.

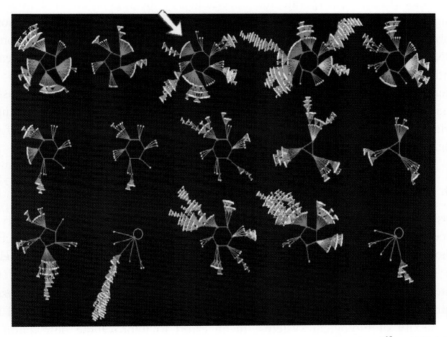

Fig. 4.3. The full state space of a random Boolean network with N = 13 nodes: 2^{13} = 4192 initial states each flow into one of 15 attractors (adapted from Wuensche (1994)). The basin of attraction marked with an arrow is the one shown in Fig. 4.2.

The attractors of gene regulatory networks can correspond to distinct cellular states such as differentiation, proliferation or to cell types; the attractors of a signal transduction network correspond to the expected response(s) to the presence or absence of a given signal (Platt and Reece 1998; Irie, Mattoo *et al.* 2004; Boczko, Cooper *et al.* 2005). Kauffman argues that life must exist on the edge of chaos such that the networks representing real genetic regulatory networks operate at the boundary between order and chaos. Phase transitions between different regimes can be analyzed by measuring the effect of perturbations, the sensitivity to initial conditions and damage spreading (Shmulevich *et al.* 2005).

The analysis of information processing through biological networks described by continuous and binary models has revealed characteristic network topologies leading to reproducible behavior in the presence of noise (Klemm *et al.* 2005). Robustness in the face of random perturbations is an important feature of biological networks. Often RBNs fail to reproduce this feature at the biologically observed connectivity. The extension of RBNs into probabilistic Boolean Networks (PBNs) incorporate the uncertainty in the regulation of a node by assigning several Boolean functions/predictors to each node, each with some probability of being chosen to advance the state of the node to which it belongs. PBNs are also more robust because the predictors are probabilistically synthesized so that each predictor's contribution is proportional to its determinative potential. Another way to introduce robustness is by using canalizing functions. This type of Boolean function in which one of the input variables is able to determine the function output regardless of the values of the other variables leads to orderly behavior (Kauffman *et al.* 2004).

4.3. Extensions of the Classical Boolean Framework

Most traditional Boolean models consider time quantized into regular intervals (time steps) and only two states, expressed/active and not expressed/ inactive (Thomas 1973; Kauffman 1993; Kauffman, Peterson *et al.* 2003). Such models are known as synchronous Boolean models because the components of the system are updated simultaneously in the algorithm; assuming that all processes require same time. Often variations of this approach are required to model the complexity of biological systems. In an effort to improve the description of variability in the durations of synthesis and decay processes, several asynchronous algorithms have been proposed. For example, each node can be assigned an individual time unit and be updated at multiples of this time unit $t_i^k = k\gamma_i$. The update function can be given as $X_i^{t_i} = B_i(X_1^{t_1}, X_2^{t_2}, ..., X_k^{t_k})$, where t_k represents the time of the last update of node k before time t_i. Known information on the timings of the processes can be included in the model by setting up inequalities of γ, e.g. in bacterial infections the fact that epithelial cells are activated before dendritic cells can be incorporated as $\gamma_{EC} < \gamma_{DC}$. A variety of asynchronous and synchronous algorithms have been studied for RBNs. Figure 4.4 shows their classification according to their updating scheme (Gershenson 2004).

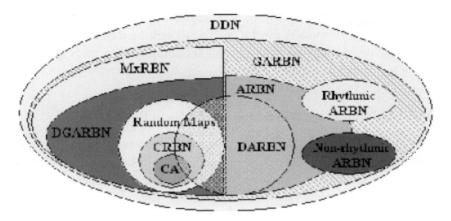

Fig. 4.4. Classification of random Boolean networks according to their updating scheme. DDNs – Discrete Dynamic Networks, MxRBNs – Mixed context RBNs, GARBNs – Generalized Asynchrnous RBNs, ARBNs – Asynchronous RBNs, DGARBN – Deterministic Generalized Asynchronous RBNs, CRBN – classical RBNs, CA – Cellular Automata, DARBNs – Deterministic Asynchronous RBNs. Figure adapted from (Gershenson 2004)

The variations to describe the activity includes using more than two states, incorporating weights to define the strengths of the input interactions and implementing independent thresholds for each components. The hybrid Boolean approach, also known as piece-wise linear formalism (Glass *et al.* 1973), combines Boolean synthesis with continuous decay. Each node is characterized by two variables, a continuous variable akin to a concentration and a discrete variable akin to an activity. The rate of change of the continuous variable is $dx_i/dt = B_i(X_1^t, X_2^t..X_k^t) - x_i$, and the Boolean variable X_i switches states when the continuous variable x_i passes a threshold θ. The limiting values 0 and 1 of the continuous variable represent, respectively, the absence of species i and maximal concentration of species i. With all this said it's important to note that all the above variations maintain the fixed points (steady states) of the traditional Boolean models.

Qualitative models such as Boolean network models have the common feature of most closely reflecting the topology of the regulatory network based on experimental data and of describing the signal processing through the network. During the network assembly perturbations can be performed to verify the network topology by comparing the outcome of the perturbed network to the experimental observations. Perturbations often include permanent silencing or activation of the nodes of the network which can be directly compared to knock-out or over-expression experiments respectively. Thus in this chapter first we will discuss network inference, mostly of gene regulatory networks. Second, we will describe how Boolean models can be used to model the dynamics of systems in plant biology, developmental biology and immunology. In each case we will discuss the methods followed by examples.

4.4. Boolean Inference Methods and Examples in Biology

Computational inference aims to extract causal relationships from experimental data and to construct a network expressing these relationships. Inference of cellular networks allows for a clearer comprehension of the inner machinery of the cell, and when combined with modeling, can also be used to make experimentally-verifiable predictions about cellular networks. A variety of computational methods for network inference exists; choosing a specific computational method depends on the nature of the data from which inferences will be made, on the type of network under consideration, on the features of the system one would most like to illuminate, and on the amount of computational time available to the researcher. Boolean methods attempt to infer causative relationships unlike empirical methods which infer associations (Fig. 4.5). Boolean methods are usually more computationally tractable than continuous (differential equation based) methods and are well suited to systems with limited known information and large size.

Deterministic Boolean methods employed for the inference of gene-regulatory networks from time-course gene expression (microarray) data seek to define a Boolean function for each gene so that the state transition tables of the corresponding synchronous Boolean network resemble the time-series pattern of the system (Smolen *et al.* 2000; Shmulevich *et al.* 2002).

Each node's logical function is found by determining the minimum set of nodes whose (changing) expression levels can explain the observed changes in state of the given node in all experimental trials. Generally, an optimization technique, such as the Coefficient of Determination (Dougherty *et al.* 2000), is employed for this inference. It is possible that more than one minimum set may be found for a particular node, and, in this case, multiple networks explain the experimental observations. A recent analysis of the attractors of multiple solutions found consistent dynamics over all solutions and relatively few fixed points (Martin *et al.* 2007). The search for the set of the nodes whose expression levels explain the observed change in the expression of a particular node can be augmented by employing probabilistic Boolean methods (Shmulevich *et al.* 2002; Dougherty *et al.* 2003) which incorporate uncertainty by assigning several

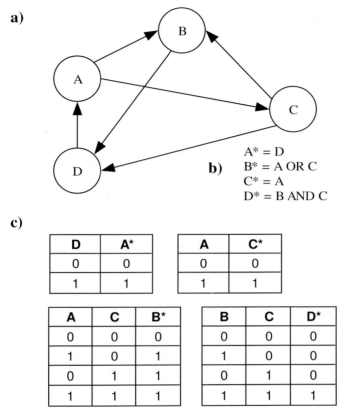

a)

b)
A* = D
B* = A OR C
C* = A
D* = B AND C

c)

D	A*
0	0
1	1

A	C*
0	0
1	1

A	C	B*
0	0	0
1	0	1
0	1	1
1	1	1

B	C	D*
0	0	0
1	0	0
0	1	0
1	1	1

Fig. 4.5. A simple Boolean network. a) Wiring diagram. The edges represent regulatory relationships, such that the state of the node at the endpoint of the edge depends on the state of the node at the beginning of the edge. b) Logical Boolean rules. The * marks the future state of the marked node. c) State look-up table for each node. The left column(s) represent the input state of the regulating node(s), and the right column represents the output state of the node which is connected by the incoming arrow to the regulating node(s).

Boolean functions to each node. The random selection of Boolean functions can be performed once at the beginning or at each time step (Derrida *et al.* 1986) by a machine learning algorithm. The dynamics of state transitions is not dependent on the methods of selection of the functions mentioned above. Algorithms such as REVEAL (REVerse Engineering Algorithm) (Liang *et al.* 1998) offer promising first steps towards large-scale network inference. In the synchronous Boolean algorithm of REVEAL, dependencies between the expressions of genes are calculated by using a quantitative information measure so that the particular rule for a gene can describe its state transitions in the time-series data.

A study of T cell proliferation in response to stimulation by the cytokine IL2 inferred the interactions among co-expressed gene clusters. The authors consider all possible networks matching the data instead of requiring enough data to infer a unique network (Martin *et al.* 2007). 99.4% of the total inferred networks had a single fixed point (steady-state). In the remaining 0.6% of the cases the cyclin-dependent kinase inhibitor was observed to fluctuate, leading to a three-step cycle. The resulting network topology elucidated the regulation between early, intermediate and late genes activated upon IL2 stimulation. A regulatory network involved in embryonic segmentation and muscle development in *D. melanogaster* was recently produced (Zhao *et al.* 2006) using an inference method combining minimum description length principle (MDL) (Rissanen 1978) with a probabilistic Boolean method. MDL reduces the search space of possible networks, which is usually high in PBNs, giving a good trade-off between modeling complexity and data-fitting. The algorithm developed in (Zhao *et al.* 2006) is useful for inferring temporal regulation and analyzing time-series datasets. Thus probabilistic Boolean networks are attractive in that they maintain the large-scale inference ability of standard Boolean methods while relaxing the determinism of the basic method.

4.6. Dynamic Boolean Models: Examples in Plant Biology, Developmental Biology and Immunology

While inference methods extract a network of interactions based on the observed data, such networks serve as an input to dynamic models. Inputs to a dynamic model include (i) the interactions and regulatory relationships between components (i.e. the network), (ii) how the strength of the interactions depends on the state of the interacting components (i.e. the state transfer functions) and (iii) the initial state of each component in the system. Given these, the model will output the time evolution of the state of the system.

To analyze the dynamics of a system of interest we can either use inferred networks or assemble a network from independent studies. In case of the latter, the topology can be verified by performing known perturbations and comparing the outcome to the result of the experimentally known mutant. In the process of network assembly inference rules need to be employed to represent the regulatory relationships concluded by various independent experimental observations since the experimentally observed relationship is often not a direct interaction (Fig. 4.6) (Li *et al.* 2006).

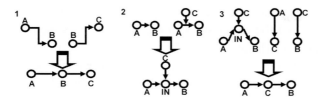

Fig. 4.6. Inference rules used in assembling component-to-component and component-to-pathway relationships in (Mendoza *et al.* 1998; Mendoza *et al.* 1999). These rules incorporate node to pathway relationships by introducing unknown intermediary nodes, then collapse the intermediary nodes with known nodes in the system if additional information is available. Figure adapted from (Li *et al.* 2006).

Experimental data is also used to design the Boolean/ transition rules of the nodes. The operator NOT is used to model an inhibition interaction between two components. The operator OR is used when two positive regulators can independently activate a target node whereas the operator AND is used when both nodes are required for the activation of the target node. Various types of interactions and regulatory relationships can be described by using these three operators. Below we will review examples of Boolean models used to simulate the dynamics of various biological systems.

Plant Biology: Signaling pathways and gene regulatory networks in plants are interesting examples of the complex regulation of genes during morphogenesis and in response to environmental signals. The development of a complex structure such as a flower has been successfully described with Boolean models. Network assembly and a Boolean model of the genes involved in *Arabidopsis thaliana* flower morphogenesis indicated six attractors, four of which correspond to the observed patterns of gene expression in four floral parts, petals, sepals, stamens and carpels (Mendoza *et al.* 1998). The fifth attractor corresponds to the cells that are not competent to flower and the sixth attractor, though not found in wild-type plants, can be induced experimentally. The model incorporates relative interaction weights and activation thresholds in addition to the Boolean rules describing regulation of each node, so that a gene is active if the weighted input of all the genes regulating it exceeds its activation threshold. A follow up study of this network identified the regulatory constraints and functional feedback circuits that lead to stable gene expression patterns (Mendoza *et al.* 1999). An extension (Espinosa-Soto *et al.* 2004) of the network published by (Li *et al.* 2006) implemented two expression levels for eight nodes and three for seven nodes. Simulations with all possible initial conditions converged to a few steady gene activity states that match gene expression profiles observed experimentally in primordial floral organ cells of wild-type and mutant plants. Furthermore, the authors also detected the differences in the network architecture of *Arabidopsis thaliana* and Petunia hybrids.

In another study the various pathways activated by abscisic acid (ABA) and leading to stomatal closure were synthesized into an asynchronous Boolean model (Li *et al.* 2006). In this model nodes depict a variety of biological molecules/components such as cytosolic Ca^{2+}, ion channels, proteins. The model identified and verified components leading to ABA hyposensitivity (Gosti *et al.* 1999; Merlot *et al.* 2001), insensitivity or impaired closure (Jacob *et al.* 1999; Wang *et al.* 2001; Coursol *et al.* 2003) as illustrated on Fig. 4.7.

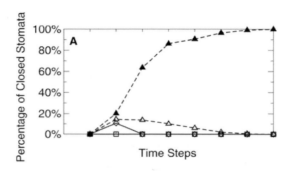

Fig. 4.7. Example output from the asynchronous Boolean model of ABA induced closure. In this model all nodes are updated in a randomly selected order during each time step. The y axis represents the percentage of the ON state of the node Closure in a large number of replicate simulations. Black triangles with dashed lines represent the normal (wild-type) response to ABA stimulus. Open triangles with dashed lines show that in wild-type, the probability of closure decays in the absence of ABA. Perturbations in depolarization (open diamonds) or anion efflux at the plasma membrane (open squares) cause total loss of ABA-induced closure. Figure adapted from (Li *et al.* 2006)

Developmental Biology: Drosophila embryonic development (Sanchez *et al.* 2001; Ghysen *et al.* 2003; Sanchez *et al.* 2003) has been studied frequently by Boolean modeling due to the well established topology of the network and the classifiable morphological effects of the network perturbations. During the initial stages of development of the fruit fly *Drosophila melanogaster*, three families of genes are successively activated (Sanson 2001): the *gap* genes, the *pair-rule* genes and the *segment polarity genes*. In two consecutive papers Sanchez and collaborators studied the *gap* and *pair-rule* cross regulatory modules respectively. The authors implemented multiple levels of activity and multiple thresholds wherever required. For example, in the case of the pair-rule module, the effect of bell shaped distribution of the concentration of the even-stripped protein was implemented by using three different thresholds levels.

Albert and Othmer (Albert *et al.* 2003) proposed a Boolean model of the segment polarity gene network whose expression is initiated by the *pair-rule* module. The synchronous model was able to reproduce the wild type gene expression patterns, as well as the ectopic expression patterns observed in over-expression experiments and various mutants, and determined that the system can have six fixed points (steady states). The model gives important insights into the network's ability to correct errors in the pre-pattern. An asynchronous model (Chaves 2005) of the same network showed that very variable synthesis and decay times can perturb the wild type pattern. Starting from the wild type initial condition, all six steady states (Albert *et al.* 2003), including mutant patterns, may occur with a certain frequency, with the incidence of the wild type steady state being 57%. On the other hand, a separation of timescales between pre- and posttranslational processes and a minimal pre-pattern ensure convergence to the wild type expression pattern regardless of fluctuations. A consequent piece-wise linear model that assumes individual activation thresholds and individual node timescales found that 100% reproducibility of the wild-type developmental process is ensured by two assumptions, (i) a separation of mRNA and protein half-lives, and (ii) an activation threshold less than 0.5. These assumptions are supported by evidence from sea urchin development (Davidson 2001).

The threshold Boolean model described in the introduction has been used to study the cell-cycle regulatory network of the budding yeast, *Saccharomyces cerevisiae*. The model can simulate the temporal evolution of protein states corresponding to the sequence of phases in the cell cycle, namely G1, S, G2 and M, before reaching a large basin of attraction corresponding to the G1 phase in which the cell grows and under appropriate conditions commits to division (Li *et al.* 2004). The states corresponding to the biological cell cycle checkpoints have large basins of attraction and the pathway through these checkpoints is an attracting trajectory. Stability analysis of this model and its comparison with random networks further support the conclusion that the cell cycle is robustly designed and is stable against perturbations. (Li *et al.* 2004). Recently (Davidich *et al.* 2008) transformed an existing continuous (ordinary differential equation based) model of the fission yeast cell-cycle (Novak *et al.* 2001) into a Boolean model in order to compare the two approaches. The Boolean model correctly reproduced the sequence of events and the steady states of the system and its robustness to perturbations of the initial conditions. The results confirm the idea that some molecular control networks are so robustly designed that timing is not a critical factor (Braunewell *et al.* 2007). The comparison between threshold

Boolean models of the fission yeast cell-cycle in two yeast species reveals their difference in circuitry and dynamics. While the *S. cerevisiae* system operates in a strongly damped mode, driven by external excitation, the *S. pombe* network represents an auto-excited system with external damping (Davidich *et al.* 2008).

Immunology: In immunology the interplay between various immune components such as cytokines and cells is crucial in deciding the outcome of the immune response against a particular pathogen. While gene regulatory networks determine cell types and signal transduction networks determine the response of particular cell types to cytokines, the interplay between different immune components decide the particular gene regulatory networks and signal transduction networks that will be activated. We will expand here on a Boolean model of the immune response against a bacterial pathogen in which diverse components describing interactions at different scales are assembled into a network. The immune response involves activation of various components such as immune cells, signaling molecules called cytokines and proteins such as complement. We have studied the activation of immune components and their regulation in response to respiratory pathogens of genus *Bordetella* (Thakar *et al.* 2007) by using an asynchronous Boolean model. We particularly compared the immune response to two pathogens, *B. bronchiseptica* and *B. pertussis*. *B. pertussis* is the causative agent of whooping cough in humans and has evolved from a *B. bronchiseptica* like progenitor. Nevertheless, *B. bronchiseptica* and *B. pertussis* have different host ranges and cause different diseases in their hosts. These pathogens can modulate the host immune response leading to their establishment and persistence in the host.

Immunologists for long time have focused on characterizing the immune responses in terms of activation of immune components in response to pathogens, for example activation of IL10 in response to the type III secretion system of *B. bronchiseptica*, or clearance of bacteria followed by activation of T helper 1 (Th1) cells. Thus, there is a lot of qualitative data available in immunology which awaits a formal mathematical description. We developed the asynchronous Boolean model using independent timescales for each process in the network. The model was augmented by using experimental information on the time scales of activation of some of the processes so that the timing of the update of a few nodes was fixed and the order of update of all other nodes was randomly chosen. We also designed two modifications of the classical Boolean functions, a threshold on the number of consecutive steps of phagocytosis in order to clear bacteria, and decay of cytokines and bacterial virulence factors, even in the presence of their regulators, after an individually chosen interval.

The model was then used to simulate time courses of wild-type infections and infections in which either host or pathogen nodes are perturbed. Despite the asynchronicity, the model results in a conserved activation pattern of key nodes, indicating three phases in the wild-type infections (Fig. 4.8). These phases correspond to the activation of innate immune responses, followed by the generation of Th2 related responses and then Th1 related responses. The clearance of bacteria was always associated with the generation of Th1 related responses, as observed experimentally. The model reproduces the early clearance of bacteria deficient in virulence factors such as the type III secretion system and the persistence of bacteria in hosts deficient in B cells and T0 cells. The model prediction that adoptive transfer of antibodies can clear *B. bronchiseptica* but not *B. pertussis* was also validated. Next, the model was used to simulate secondary infections by the same and different pathogen. The novel model prediction that the secondary invasion of the same pathogen in the third phase can be cleared faster was experimentally tested and validated. The model also indicates that Th1 related cytokines and antibodies are the most important factors controlling the bacterial numbers.

B. bronchiseptica

B. pertussis

Fig. 4.8. Activation pattern indicating the activity of the immune components at each time-step in. The two panels correspond to the two bacterial strains modeled. Colored squares represent the ON state of the given node, while white boxes mean the OFF state of the node. The white box in the lower right corner of each pattern represents the clearance of bacteria on the last time-step. The figure is adapted from (Thakar *et al.* 2007).

Antigen specific immune responses are regulated by the T helper cell types, thus the differentiation of T cells has been the focus of many studies. Mendoza and Xenarios (Mendoza *et al.* 2006) developed a generalized formulation to describe any static network using discrete and hybrid dynamic models. The hybrid formalism is similar to the piecewise linear formalism described in the previous section except the activation term is a sigmoidal function of activating and inhibiting interactions. This method has been applied to the cytokine interaction network implicated in T cell differentiation. The authors have studied the steady states leading to the activation of IL12 or IL4 leading to the differentiation of naïve T helper cells into their subtypes, T helper 1 and T helper 2 respectively. Smaller systems have also been successful in integrating known information into Boolean models to formulate a constructive hypothesis. A five component model for T cell receptor mediated signaling was studied to reveal the mechanisms by which the receptor can induce alternative cellular responses (Kaufman *et al.* 1999). The authors show that different signaling conditions can occur for a single ligand-binding event per receptor in time-dependent manner so that the relative time of binding of the ligand and the tyrosine kinase activity decides the outcome. The model predicts two different stable states can be reached after ligand dissociation, one corresponding to immunocompetence and another to unresponsiveness (anergy). Another interesting example is a multi-scale model of the genetic and molecular features of the evolution of colorectal cancer (Ribba *et al.* 2006). A Boolean model of genes involved in cell proliferation and death pathways that are frequently mutated in colorectal cancer patients is integrated with a discrete model of the cell cycle and a fluid dynamics tissue model. Such multi-scale model can incorporate regulating factors such as hypoxia, tumor geometry and tissue dynamics in predicting and improving radio-therapeutic efficacy.

4.7. Conclusions

Theoretical models developed in close collaborations with experimentalists are required to analyze the implications of a set of observations and also to design novel explanatory hypotheses. We reviewed here applications of the Boolean approach in inferring the network of interactions from expression data and in modeling the dynamics of a network of interactions. The Boolean approach is a coarse grained method which is usually used to analyze systems with limited time course and kinetic information. Boolean networks thus show a way to start modeling dynamics of molecular networks at an earlier stage than we are used to today. The success of these models in describing specific systems and processes indicates that the topology of regulatory networks has a significant role in restricting their dynamical behavior.

Boolean modeling approaches rely on the network topology and provide a backbone to integrate different studies. Network topology can be improved by perturbation studies and of course by additional data. In the current state of knowledge in many systems, even the published data cannot be easily wired due to the lack of causal information (refer to Fig. 4.5). Thus the network assembly is already a step forward in defining a system and interpreting the experimental data. Often there are debates in interpreting the model predictions. The dynamical output of the network is a collective behavior of all the components. It is impossible to use experimental outcomes such as "bacteria deficient in the type III secretion system are cleared faster by the immune system" as inputs to the model. Instead, specific information on the function of the type III secretion system can be used to infer interactions that are then used during the construction of the network. While there is no guarantee that the dynamic behavior of the network will reflect the original outcome, the correctly predicted dynamic behavior of the network serves as a validation of the wiring.

The second important characteristic of the Boolean approach is its usability in analyzing large systems. The Boolean simplification into two states is often unfavorably compared to continuous approaches. However, the latter approach is only practical for small subsystems. Though properly chosen subsystem models have been successful, they are inadequate to study the emergence of holistic properties. Instead, it is often advantageous to employ a system-level network-based modeling approach. The resulting network models not only provide holistic insight but also predict the outcome of subsystems. The input to the network is from completely independent studies and the model is verified by comparing its dynamic (time-elapsed) behavior with experimental time-course observations.

The Boolean approach helps us understand the dynamics of the system by giving an activation pattern where the sequence of dynamic events can be compared between simulated and experimental data. The reproduction of such a sequence indicates the components important in various stages of the dynamics. Some of the Boolean approaches also give insight about the timescale of the processes (e.g. asynchronous Boolean models) and the activation thresholds (e.g. threshold models) of the nodes. In a sense we can view them as coarse-graining of the differential equation models that keeps the same steady states.

As we have illustrated in this chapter, Boolean modeling can successfully describe a variety of networks from the molecular to the physiological level. This does not mean that all networks are suitably described by Boolean models. For example, existing Boolean approaches may not be successful in systems in which stochastic effects propagate from the micro- to the macro-level. Probabilistic Boolean Networks introduce stochasticity in defining the causal relationships. Though this is useful to infer a network; dynamic stochasticity such as in the transport of molecules cannot be easily implemented in the current Boolean models. One can argue that the final outcome of most of the biological systems is rather deterministic and robust and thus, stochasticity is a rather exotic phenomenon in regulatory networks (Acar *et al.* 2008). Nevertheless, it may be relevant in specific circumstances, for example, cell differentiation. We expect that exploring the incorporation and effects of stochasticity will play an increasing role in discrete modeling. Still, in the current state of knowledge on biological networks the most frequent model failure is also the simplest: insufficient information of the network architecture. We hope that increasing communication between modelers and experimentalists will decrease this deficiency.

References

Albert, R. and H. G. Othmer (2003). "The topology of the regulatory interactions predicts the expression pattern of the segment polarity genes in Drosophila melanogaster." *J Theor Biol* 223(1): 1-18.

Bower, J. M., Bolouri, H. (2001). *Computational Modeling of Genetic and Biochemical Networks*. Cambridge, MA, MIT Press.

Braunewell, S. and S. Bornholdt (2007). "Superstability of the yeast cell-cycle dynamics: ensuring causality in the presence of biochemical stochasticity." *J Theor Biol* 245(4): 638-43.

Chaves, M., Albert, R., Sontag, E.D. (2005). "Robustness and fragility of Boolean models for genetic regulatory networks." *Journal of Theoretical Biology* 235(3): 431-449.

Coursol, S., L. M. Fan, *et al.* (2003). "Sphingolipid signalling in Arabidopsis guard cells involves heterotrimeric G proteins." *Nature* 423(6940): 651-4.

Davidich, M. and S. Bornholdt (2008). "From differential equations to Boolean networks: A case study in modeling regulatory networks." *J Theor Biol.*

Davidich, M. I. and S. Bornholdt (2008). "Boolean network model predicts cell cycle sequence of fission yeast." *PLoS ONE* 3(2): e1672.

Davidson, E. H. (2001). *Genomic Regulatory Systems: Development and Evolution*. San Diego, Academic Press.

Derrida, B. (1987). "Dynamical phase transition in nonsymmetric spin glasses." *J Phys A: Math Gen* 20: L721-L725.

Derrida, B. and Y. Pomeau (1986). "Random networks of automata: a simple annealed approximation." *Europhysics Letters.*

Dougherty, E. R., S. Kim, *et al.* (2000). "Coefficient of determination in nonlinear signal processing." *Signal Processing* 80: 2219-2235.

Dougherty, E. R. and I. Shmulevich (2003). "Mappings between probabilistic Boolean networks." *Signal Processing* 83: 799-809.

Espinosa-Soto, C., P. Padilla-Longoria, *et al.* (2004). "A gene regulatory network model for cell-fate determination during Arabidopsis thaliana flower development that is robust and recovers experimental gene expression profiles." *Plant Cell* 16(11): 2923-39.

Fall, C. P., Marland, E. S., Wagner, J. M. and Tyson, J.J. (2002). *Computational Cell Biology*. New York, Springer.

Gershenson, C. (2004). Introduction to Random Boolean Networks. Workshop and Tutorial *Proceedings, Ninth International Conference on the Simulation and Synthesis of Living Systems*. M. Bedau, P. Husbands, T. Hutton, S. Kumar and H. Suzuki: 160-173.

Ghysen, A. and R. Thomas (2003). "The formation of sense organs in Drosophila: a logical approach." *Bioessays* 25(8): 802-7.

Glass, L. and S. A. Kauffman (1973). "The logical analysis of continuous, non-linear biochemical control networks." *J Theor Biol* 39(1): 103-29.

Gosti, F., N. Beaudoin, *et al.* (1999). "ABI1 protein phosphatase 2C is a negative regulator of abscisic acid signaling." *Plant Cell* 11(10): 1897-910.

Heinrich, R. and S. Schuster (1996). *The Regulation of Cellular Systems*. New York, Chapman & Hall.

Jacob, T., S. Ritchie, *et al.* (1999). "Abscisic acid signal transduction in guard cells is mediated by phospholipase D activity." *Proc Natl Acad Sci USA* 96(21): 12192-7.

Kauffman, S., C. Peterson, *et al.* (2004). "Genetic networks with canalyzing Boolean rules are always stable." *Proc Natl Acad Sci USA* 101(49): 17102-17107.

Kauffman, S. A. (1993). *The Origins of Order : Self Organization and Selection in Evolution*. New York, Oxford University Press.

Kaufman, M., F. Andris, *et al.* (1999). "A logical analysis of T cell activation and anergy." *Proc Natl Acad Sci USA* 96(7): 3894-9.

Klemm, K. and S. Bornholdt (2005). "Topology of biological networks and reliability of information processing." *Proc Natl Acad Sci USA* 102(51): 18414-9.

Kuerten, K. E. (1988). "Critical phenomena in model neural networks." *Phys Lett A* 129(157-160).

Li, F., T. Long, *et al.* (2004). "The yeast cell-cycle network is robustly designed." *Proc Natl Acad Sci USA* 101(14): 4781-6.

Li, S., S. M. Assmann, *et al.* (2006). "Predicting Essential Components of Signal Transduction Networks: A Dynamic Model of Guard Cell Abscisic Acid Signaling." *PLoS Biol* 4(10).

Liang, S., S. Fuhrman, *et al.* (1998). "Reveal, a general reverse engineering algorithm for inference of genetic network architectures." *Pac Symp Biocomput*: 18-29.

Martin, S., Z. Zhang, *et al.* (2007). "Boolean dynamics of genetic regulatory networks inferred from microarray time series data." *Bioinformatics* 23(7): 866-74.

Mendoza, L. and E. R. Alvarez-Buylla (1998). "Dynamics of the genetic regulatory network for Arabidopsis thaliana flower morphogenesis." *J Theor Biol* 193(2): 307-19.

Mendoza, L., D. Thieffry, *et al.* (1999). "Genetic control of flower morphogenesis in Arabidopsis thaliana: a logical analysis." *Bioinformatics* 15(7-8): 593-606.

Mendoza, L. and I. Xenarios (2006). "A method for the generation of standardized qualitative dynamical systems of regulatory networks." *Theor Biol Med Model* 3: 13.

Merlot, S., F. Gosti, *et al.* (2001). "The ABI1 and ABI2 protein phosphatases 2C act in a negative feedback regulatory loop of the abscisic acid signalling pathway." *Plant J* 25(3): 295-303.

Novak, B., Z. Pataki, *et al.* (2001). "Mathematical model of the cell division cycle of fission yeast." *Chaos* 11(1): 277-286.

Ribba, B., T. Colin, *et al.* (2006). "A multiscale mathematical model of cancer, and its use in analyzing irradiation therapies." *Theor Biol Med Model* 3: 7.

Rissanen, J. (1978). "Modeling by shortest data description." *Automatica* 14: 465-471.

Sanchez, L. and D. Thieffry (2001). "A logical analysis of the Drosophila gap-gene system." *J Theor Biol* 211(2): 115-41.

Sanchez, L. and D. Thieffry (2003). "Segmenting the fly embryo: a logical analysis of the pair-rule cross-regulatory module." *J Theor Biol* 224(4): 517-37.

Sanson, B. (2001). "Generating patterns from fields of cells. Examples from Drosophila segmentation." *EMBO Rep* 2(12): 1083-8.

Shmulevich, I., E. R. Dougherty, *et al.* (2002). "Probabilistic Boolean Networks: a rule-based uncertainty model for gene regulatory networks." *Bioinformatics* 18(2): 261-74.

Shmulevich, I., S. A. Kauffman, *et al.* (2005). "Eukaryotic cells are dynamically ordered or critical but not chaotic." *Proc Natl Acad Sci USA* 102(38): 13439-44.

Smolen, P., D. A. Baxter, *et al.* (2000). "Mathematical modeling of gene networks." *Neuron* 26(3): 567-80.

Thakar, J., M. Pilione, *et al.* (2007). "Modeling Systems-Level Regulation of Host Immune Responses." *PLoS Comput Biol* 3(6): e109.

Voit, E. O. (2000). *Computational Analysis of Biochemical Systems*. Cambridge, Cambridge University Press.

von Dassow, G., E. Meir, *et al.* (2000). "The segment polarity network is a robust developmental module." *Nature* 406(6792): 188-92.

Wang, X. Q., H. Ullah, *et al.* (2001). "G protein regulation of ion channels and abscisic acid signaling in Arabidopsis guard cells." *Science* 292(5524): 2070-2.

Zhao, W., E. Serpedin, *et al.* (2006). "Inferring gene regulatory networks from time series data using the minimum description length principle." *Bioinformatics* 22(17): 2129-35.

Chapter 5

Complexity of Boolean Dynamics in Simple Models of Signaling Networks and in Real Genetic Networks

Albert Díaz-Guilera[1,2] and Elena R. Álvarez-Buylla[3]

[1] *Departament de Física Fonamental, Universitat de Barcelona,*
Barcelona, 08028, Spain
[2] *Department of Chemical and Biological Engineering, Northwestern University*
Evanston, IL 60202, USA
[3] *Instituto de Ecología, Departamento de Ecología Funcional,*
Universidad Nacional Autónoma de México, México, D.F. 04510, Mexico

5.1. Introduction

Complex systems are composed of interacting units that communicate among themselves and process external environmental stimuli [Buchman (2002)]. Generally, the dynamical properties of such systems are analyzed in terms of the time series of some relevant property of the system. In this sense, the complexity of the dynamics is understood as the existence of non-trivial correlations of the fluctuations. The Fourier transform of the correlation function of the fluctuations (the spectrum) measures the degree of complexity of the system. When correlations span across multiple time scales the spectrum has a power-law behavior and it is the exponent of the power-law that quantifies the complexity of the generated dynamics [Bassingthwaighte *et al.* (1994); Malik and Camm (1995); Goldberger *et al.* (2002)].

One of the crucial findings in the last decade concerning the structure of many complex systems is the fact that the topology of the interactions between single units plays a crucial role. Real topologies are far from being regular or completely random and the topological correlations add new ingredients to the dynamical models considered so far [Watts and Strogatz (1998); Albert and Barabási (2002)].

Thus, trying to understand how complex systems dynamically evolve and process information has stimulated the development of network theory [Boccaletti *et al.* (2006)]. In Biology, the concerted action of multiple interconnected genes and proteins underlie developmental processes and phenotypical traits. Several years ago, Kauffman (1969) proposed that cellular states or types corresponded to steady states or attractors of such dynamic complex networks and recent experimental evidence supports this hypothesis [Huang *et al.* (2005)].

As detailed molecular, as well as high throughput genomic data, accumulate, a repertoire of gene regulatory networks (GRN) grounded on experimental data is being built [Mendoza and Alvarez-Buylla (1998); Albert and Othmer (2003); Espinosa-Soto *et al.* (2004)]. Furthermore, some of the studies are suggesting that qualitative models such as Boolean networks recover the fundamental structural characteristics (eg., topology of GRN interconnections) and dynamical behaviors (eg., attractors attained that correspond to multigene expression profiles characteristic of certain cell types, and robustness to transient and genetic perturbations). Thus, a continuous feedback from theoretical studies exploring the behavior of simple model GRN, to those analyzing the behavior of experimentally grounded biological GRNs is useful. Such interplay between theoretical and experimental GRNs is providing guidelines to understanding both generic and specific structural and dynamical characteristics of biological GRNs, as well as more general issues concerning the relationship between network structure and dynamics, or network behavior under noisy environments.

Recent studies are demonstrating that the coordinated expression of multiple molecular components is essential to sustain functionality in noisy environments [Aldana *et al.* (2007)]. The global dynamics of biological GRNs must be robust in order to guarantee stability under a broad range of external conditions. But, at the same time, GRNs should be sufficiently flexible to be able to recognize and integrate external signals that specifically elicit adaptive mechanisms to the environmental challenges met by each organism.

Interestingly, the so called critical dynamical systems, that operate at the brink of a phase transition between order and chaos, have such a compromised behavior between robustness and adaptability. Precisely, critical dynamics is related to the appearance of long-range time correlations observed in many natural systems, as described at the beginning of the section. A recent study showed that biological GRNs of a wide range of sizes (eg., large genomic as well as small modules), and inferred with very different methods for contrasting organisms (eg., bacteria, yeast, plants and animals), all operate close to a critical dynamics [Balleza *et al.* (2008)]. This result suggests that the capacity to compromise between being able to recover previous stable states, as well as exploring all of them or new ones, might have being fundamental during the evolutionary history of GRN assemblage. At the same time, such a global trait of GRN could have been subject to natural selection during the early evolution of life, and at the same time may provide an explanation for the great diversity and dynamically robust forms that have evolved among living beings.

Cells that function with a critical dynamics are able to keep previous stable states and at the same time explore among them or even among new ones, are able to bind past discriminations to future reliable actions. Systems that function under a critical regime have a near parallel flow of information that optimizes the capacity to bind past and future decisions. In contrast, ordered dynamical systems converge in state space and forgets past distinctions. Finally, in a chaotic system, even small stochastic fluctuations may cause wide divergences in state space trajectories thus precluding reliable action [Kauffman (1969); Aldana *et al.* (2007)].

Previous studies have also demonstrated with simulated transient or permanent mutations that biological networks are robust to noisy perturbations [von Dassow *et al.* (2000); Albert and Othmer (2003); Espinosa-Soto *et al.* (2004)]. However, besides recent studies demonstrating that a wide array of biological GRN are critical, no study has shown that biological networks are robust to noisy perturbations and at the same time optimize their capacity to explore alternative states in time and space. Little is known as well as to which structural traits (topologies of interconnections and structure of logical functions) are linked to the optimized capacity of GRN to explore alternatives, and at the same time are able to retain previous attractors.

In this contribution we first review our own work on simple model networks in which we have particularly searched for generic dynamical behaviors in the face of noise. Then we review our work on an experimentally grounded GRN, that constitutes a model for the logic of necessary and sufficient gene regulatory interactions required for primordial cell fate determination during early stages of flower development. We specifically address if such network has a peculiar behavior in response to noise in comparison to the simple model networks analyzed, and if such behavior maximizes its capacity to explore the state space at the same time that it retains the observed steady states under different noisy regimes.

5.2. Noisy Dynamics of Boolean Networks

Both model and real networks are constituted by units that process information and are subject to some noise regime. Concerning the unit's dynamics, we assume that the state of the units are Boolean variables. Boolean variables, which can take one of two values, 0 (OFF) or 1 (ON), and Boolean functions have been extensively used to model the state and dynamics of complex systems — see [Kaplan and Glass (1997)] for an introduction. Even though Boolean networks are oversimplifications of real systems, it has been shown that Boolean functions are good approximations to the nonlinear functions encountered in many control systems [Kauffman (1993); Weng *et al.* (1999); Aldana *et al.* (2003); Wolfram (2002)]. For instance, Random Boolean networks (RBNs) were proposed by Kauffman (1993) as models of genetic regulatory networks, and have also been studied in a number of other contexts [Weng *et al.* (1999); Aldana *et al.* (2003)]. Wolfram (2002), in contrast, proposed that cellular automata (CA) models — a class of ordered Boolean networks with *identical* units — may explain the real-world's complexity.

In general, it is assumed that the state $\sigma_i(t+1)$ of unit i at time $t+1$ depends on the state of the set of its neighbors — including itself — at time t as

$$\sigma_i(t+1) = \mathcal{F}_i\left[\sigma_{i_1}(t), \sigma_{i_2}(t), \ldots, \sigma_{i_{k_i}}(t)\right], \tag{5.1}$$

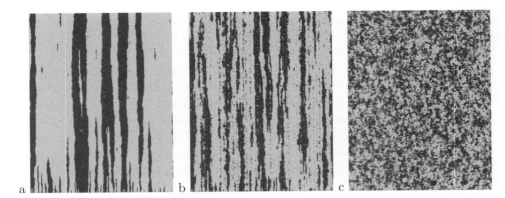

Fig. 5.1.   Time evolution of the state of a set of units placed on a circle, subject to periodic boundary conditions (see Fig. 5.2 for details of the model). Black (gray) squares correspond to nodes with 0 (1) states. Time goes from top to bottom and unit labels from left to right. All units process information according to the majority rule (rule 232, see text) and noise intensity is fixed. The three panels differ in the topology; whereas all of them have the same initial circular pattern they vary in the number of excess connections (short-cuts). They have in common the initial state and the sequence of random numbers mimicking noise. For a low random connectivity (left) the initial configuration is rapidly stabilized and patterns keep their shapes, only affected by the noise from time to time. For larger number of short-cuts, initial patterns develop and can spread over different time scales. Finally, for a larger random connectivity the dynamics becomes dominated by the external fluctuations and correlations are rapidly lost.

where k_i is the connectivity of unit i — i.e., the number of inputs that unit i receives — and $\mathcal{F}_i$ is a Boolean function (or rule), that completely specifies the way that unit i is processing the information that arrives to it and hence is fixed in time. Note that for a number K of inputs there are 2^{2^K} different functions: for example, for $K = 3$ there are 256 different Boolean functions.

Generally, the dynamics in real systems is subject to noise effects. In our case, the noise is intended to mimic two effects that are always present in living processes: (*i*) communication errors due to the intrinsic noise from *in vivo* conditions, and (*ii*) external stimuli affecting the transmission of signals to a unit [Mar *et al.* (1999)]. We introduce noise in the model dynamics by assuming that a unit has a probability η of "reading" a random Boolean variable instead of the "true" state of the neighbor. It is important to note that in our model the noise acts only on the inputs of the neighbors of a unit. This implies that the state of a unit is not changed because of noise only.

As an example of this type of dynamics we present in Fig. 5.1 the evolution of the state of a particular model (to be described in detail in Sect. 5.4). This example illustrates the great importance of the topology, apart from the dynamical rules of the units and the intrinsic noise.

5.3. Correlation Functions of the Fluctuations

In discrete models it is usual to define the state $S(t)$ of the system as the sum of the states of all the units

$$S(t) = \sum_i \sigma_i(t) \,. \tag{5.2}$$

We start all of our numerical simulations with a random initial configuration and let the system evolve according to the rules of the individual units. We record the state of the system at each time step according to (5.2) and quantify the complexity of the series generated in terms of the auto-correlation function of $S(t)$ [Goldberger *et al.* (2002)].

To analyze the character of the fluctuations it is useful to apply the detrended fluctuation analysis (DFA) method to the time series generated by the model [Peng *et al.* (1995); Taqqu *et al.* (1995)]. In this method, the first step is to integrate the original time series. Then, the integrated time series is divided into boxes of size n and, for each box, a least-square linear fit is performed. Next, the root-mean-square deviation of the integrated time series from the fit, $F(n)$, is calculated. This process is repeated over different box sizes or time scales.

For self-similar signals one finds that $F(n)$ satisfies a power law relation with the size of the box n, that is $F(n) \sim n^\alpha$, with α the scaling correlation exponent. This method quantifies long-range time-correlations in the dynamical output of a system by means of a single scaling exponent α. The exponent α is related to the exponent β of the power spectrum of the fluctuations, $\mathcal{S}(f) \sim 1/f^\beta$, through the relation $\beta = 2\alpha - 1$.

Brownian noise yields $\alpha = 1.5$, while uncorrelated white-noise yields $\alpha = 0.5$. For a number of physiological signals from free-running, healthy, mature systems, the scaling exponent α takes values close to one, so-called $1/f$-behavior, which can be seen as a "trade-off" between the two previous limits [Goldberger *et al.* (2002)]. However, different types of correlations (given by different values of the exponent α) are encountered in some immature, diseased, or aged physiological systems [Buchman (2002); Lipsitz (2002)].

5.4. Model Signalling Networks

In the previous section we outlined some results about the complexity of physiological signals measured experimentally. Basically, we highlight the fact that for healthy mature individuals some of the signals show the $1/f$ behavior, which in terms of the exponent of the DFA means that $\alpha \simeq 1$ [Goldberger *et al.* (2002)]. Additional sets of measures showed that this exponent changes with disease and age. On the other hand, it is also well known [Lipsitz (2002); Buchman (2002)] that disease and age produce degradation in the communication between subunits in a physiological system. For this reason we proposed a stylized model in which, with

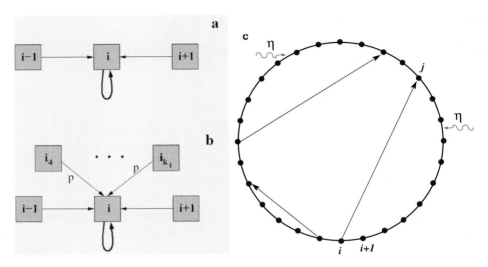

Fig. 5.2. The units forming the simple model are placed on the nodes of a one-dimensional lattice and establish bi-directional nearest-neighbor connections. With some probability we add long-range unidirectional connections until there are $k_e N$ of such excess links, k_e being the average excess connectivity, and N the number of units. Each unit processes a set of input signals in the way that it is described in the text. The signal received from one of its neighbors is replaced with a random value with probability η.

a minimal set of ingredients, we can show the relationship between age or disease, lost of links, noise, and the spectrum of the resulting fluctuations.

5.4.1. *Topology*

Concerning the system's topology of simple signalling model networks, we assume that the units are placed on the nodes of a one-dimensional lattice and that each unit is bidirectionally connected to its two nearest neighbors. We then increase the topological complexity of the system by adding, with probability k_e, an incoming connection to each node [Newman and Watts (1999)] (Fig. 5.2). The basic topology we considered consists in placing units on a circle with connections to the nearest-neighbors. This is a very regular, ordered (and artificial) network structure. To consider a more general topology we first allow for "connection errors," which we implement through long-range connections to randomly-selected units on the circle ("short-cuts"). This corresponds to the "small-world" topology proposed by Watts and Strogatz (1998); Strogatz (2001). Although quite stylized, this topology captures some aspects of real-world networks such as (i) the small number of "degrees-of-separation" between the units and (ii) the local order through the clustering coefficient.

5.4.2. *Boolean rules*

Without short-cuts, the units of our model have exactly three inputs. In this case, we can write the truth table of Boolean functions (or rules) of three-inputs in the following form [Wolfram (1994)]:

$\sigma_{i_1}(t)$	1	0	1	0	1	0	1	0
$\sigma_i(t)$	1	1	0	0	1	1	0	0
$\sigma_{i_2}(t)$	1	1	1	1	0	0	0	0
$\sigma_i(t+1)$	b_7	b_6	b_5	b_4	b_3	b_2	b_1	b_0

$$(5.3)$$

where b_j is the output for each of the eight possible combinations of inputs, and can take values 0 or 1. The rules are designated by the decimal number that corresponds to the binary number $b_7 b_6 b_5 b_4 b_3 b_2 b_1 b_0$; for example, the Boolean function 11101000 is rule 232 [Wolfram (1994); Kaplan and Glass (1997)]. This rule returns as an output the value in the majority among the inputs, and hence it is also named the "majority" rule.

In the following, we focus our attention on rules that can be generalized to an arbitrary number of inputs. In general, a three-input rule cannot be generalized to any number of inputs. An important class of Boolean rules that is mostly excluded by the generalizable Boolean rules considered above is the class of canalizing functions [Wolfram (1994)]. In a canalizing rule, the output value of the rule is solely determined by one input, the canalizing variable. Three of the rules we study here are canalizing: 1, 19, and 50. As we have reported [Amaral *et al.* (2004)], two of these rules lead to a broad range of dynamical behaviors. However, the majority rule, which is not canalizing, also displays a broad range of dynamical behaviors, suggesting that canalizing variables are not necessary to obtain such diverse dynamics.

In order to define the generalizable Boolean functions, we replace the inputs of the neighbors in a 3-input rule by an average input $\sigma_n(t)$:

$$\sigma_n(t) = \frac{1 + \text{sign}(\sigma_{i_1} + \sigma_{i_2} - 1)}{2},$$

$$(5.4)$$

where we define $\text{sign}(0) \equiv 0$; a value of $1/2$ for the state σ_n is what we call a Tie. The average input $\sigma_n(t)$ can be generalized to include an arbitrary number of inputs:

$$\sigma_n(t) = \frac{1 + \text{sign}\left(2\sum_{j=1}^{j=k_i} \sigma_{i_j} - k_i\right)}{2}.$$

$$(5.5)$$

The Boolean rules of three inputs that are generalizable are symmetric rules that have $b_1 = b_4$ and $b_3 = b_6$:[1]

---

[1] Only in this case it does not matter where the Tie comes from, a 1 on the right and a 0 on the left or just the other way around.

$\sigma_n(t)$	1	T	1	T	T	0	T	0
$\sigma_i(t)$	1	1	0	0	1	1	0	0
$\sigma_i(t+1)$	b_7	b_6	b_5	b_4	b_3	b_2	b_1	b_0

$$(5.6)$$

For the sake of clarity we will keep using the 8-bit decimal representation used for Boolean rules with 3-inputs to label the corresponding generalizable rule of arbitrary number of inputs [Wolfram (1994)]. However, the truth table of a generalizable rule can be simplified as follows:

$\sigma_n(t)$	1	T	0	1	T	0
$\sigma_i(t)$	1	1	1	0	0	0
$\sigma_i(t+1)$	b_5'	b_4'	b_3'	b_2'	b_1'	b_0'

$$(5.7)$$

where b_i' is the output of the function in each of the six different input conditions. This representation, in terms of six possible input combinations enables a better visual identification of the rules. In the top of Fig. 5.5 we show the truth tables in this representation of a small subset of the generalizable Boolean functions.

Because of symmetries, there are only $2^6 = 64$ independent generalizable rules. However, some of these rules are conjugate rules and one does not need to investigate the whole set of rules. Two conjugate rules will have identical dynamics if the zeros and ones are switched for one of them [Wolfram (1994)]. Additionally, some rules are self-conjugates. These are rules 23, 51, 77, 105, 150, 178, 204, and 232. Self-conjugate rules have an inverse rule. Two rules are said to be inverse if, when starting both with the same initial condition, they will be in the exact same state every other time step and in the inverse state the other time steps. An example of inverse rules is given by rules 204 (the "identity" rule) and 51 (the "negation" rule). The identity rule will keep the initial state, whereas the negation rule will switch between the initial state and its inverse. Other pairs of inverse rules are {23, 232}, {77, 178}, and {105, 150}.

5.4.3. *Noisy dynamics*

Even in the presence of noise, not all of the 64 generalizable rules that we may consider display fluctuating time series. For example, rules 0, 4, 72, 76, 128, 132, and 200 converge to a fixed state, in which all units are in state zero (see Amaral *et al.* (2004)). Additionally, the output of rules 51 and 204 depends only on the state of σ_i, so it will not be affected by noise and, as such, will not display fluctuations.

Because of these facts and because conjugate rules have identical dynamics, we finally investigated only 24 of the 64 generalizable rules. These rules are 1, 5, 18, 19, 22, 32, 33, 36, 37, 50, 54, 73, 77, 90, 94, 104, 105, 108, 122, 126, 146, 160, 164, and 232. This process of elimination of irrelevant rules is schematized in Amaral *et al.* (2004).

As an example of the noisy dynamics, the analysis and the different types of behaviors we present some results in Figs. 5.1-5.4 for units operating with rule 232.

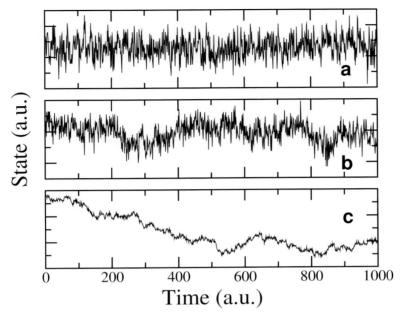

Fig. 5.3. We show $S(t)$ for a system with $N = 4096$ units, $\eta = 0.1$, $\mathcal{F}_i = 232$, and $k_e = 0.90$, $k_e = 0.45$, $k_e = 0.15$ (from top to bottom corresponding to the three panels of Fig. 5.1). The three values of k_e lead to quite different dynamics of the system. **(a)** For a large number of random links, $k_e = 0.90$, the dynamics are very poorly correlated. **(b)** For an intermediate value of k_e, long-range correlations emerge and the power spectrum displays a power-law behavior, $\mathcal{S}(f) \propto 1/f^{\beta}$, with $\beta \approx 1$. **(c)** For a small number of random links, the time correlations display trivial long-range correlations such as found for Brownian noise.

In Fig. 5.1 we show the evolution of single nodes (it displays the specific state of each unit) in time. There we can see how a system with a low number of shortcuts (left) evolves quasi-deterministically since the majority rule (232) is quite stable in a ring, even with the effect of external noise. On the other hand, when the number of shortcuts is very large the system becomes quite disordered and information about the initial state is rapidly lost. Interestingly from a biological point of view, we notice that for an intermediate number there is a tradeoff between these two effects, since a clear mixing of robustness and flexibility is displayed.

A more global picture is given by the state of the system, as defined by Eq. (5.2). This is what we plot in Fig. 5.3. Here we can notice again the different behaviors already shown in the previous figure. A low number of shortcuts makes the system to keep trapped in some of the attractors of the dynamics whereas a large number of extra connections makes the system to wander around not being able to stabilize in any of the deterministic configurations. Finally, for an intermediate value the system is able to jump in a reasonable time from one attractor to the other showing effects that are important at all time scales.

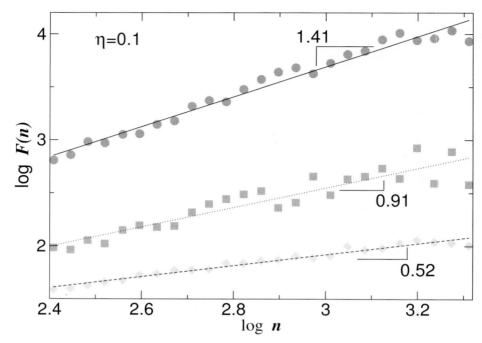

Fig. 5.4. Estimation of temporal auto-correlations of the state of the system by the detrended fluctuation analysis (DFA) method [Peng et al. (1995)]. We show the log-log plot of the fluctuations $F(n)$ in the state of the system, versus time scale n for the time series shown in Figs. 5.3a–c. In such a plot a straight line indicates a power-law dependence $F(n) \propto n^\alpha$. The slope of the lines yields the scaling exponent α. For panel **a** in Fig. 5.3 the value of the exponents is 0.52, for panel **b** is 0.91, and for panel **c** is 1.41.

Finally, in Fig. 5.4 we show the DFA of the time series of Fig. 5.3. As explained in Sect. 5.3 a straight line in this plot corresponds to self-similar dynamics. In terms of the noise correlations we can say in this case that the top set of points (left in Fig. 5.1, bottom in Fig. 5.3) is very close to Brownian noise ($\alpha = 1.41$), the bottom set (right in Fig. 5.1, top in Fig. 5.3) to white noise ($\alpha = 0.52$) whereas the middle set (center in Fig. 5.1, middle in Fig. 5.3) corresponds to $1/f$ noise ($\alpha = 0.91$).

5.4.4. *Complex dynamics from simple models*

The previous paragraphs show a way in which the system can be analyzed systematically. Actually, what we want to study is how a single characterization of the system (in our case the exponent α) depends on the details, the topology, the noise intensity, the Boolean rules, and so. To do so we systematically run simulations for a fixed time, and compute the exponent for a single run with a fixed box size. Afterwards, we perform averages over different initial conditions. Thus a given rule, topology, and noise intensity give rise to and exponent value, and comparison between different effects can be performed.

Prior to presenting our results it is worthwhile to note that both the RBN and CA models can be seen as limiting cases of the present model in the absence of noise: A RBN model corresponds to a completely random network with different Boolean rules for the units, while a CA model corresponds to $k_e = 0$ and all units evolving according to the same rule.

In our systematic analysis, we first calculated the exponent α for systems whose units are randomly assigned a Boolean function from the set of 256 functions of three inputs — as in the classical RBN model. We found white-noise dynamics for essentially any pair of values of k_e and η within the ranges considered, suggesting that a system of random Boolean functions cannot generate complex dynamics. This result is not unexpected, since the random collection of Boolean functions comprising the system prevents the development of any order or predictability in its dynamics.

In order to prevent this fact, we considered systems whose units all evolve according to the same rule, as in CA models (but noisy dynamics in our case). We systematically studied the 24 different Boolean functions of three inputs which do display fluctuations, for different pairs of values of k_e and η.

In our previous works [Amaral *et al.* (2004); Díaz-Guilera *et al.* (2007)] we showed the phase spaces for different rules displaying different behaviors. We classified them in three different sets depending on the diversity of ranges of dynamics. In the first and more interesting set, shown in Fig. 5.5, three different types of dynamical behavior are displayed, depending on both the intensity of the noise and on the excess connectivity. In a second set, complex fluctuations are only obtained for small intensities of the noise without dependence on the excess connectivity. Finally, there is a collection of rules for which fluctuations are always uncorrelated. Only the first case is then worth of consideration. In particular, rule 232 (the majority rule) which is encountered in many biological or physical models, plays the most important role in our further analysis. In our case the most interesting result is the dependence of the exponent on the topology for a fixed noise intensity, as has been discussed along the previous paragraphs. This simple model (rule 232 in Fig. 5.5) can hence explain the experimentally observed dependence of the scaling properties of some physiological signals on age and failure. Age and failure can be related, respectively, to missing physiological "links" and to misscommunication among some physiological subunits [Lipsitz (2002); Goldberger *et al.* (2002)].

5.4.5. *Robustness*

In order to determine the generality of the results presented above, one needs to address the questions of how these findings are affected by (i) changes in the topology of the network or (ii) "errors" in the units' implementation of the rules. Since from both a biological and a physical point of view, the majority rule has a meaningful interpretation we take it as a starting point, but it is worth to note that it could be performed for any of the "complex" rules represented in Fig. 5.5 (rules 19 and 1).

A. Díaz-Guilera & E. R. Álvarez-Buylla

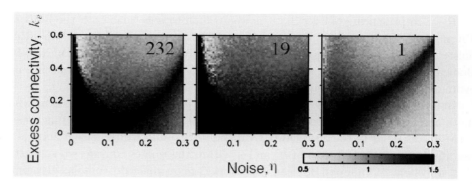

Fig. 5.5. Phase space for signaling networks with units obeying uniformly rules 232, 19 and 1. In the phase spaces, we use the color scheme shown in the bar to indicate the value of the scaling exponent α characterizing the auto-correlations in the dynamics of the system by means of the DFA method [Peng et al. (1995)]. The exponent is systematically estimated for time-scales $40 < n < 4000$. We show α for 61×61 pairs of values of k_e and the noise η in the communication between the units comprising the network. For all simulations, we follow the time evolution of systems comprising 4,096 units for a transient period lasting 8,192 time steps, and then record the time evolution of the system for an additional 10,000 time steps. The rules shown in this figure display different types of noisy fluctuations depending on the values of η and k_e. For all three rules, the system generates $1/f$-noise for a broad range of noise intensities.

Concerning (i) we note that the topology considered so far takes as initial setup, before adding the extra long-range connections, a one-dimensional ring of nearest-neighbor links. One of the first questions one asks itself is whether this particular setup is relevant in the development of the complexity of the signals discussed in the previous section. We have tackled this problem, from two different points: a) considering next-nearest neighbors connections, and b) connectivity distributions other that the homogeneous ones considered so far [Díaz-Guilera et al. (2007)].

In the first case, we considered as initial setup a one-dimensional array in which each unit i has bidirectional connections to its nearest neighbors ($\pm$ 1) and to its next-nearest neighbors ($\pm$ 2). We noticed that the inclusion of two additional neighbors before we add the extra connections leads to a similar structure of the phase space and the three different regimes are still observed. It is noted, however, that more noise intensity is needed to obtain the regimes observed in the original system with only nearest neighbors.

In the second case, we note that the network topologies considered so far span the cases of ordered one-dimensional lattices, small-world networks, and random graphs [Amaral et al. (2000)]. However, all networks considered are comprised of units with approximately the same degree, i.e., the same number of connections. To investigate the role of the distribution of number of connections, we also studied networks which span the range of empirically observed degree distribution: a delta-distribution, an exponential distribution, and a power law distribution.

In particular, we considered different distributions of number of incoming and outgoing connections for networks with exponential distribution of local links with different means. Allowing the number of incoming connections to fluctuate also changes the distribution of number of outgoing connections, which becomes a Poisson distribution. The change in the increased number of local connections leads to no significant change in the results. The reason is that the complex dynamics are generated at the boundary between domains: when one allows some units to have more local connections, these units still have the same number of units to the left and to the right, so the existence of a single long-range connection is enough to destabilize the boundary.

Additionally, we considered networks with power-law distributions either of incoming or outgoing links. Taking the one-dimensional ring of Fig. 5.2 again as initial setup, we add the additional connections according to the preferential attachment rule for outgoing units and incoming units [Albert and Barabási (2002)]. The former case gives rise to a network with a broad distribution of outgoing links while the latter gives rise to a network with broad distribution of incoming links. As one might expect, a power law distribution of outgoing links leads to no significant change in the phase-space describing the dynamic behaviors since it only makes the network a small-world more efficiently than random long-range connections [Cohen and Havlin (2003)]. In contrast, a power law distribution of incoming links *does* lead to a change in the phase-space of dynamical behaviors. The reason may be that since information travels only one way on the connections, the fact that some units are receiving so many of the long-distance connections will make it harder for the system to reach the small-world regime. Interestingly, this asymmetrical distribution of incoming/outgoing links has been observed in genetic regulatory networks for species across four different kingdoms [Balleza *et al.* (2008)].

The second class of change we considered is the way in which the units implement the rules. We looked at this from two different perspectives: one consists in restricting the majority rule to need more than the simple majority and the second in allowing a subset of units to operate according to another rule, which accounts for the effect of "errors" in the units implementation.

The first attempt was to construct a more restrictive majority rule. To do this we considered a clear majority rule by demanding the majority (half plus one) plus one to adopt the state. In this way we have a more restrictive majority rule, that is, a unit would remain in its initial state unless a clear majority operates. We perform simulations to evaluate the correlation exponent for systems with a fraction of units operating according either to a clear majority or simple majority rule. The use of a clear majority rule forces the units to remain in its state for more extended periods before switching to the opposite state. The phase space shows the same regimes as in Fig 5.5 (left) but a more extended region corresponding to Brownian dynamics is observed, due to the fact that the dynamics is more persistent in the present case.

We also investigated the effect of allowing the co-existence in the system of distinct Boolean rules. To this end, we first explored systems composed of units operating according either to rule 232 or according to a randomly selected rule. The inclusion of such randomness leads to a progressive decrease in the richness of the phase space of the system. We found that with as many as 1/4th of all units operating according to random Boolean functions the model still displays a rich phase space, including white, $1/f$, and Brownian noise. But, if more than 1/4th of all the units operate according to randomly selected Boolean functions, then the phase-space displays mostly white-noise dynamics, due to an increase in the randomness of the system. Additionally, we explored the existing dynamical behaviors for systems composed of units operating according to either rule 232 or rule 50. When both rules are present in the system (and at least 50% of the units operate according to rule 232) we still found several distinct classes of dynamical behaviors, including a wide range of parameter values that generate $1/f$ noise.

5.5. Dynamics in Real Genetic Regulatory Networks

Given the rich dynamical behavior in response to noise that depends on the topology and logical rules assumed, it becomes then very interesting to address the behavior of real GRN. Thus, in this section we present the same type of analysis done above for simple model networks, for the GRN underlying cell-fate determination during early stages of flower development (herein: floral GRN) in *Arabidopsis thaliana* (Fig. 5.6). This plant is the experimental system for molecular genetic studies. Plants share many more aspects of their basic molecular components and regulatory circuits with animals than thought before (Jones *et al.*, 2008). But their cellular structure is simpler than that of animals and are thus useful systems to scale from the structure and function of GRN, to cell-fate determination and the emergence of multicellular and morphogenetic patterns and dynamics *in vivo*. We focus here on a GRN that underlies cell fate determination at early stages of flower development.

Flowers are the reproductive and most complex structures of plants. They are characteristic of the most recently evolved lineage of plant species: the flowering plants or angiosperms. In contrast to animals, plant structures are formed along their complete life cycles from groups of undifferentiated cells, called meristems. These are useful structures for *in vivo* studies of the dynamics of cell-fate determination. Upon induction to flowering, the shoot apical meristem, from which the aerial parts of plants are formed, turns into an inflorescence meristem, from which flanks, flower meristems are produced. Adult floral morphology originates from these floral meristems or buds originally constituted of undifferentiated cells. During the three first days after emergence from the inflorescence flanks, floral buds are partitioned into four regions. Each one comprises the primordial cells of sepals, petals, stamens and carpels, that are the four types of organs which characterize the great majority of flowers (Fig. 5.6). In fact, these organs are formed also in a spatio-temporal stereotypical fashion from the outermost to the flower center: first sepals, then petals, stamens and finally carpels in the center [Bowman (1994)].

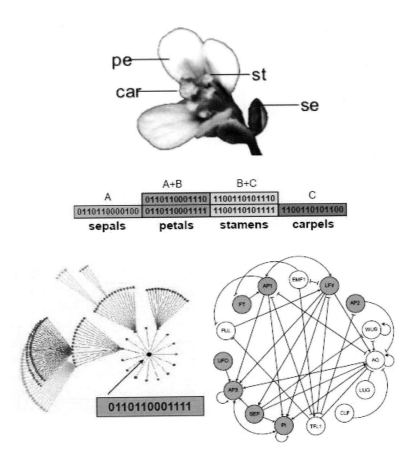

Fig. 5.6. *Arabidopsis thaliana*. ABC and GRN model for the specification of floral organ primordial cells. This plant species has the stereotypical floral morphology with sepals, petals, stamens and carpels that emerge sequentially from the periphery to the central part of the flower. A GRN model that integrates available experimental data for fifteen genes or proteins (or protein complexes), represented by nodes, and their regulatory interactions (activations - arrows; and repressions - blont-end edges). This GRN was modeled as a Boolean system and it converges to ten fixed-point attractors. Four of them correspond to cells in the inflorescence meristem. The floral meristems (buds) form from the flanks of the inflorescence meristem (see text). Within three days of their formation, the floral buds are partitioned into four regions, each comprised of the primordial cells that will form the floral organs. The other six attractors of the GRN are shown here: one corresponds to the combinations of gene activation states characteristic of sepal primordial cells (green), two correspond to petal primordial cells (red; one with the gene UFO turned "on" and the other one with this gene turned "off"), two to stamen primordial cells (yellow; one with the gene UFO turned "on" and the other one with this gene turned "off"), and one to carpel primordial cells (purple). The basin of attraction that corresponds to one of the petal attractors is illustrated and the corresponding attractor shown. The genes LUG and CLF are not listed in the attractors because they remain ON (1) at all times in all types of primordial floral cell types modeled here. In the network, the nodes that are ON in the exemplified petal attractor are in grey, all others are in white.

In previous works, we have integrated the experimental data on the molecular components that are necessary and sufficient for early floral organ cell-fate determination. We have used this data to propose a GRN of fifteen nodes (proteins or genes) and their interactions [Mendoza and Alvarez-Buylla (1998); Espinosa-Soto *et al.* (2004); Chaos *et al.* (2006)]. The wealth of data on which this GRN is grounded have been accumulated during the last 15 years in many different laboratories and gave rise to the postulation of the ABC model of flower development, that was derived from genetic analyses of floral organ homeotic mutants in two plant study systems: *Antirrhinum majus* L. and *Arabidopsis thaliana* (L.) Heynh [Coen and Meyerowitz (1991)]. The ABC model states that the identities of the floral organ types are established by combinations of genes grouped in three main classes: the A, B and C. A genes alone determine sepal identity, A plus B petal identity, B plus C stamen identity, and C alone carpel identity (Fig. 5.6).

However, the ABC combinatorial model does not provide an explanation for how such combinatorial selection of gene activity is established during floral organ primordia specification, and how does the spatio-temporal pattern of ABC and non-ABC gene expression is dynamically established, or which other genes interact with the ABC and together are sufficient for the specification of the four types of primordial cell types in a floral bud. Furthermore, the conserved pattern of floral organ determination and also the overall conservation of the ABC genes patterns of expression among eudicotyledoneous (more recently evolved angiosperms) species suggest a robust mechanism underlying such combinatorial selection of gene activities. Such mechanism recreates the conserved floral pattern in every individual of most angiosperms irrespective of the environmental conditions encountered, and it has done so during evolution in many flowering plant lineages, also independently of the genetic backgrounds of the species involved. The ABC model by itself does not provide an explanation to such robustness either. We have been developing qualitative GRN models for tackling these questions.

Discrete models of real GRN are abstractions of the regulatory interactions among genes. These undergo a network of mRNA transcription, translation, protein modifications and transport until its final functional destiny. Most data available for plant development is qualitative and at the transcriptional regulatory level, although recent data are suggesting that miRNA regulation are important during plant development [Chen (2004)]. In any case, the temporal scales of the molecular processes of the pathway that goes from DNA transcription to protein function are relatively short in comparison to those of the processes of pattern formation. Therefore, it is reasonable to focus only on the qualitative regulatory interactions. Also, formal analyses of equivalent continuous and discrete models have analytically shown that both yield equivalent results [Thomas *et al.* (1995)]. Moreover, in systems of many components with many non-linearities the behavior of the system depends mostly on the qualitative aspects of the GRN topology rather than on the kinetic details for each interaction and component.

5.5.1. *A GRN underlying cell-fate determination during early stages of flower development*

We postulate Boolean models in which genes or proteins (GRN nodes) may attain two values: 0 (OFF) or 1 (ON). The activity of each gene or protein depends on updating logical Boolean functions grounded on experimental data. Such models have been successful at recovering observed gene expression arrays in animal and plant cells [von Dassow *et al.* (2000); Albert and Othmer (2003); Espinosa-Soto *et al.* (2004)]. Discrete models are further justified because recent experimental evidence is suggesting that gene expression is digital and stochastic at the individual cell level, rather than continuous [Hume (2000); Ozbudak *et al.* (2002); Elowitz *et al.* (2002); Blake *et al.* (2003); Paulsson (2004); Walters *et al.* (1995); Fiering *et al.* (2000); Ho *et al.* (1996)]. As discussed in Sect. 5.1, random Boolean networks (RBN) have successfully described in a qualitative way several important aspects of gene regulation and cell differentiation. But, in any case, it is important to work with experimentally grounded GRNs.

In this particular case, the rules are derived from experimental data, see Espinosa-Soto *et al.* (2004); Chaos *et al.* (2006). A particular Boolean GRN has a 2^N possible gene expression configurations or states, where N is the number of nodes. Each state leads to another state or to itself in which case such state is a point attractor (steady state). A system may have more than one attractor and, eventually, all initial states will reach an attractor. The set of all states that lead to a specific attractor correspond to the basin of that attractor. The basins of attraction and attractors of a GRN depend on the number of elements, the number of possible states of each element, the topology of the network, and on the logical rules of each gene. Consequently, the dynamics of the system is deterministic, in which the fates of all states are known.

We have proposed a 15-node GRN that includes the ABC and non-ABC genes. This model attains only ten fixed-point attractors: four correspond to configurations characteristic of the inflorescence meristem, and the rest correspond to configurations of primordial sepal (one attractor), petal (two attractors), stamen (two attractors) and carpel (one attractor). These results suggested that the proposed GRN model incorporates the key components of a developmental module or subnetwork that underlies the combinatorial gene activities predicted in the ABC model (Fig. 5.6).

In conclusion the proposed GRN provides a dynamical explanation for the ABC model. It also shows that precise signaling pathways are not required to restrain organ primordia cell types during *A. thaliana* flower development, but rather that cell-fate are determined by overall gene network topology and dynamics. This was confirmed by robustness analyses of random perturbations of gene interaction parameters and by the fact that a three-state and the Boolean GRN yielded the same attractors [Espinosa-Soto *et al.* (2004); Chaos *et al.* (2006)]. It is likely that new components are part of the uncovered module. But new components of mechanistic details will likely have a similar qualitative behavior as the 15-gene regulatory network that has been uncovered. This is supported by the fact that the present model has been overall validated with simulations of several mutants and it has been able to make predictions that have been later confirmed with additional experimental data. For example, we had predicted that *AG*, one of the GRN nodes, should have a positive direct or indirect feedback loop and this has been confirmed in a recent paper [Gomez-Mena *et al.* (2005)].

5.5.2. *Noisy dynamics*

In this paper, we use this GRN to explore the response of a biological network to random perturbations. This is a follow up of a study, that we have recently published which demonstrates that when this GRN is subject to noise the sequence of attractor visitation recovered mimics that observed in nature. In particular, is the same as that observed for the ABC functions. The A genes are turned first, then the B ones defining the A to AB transition, and then the C, defining the AB to BC and C transitions [Alvarez-Buylla *et al.* (2008)]. We explore here in a systematic way the complex response of the network to noise. We also relate the complexity of the dynamical output signal to the flexibility that the system has to explore the set of deterministic attractors.

First of all, we look at the noisy dynamics of the original network with the rules experimentally obtained [Alvarez-Buylla *et al.* (2008)]. However, in this contribution we do so from the point of view of the complexity of the generated time series. We plot the time series of the state of the system, as defined by Eq. (5.2), and we can observe the different behavior for different noise intensities in Fig. 5.7. For a low noise intensity, the system stays most of the time in a subset of the available attractors. On the other hand, for a large noise intensity the system is not able to reside for a long a time close to any of the attractors and the dynamics is clearly dominated by noise. However, for an intermediate value (around 3%) the system jumps between the different attractors and stays for a reasonable time in most of them. The latter behavior is the required mixture of robustness and flexibility in many biological system. When applying the DFA to the time series we see in Fig. 5.8 how the dynamical behavior turns into precise values of the α exponent. Thus, for low noise intensity we obtain Brownian noise, for a large noise intensity white noise, and for the intermediate value we get $1/f$ noise. This is in fact a very remarkable result that relates the correlations of the fluctuations with the observed robustness and flexibility in the dynamics of the model around the attractors. In Fig. 5.9 (top set of points) we have plotted the dependence of the exponent on the intensity of the noise, and we can see that only for intermediate values this exponent is close to 1.

In order to check the biological relevance of the observed rules, we have repeated the previous analysis on the network preserving the topology but allowing for changes in the way the units process the information, i.e. altering the logical

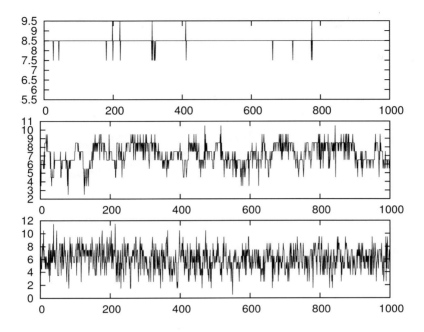

Fig. 5.7.   Time series of the noisy dynamics of the GRN of floral development in *A. thaliana* for three values of noise intensity. From top to bottom: 0.001, 0.03, and 0.2.

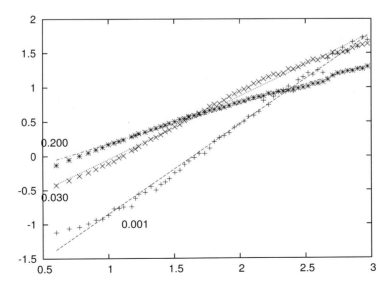

Fig. 5.8.   DFA analysis of the three series of data from Fig. 5.7. The noise values are indicated inside the panel. The corresponding exponents fitting each series of points to a straight line in the whole interval are, respectively, 0.577, 0.917 and 1.317, for each of the noise levels (0.200, 0.030, and 0.001).

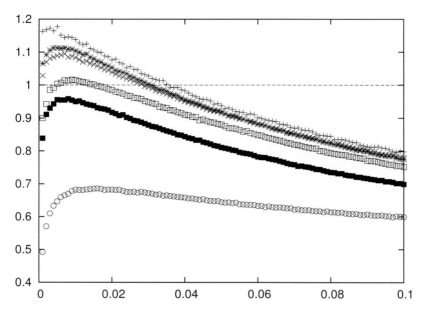

Fig. 5.9.   Value of the exponent α as a function of the noise intensity for the GRN for the floral development of Arabidopsis thaliana. We have preserved the topological structure but muted the Boolean rules with a certain probability. Each set of points corresponds to different probabilities (from top to bottom 0.00, 0.02, 0.04, 0.10, 0.20, and 0.50). Even with the same probability, each single mutation can give rise to very different outputs. For this reason, we have averaged over initial conditions (as usual) and over mutations with the same probability. Fluctuations are hence quite remarkable in this plot, but we have preferred to removed them to preserve the clarity of the results. The horizontal line corresponds to the $1/f$ noise.

rules. We allow for violations of the rules with certain probability and look at the exponent of the time series of the generated fluctuations. Mutations are performed with certain probability p; different realizations for the same value of p gives rise obviously to different rules, but these precise implementations of the rules can make huge changes in the number of attractors and in the structure of their basins of attraction. Then the response of the system depends not only on the value of p but also on the single realization of the mutation. For this reason, when plotting in Fig. 5.9 the exponent as a function of p, every single point is an average over different realizations (apart from an average over different initial conditions as we have done in the simple model), but with large fluctuations (not shown in the plot).

In Fig. 5.9 it is clear that as the mutation probability increases the response of system moves away from the $1/f$ noise behavior, and the resulting systems are not able to show the variety of behaviors we have observed in the network with the original rules or in the simple model of the first part of our work. A more detailed analysis of how the alteration of logical rules affects the visitation of the attractors and the noise spectra is still needed.

5.6. Conclusions

We have presented here analysis of Boolean dynamics on both a simplified model of signalling networks and a real genetic regulatory network model (in our particular case the GRN underlying cell-fate determination during early stages of flower development of *Arabidopsis thaliana*. In both cases we have observed through computer simulations that different rules, noise intensities, and topologies can give rise to very different complex dynamical behaviors. When the complexity of the dynamical behavior is quantified in terms of the spectrum of the correlation function of the variable describing the state of the system, we can say that the observed behaviors range from white noise to Brownian noise. These are two limiting cases very well known in linear dynamical systems. But complex non-linear systems also show, for intermediate noise levels, a very interesting $1/f$ behavior.

We have explored if such behavior is also observed in biological systems. In particular, we have observed that the correlations of the fluctuations are related to the rate of visitation to the different deterministic attractors that the static network and rules generate. Robustness and flexibility are necessary in biological systems, and thus the observed behavior may be precisely interpreted as the ability of the system to visit all the available attractors. Therefore, relating such compromise between robustness and flexibility to the complexity of the time signal is a remarkable finding. In this context, it is remarkable that the wild type experimentally grounded GRN studied, also showed a similar compromised behavior. Such behavior is lost when the probability of violating the documented rules is increased. The preliminary analysis provided here concerning the *Arabidopsis thaliana* GRN in response to noise, adds up to previous works showing that this, as well as other biological GRNs, show a critical behavior at the brisk between ordered and chaotic dynamics.

References

Albert, R. and Barabási, A.-L. (2002). Statistical mechanics of complex networks, *Rev. Mod. Phys.* **74**, pp. 47–96.

Albert, R. and Othmer, H. G. (2003). The topology of the regulatory interactions predicts the expression pattern of the segment polarity genes in drosophila melanogaster, *Journal of Theoretical Biology* **223**, pp. 1–18.

Aldana, M., Balleza, E., Kauffman, S. and Resendiz, O. (2007). Robustness and evolvability in genetic regulatory networks, *J. Theor. Bio.* **245**, pp. 433–448.

Aldana, M., Coppersmith, S. and Kadanoff, L. P. (2003). Boolean dynamics with random coupling, in E. Kaplan, J. E. Marsden and K. R. Sreenivasan (eds.), *Perspectives and Problems in Nonlinear Science. A celebratory volume in honor of Lawrence Sirovich*, Springer Applied Mathematical Sciences Series (Springer, Berlin), pp. 23–90.

Alvarez-Buylla, E., Benitez, M., Chaos, A., Cortes-Poza, Y., Espinosa-Soto, C., Beau-Lotto, R., Malkin, D., Escalera-Santos, G. and Padilla-Longoria, P. (2008). Floral morphogenesis: Stochastic explorations of a gene network epigenetic landscape, PLoS ONE (in press).

Amaral, L. A. N., Díaz-Guilera, A., Moreira, A. A., Goldberger, A. L. and Lipsitz, L. A.

(2004). Emergence of complex dynamics in a simple model of signaling networks, *Proc. Nat. Acad. Sci. USA* **101**, pp. 15551–15555.

Amaral, L. A. N., Scala, A., Barthélémy, M. and Stanley, H. E. (2000). Classes of small-world networks, *Proc. Nat. Acad. Sci. USA* **97**, pp. 11149–11152.

Balleza, E., Alvarez-Buylla, E. R., Chaos, A., Kauffman, S., Shmulevich, I. and Aldana, M. (2008). Critical dynamics in genetic regulatory networks: Examples from four kingdoms, *PLoS ONE* **3**, pp. e2456+.

Bassingthwaighte, J. B., Liebovitch, L. S. and West, B. J. (1994). *Fractal Physiology* (Oxford University Press, New York).

Blake, W. J., KAErn, M., Cantor, C. R. and Collins, J. J. (2003). Noise in eukaryotic gene expression. *Nature* **422**, pp. 633–637.

Boccaletti, S., Latora, V., Moreno, Y., Chavez, M. and Hwang, D.-U. (2006). Complex networks: Structure and dynamics, *Phys. Rep.* **424**, pp. 175–308.

Bowman, J. (1994). *Arabidopsis. An Atlas of Morphology and Development* (Springer-Verlag, New York, NY).

Buchman, T. G. (2002). The community of the self, *Nature* **420**, 6912, pp. 246–251.

Chaos, A., Aldana, M., Espinosa-Soto, C., Ponce de Leon, B. G., Arroyo, A. G. and Alvarez-Buylla, E. R. (2006). From genes to flower patterns and evolution: Dynamic models of gene regulatory networks, *J. Plant Growth Regul.* **25**, pp. 278–289.

Chen, X. (2004). A microRNA as a translational repressor of APETALA2 in Arabidopsis flower development. *Science* **303**, pp. 2022–2025.

Coen, E. S. and Meyerowitz, E. M. (1991). The war of the whorls: genetic interactions controlling flower development. *Nature* **353**, pp. 31–37.

Cohen, R. and Havlin, S. (2003). Scale-free networks are ultrasmall, *Phys. Rev. Lett.* **90**, p. art. no. 058701.

Díaz-Guilera, A., Moreira, A. A., Guzman, L. and Amaral, L. A. N. (2007). Complex fluctuations and robustness in stylized signalling networks, *Journal of Statistical Mechanics: Theory and Experiment* **2007**, 01, p. P01013.

Elowitz, M. B., Levine, A. J., Siggia, E. D. and Swain, P. S. (2002). Stochastic gene expression in a single cell. *Science* **297**, pp. 1183–1186.

Espinosa-Soto, C., Padilla-Longoria, P. and Alvarez-Buylla, E. (2004). A gene regulatory network model for cell-fate determination during Arabidopsis thalianal flower development that is robust and recovers experimental gene expression profiles, *Plant Cell* **16**, pp. 2923–2939.

Fiering, S., Whitelaw, E. and Martin, D. (2000). To be or not to be active: the stochastic nature of enhancer action, *Bioessays* **22**, pp. 381–387.

Goldberger, A. L., Amaral, L. A. N., Haussdorf, J. M., Ivanov, P. C., Peng, C.-K. and Stanley, H. E. (2002). Fractal dynamics in physiology: Alterations with disease and aging, *Proc. Nat. Acad. Sci. USA* **99 Sup 1**, pp. 2466–2472.

Gomez-Mena, C., de Folter, S., Costa, M., Angenent, G. and Sablowski, R. (2005). Transcriptional program controlled by the floral homeotic gene AGAMOUS during early organogenesis, *Development* **132**, pp. 429–438.

Ho, S., Biggar, S., Spencer, D., Schreiber, S. and Crabtree, G. (1996). Dimeric ligands define a role for transcriptional activation domains in reinitiation, *Nature* **382**, pp. 822–826.

Huang, S., Eichler, G., Bar-Yam, Y. and Ingber, D. E. (2005). Cell fates as high-dimensional attractor states of a complex gene regulatory network, *Phys. Rev. Lett.* **94**, p. 128701.

Hume, D. (2000). Probability in transcriptional regulation and its implications for leukocyte differentiation and inducible gene expression, *Blood* **96**, pp. 2323–2328.

Kaplan, D. and Glass, L. (1997). *Understanding Nonlinear Dynamics* (Springer-Verlag, Berlin).

Kauffman, S. A. (1969). Metabolic stability and epigenesis in randomly constructed genetic nets, *J. Theor. Biology* **22**, pp. 437–467.

Kauffman, S. A. (1993). *The Origins of Order: Self-Organization and Selection in Evolution* (Oxford University Press, Oxford).

Lipsitz, L. A. (2002). Dynamics of stability: The physiologic basis of functional health and frailty, *J. Gerontology* **57A**, 3, pp. B115–B125.

Malik, M. and Camm, A. J. (eds.) (1995). *Heart Rate Variability* (Futura, Armonk NY).

Mar, D. J., Chow, C. C., Gerstner, W., Adams, R. and Collins, J. J. (1999). Noise shaping in populations of coupled model neurons, *Proc. Nat. Acad. Sci. USA* **96**, pp. 10450–10455.

Mendoza, L. and Alvarez-Buylla, E. (1998). Dynamics of the genetic regulatory network for Arabidopsis thaliana flower morphogenesis, *J. Theor. Biol.* **193**, pp. 307–319.

Newman, M. E. J. and Watts, D. J. (1999). Renormalization group analysis of the small-world network model, *Phys. Lett.* **263A**, pp. 341–346.

Ozbudak, E. M., Thattai, M., Kurtser, I., Grossman, A. D. and van Oudenaarden, A. (2002). Regulation of noise in the expression of a single gene. *Nat. Genet.* **31**, pp. 69–73.

Paulsson, J. (2004). Summing up the noise in gene networks. *Nature* **427**, pp. 415–418.

Peng, C.-K., Havlin, S., Stanley, H. E. and Goldberger, A. L. (1995). Quantification of scaling exponents and crossover phenomena in nonstationary heartbeat time series, *Chaos* **5**, pp. 82–87.

Strogatz, S. H. (2001). Exploring complex networks, *Nature* **410**, pp. 268–276.

Taqqu, M. S., Teverovsky, V. and Willinger, W. (1995). Estimators for long-range dependence: an empirical study, *Fractals* **3**, pp. 785–798.

Thomas, R., Thieffry, D. and Kaufman, M. (1995). Dynamical behaviour of biological regulatory networks-I. Biological role of feedback loops and practical use of the concept of the loop-characteristic state, *Bull Math Biol* **57**, pp. 247–276.

von Dassow, G., Meir, E., Munro, E. M. and Odell, G. M. (2000). The segment polarity network is a robust developmental module. *Nature* **406**, pp. 188–192.

Walters, M., Fiering, S., Eidemiller, J., Magis, W., Groudine, M. and Martin, D. (1995). Enhancers increase the probability but not the level of gene-expression, *Proc. Natl. Acad. Sci. U. S. A.* **92**, pp. 7125–7129.

Watts, D. J. and Strogatz, S. (1998). Collective dynamics of "small-world networks, *Nature* **393**, pp. 440–441.

Weng, G., Bhalla, U. S. and Iyengar, R. (1999). Complexity in biological signaling systems, *Science* **284**, pp. 92–96.

Wolfram, S. (1994). *Cellular Automata and Complexity: Collected Papers* (Westview).

Wolfram, S. (2002). *A New Kind of Science* (Wolfram Media, Champaign, IL).

Chapter 6

Geometry and Topology of Folding Landscapes

Lorenzo Bongini[1] and Lapo Casetti[1,2]

[1] *Dipartimento di Fisica and Centro per lo Studio delle Dinamiche Complesse (CSDC), Università di Firenze, via G. Sansone, 1, 50019 Sesto Fiorentino (FI), Italy*

[2] *INFN, Sezione di Firenze, Italy*

6.1. Introduction

Protein folding is one of the most fundamental and challenging open questions in molecular biology. Proteins are polymers made of aminoacids and since the pioneering experiments by Anfinsen and coworkers [1] it has been known that the sequence of aminoacids—also called the primary structure of the protein—uniquely determines its native state, or tertiary structure, i.e., the compact configuration the protein assumes in physiological conditions and which makes it able to perform its biological tasks [2]. To understand how the information contained in the sequence is translated into the three-dimensional native structure is the core of the protein folding problem, and its solution would allow one to predict a protein's structure from the sole knowledge of the aminoacid sequence: being the sequencing of a protein much easier than experimental determination of its structure, it is easy to understand the impact such a solution would have on biochemistry and molecular biology. Despite many remarkable advances in the last decades, the protein folding problem is still far from a general solution [2].

A polymer made of aminoacids is referred to as a polypeptide. However, not all polypeptides are proteins: only a very small subset of all the possible sequences of the twenty naturally occurring aminoacids have been selected by evolution. According to our present knowledge, all the naturally selected proteins fold to a uniquely determined native state, but a generic polypeptide does not (it rather has a glassy-like behavior). Then the following question naturally arises: what makes a protein different from a generic polypeptide? or, more precisely, which are the properties a polypeptide must have to behave like a protein, i.e., to fold into a unique native state regardless of the initial conditions, when the environment is the correct one?

The *energy landscape* picture has emerged as a promising approach to answer this question. Energy landscape, or more precisely potential energy landscape, is the name commonly given to the graph of the potential energy of interaction between the microscopic degrees of freedom of the system [3]; the latter is a high-dimensional surface, but one can also speak of a free energy landscape when only its projection on a small set of collective variables (with a suitable average over all the other degrees of freedom) is considered [3]. Before having been applied to biomolecules, this concept has proven useful in the study of other complex systems, especially of supercooled liquids and of the glass transition [4]. The basic idea is very simple, yet powerful: if a system has a rugged, complex energy landscape, with many minima and valleys separated by barriers of different height, its dynamics will experience a variety of time scales, with oscillations in the valleys and jumps from one valley to another. Then one can try to link special features of the behavior of the system (i.e., the presence of a glass transition, the separation of time scales, and so on) to special properties of the landscape, like the topography of the basins around minima, the energy distribution of minima and saddles connecting them and so on. Anyway, a complex landscape yields a complex dynamics, where the system is very likely to remain trapped in different valleys when the temperature is not so high. This is consistent with a glassy behavior, but a protein does not show a glassy behavior, it rather has relatively low frustration. This means that there must be some property of the landscape such to avoid too much frustration. This property is commonly referred to as the *folding funnel* [5]: though locally rugged, the low-energy part of the energy landscape is supposed to have an overall funnel shape so that most initial conditions are driven towards the correct native state. The dynamics must then be such as to make this happen in a reasonably fast and reliable way, i.e., non-native minima must be efficiently connected to the native state so that trapping in the wrong configuration is unlikely.

However, a direct visualization of the energy landscape is impossible due to its high dimensionality, and its detailed properties must be inferred indirectly. In the following, we will describe two strategies to analyze the energy landscape of model proteins: a *local* one and a *global* one. The former strategy, addressed to in Sec. 6.2, is essentially topological in character and amounts to define a network (a graph) whose nodes are the minima of the potential energy and whose edges are the saddles connecting them. This network has, however, to be properly renormalized in order to yield significant results: the renormalization procedure is described in Sec. 6.2.1. The latter strategy (Sec. 6.3) is instead a geometrical one, and is based on defining global geometric quantities able to characterize the folding landscape as a whole. In both cases interesting information about the differences between the landscapes of proteinlike systems and those of generic polymers can be obtained.

6.2. The Connectivity Graph of the Energy Landscape

We can think of the energy landscape as the collection of the basins of attraction of several local minima of the potential. A basin of attraction of a local minimum is the collection of points in phase space, whose overdamped dynamics converges to that minimum. Accordingly, any trajectory in phase space evolves by going through contiguous basins of attractions of different local minima. In this sense we can think of the dynamics as a sequence of transitions between different minima separated by energy barriers corresponding to saddles. This essentially amounts to approximate the dynamics as a sequence of thermally activated transitions between different local minima. We expect that such an approximation is a suitable one when the temperature of the system (expressed in units of the Boltzmann constant) is small enough to be comparable with the typical height of the saddles separating nearby minima. Relying upon these considerations, we can assume that for sufficiently (but no too) small temperatures the molecular dynamics can be effectively replaced by a stochastic dynamics defined onto a connected graph. This approach has been put forward in [6], where the graph has been defined such as its nodes are the local minima and their local connectivity is determined by the existence of a saddle separating them from other local minima. The transition rate between connected minima can be determined by purely geometric features of the energy landscape as a suitable generalization of the Arrhenius law to a high–dimensional space, i.e., Langer's formula, see Eq. (6.1).

Reconstructing the energy landscape of a protein model amounts then to first identify all the local minima of the potential energy. Since the number stationary points of the potential energy typically grows exponentially with the number of degrees of freedom, such a task is practically unfeasible for accurate all-atom potential energies, but a reasonable sampling of the minima may become accessible for minimalistic potentials. Minimalistic models are those where the polymer is described at a coarse-grained level, as a chain of N beads where N is the number of aminoacids; no explicit water molecules are considered and the solvent is taken into account only by means of effective interactions among the monomers. Minimalistic models can be relatively simple, yet in some cases yield very accurate results which compare well with experiments [7, 8]. The local properties of the energy landscape of minimalistic models have been recently studied (see e.g. Refs. [9–13]) and very interesting clues about the structure of the folding funnel and the differences between protein-like heteropolymers and other polymers have been found: in particular, it has been shown that a funnel-like structure is present also in homopolymers, but what makes a big difference is that in protein-like systems jumps between minima corresponding to distant configurations are much more favoured dynamically [14].

As stated above, at sufficiently low temperatures, the dynamics of many systems characterized by a rough energy landscapes can be summarized as a quick random wandering inside metastable states intertwined by thermally activated jumps to a new state. Typically a metastable state consists in a set of minima of the potential energy whose basins of attraction are separated from the rest of the configuration space by energy barriers sufficiently high to determine a separation of time scales between the exploration of the set itself and the average time necessary to leave it. In the limit of extremely low temperature each minimum of the potential corresponds to a single metastable state. The rate of the transition between two potential energy minima is dictated by the energy and the shape of the highest energy point on the minimal energy path connecting them, which, for C^2 potentials, can be shown to be a saddle of the first order of the potential. The rate of the passage from the i-th to the j-th minimum can be approximated, for realistic friction coefficients and sufficiently low temperatures, by the Langer estimate [15] which corrects the standard Arrhenius term by an entropic factor depending on the curvature of the potential both in the saddle and the starting minimum:

$$\Gamma_{i,j} = \frac{\omega_{\parallel\, i,j}}{\pi\gamma} \frac{\prod_{k=1}^{N'} \omega_i^{(k)}}{\prod_{k=1}^{N'-1} \omega_{\perp\, i,j}^{(k)}} \exp\left(-\frac{V_{s\, i,j} - V_i}{k_B T}\right) \tag{6.1}$$

where the $\omega_i^{(k)}$'s are the $N' = dN - 3$ non zero eigenfrequencies of the minimum i (d is the spatial dimension), $\omega_\perp^{(k)}$'s are the $N' - 1$ non zero frequencies of the saddle lying on the border between the basins of attraction of i and j, while $\omega_\parallel$ is associated with the only expanding direction. Finally, γ is the dissipation rate, while the exponential factor depends on the height of the energy barrier, $V_{s\, i,j} - V_i$, normalized to the reduced temperature $k_B T$, k_B being the Boltzmann constant. The analysis of saddles and their crossing rates becomes then essential to the study of protein folding, because they summarize the only relevant dynamic contribution as far as large conformational changes are concerned. In [16] it has been shown that distant minima are connected with saddles characterized by higher jumping rates in the case of a fast-folding hetero-polymer. This observation nicely fits with the scenario of the diffusion on the energy landscape of a fast-folder being helped by the existence of fast connections between distant configurations.

Given this scenario it is then natural to study the folding dynamics of a protein as a diffusion process on a *connectivity graph*, that is a graph whose $N_{\min}$ nodes represent the basins of attraction of the minima of the energy landscape of the protein and whose N_{sad} edges represent connections between such basins, i.e., first order saddles. For a realistic description connections must be weighted according to their dynamical relevance, which is well represented by the corresponding rate of barrier crossing. In such a framework the entire system can be summarized by a non–symmetric $N_{\min} \times N_{\min}$ connectivity matrix Γ whose element $\Gamma_{i,j}$ equals the jumping rate from the the i-th to the j-th minimum, if the two are directly connected, and is 0 otherwise.

According to this formulation the probability that the protein resides in the basin of attraction of the minimum i obeys the following master equation:

$$\dot{P}_i = \sum_{j=1}^{N_{\min}} P_j \Gamma_{j,i} - P_i \sum_{j=1}^{N_{\min}} \Gamma_{i,j} \tag{6.2}$$

which can be cast into the matrix form $\dot{P} = WP$ by defining an evolution matrix W such that

$$W_{i,j} = \Gamma_{j,i} - \delta_{i,j} \sum_{j=1}^{N_{min}} \Gamma_{i,j}. \tag{6.3}$$

In graph theory literature $L = -W$ is often referred to as the Laplacian matrix. The solutions of the master equation can be straightforwardly derived by diagonalizing the evolution W; unfortunately, the latter operation is often computationally quite expensive. One specific eigenstate can nonetheless be easily computed without need of any diagonalization: the kernel P_0 which corresponds to the stationary condition on the graph ($\dot{P}_0 = 0$). By setting the left hand side of equation (6.2) equal to 0 one finds that the components of P_0 are:

$$P_{0,i} = \alpha \frac{e^{-\frac{V_i}{k_B T}}}{\prod_{k=1}^{N'} \omega_i^{(k)}}, \tag{6.4}$$

where α is a normalization constant such that $\sum_i P_{0,i} = 1$. Note that the stationary probability only depends on minima and any information about the saddles is lost. Actually the $P_{0,i}$'s correspond to the basin occupation probabilities that one would obtain by approximating the potential at the second order in the minima and by computing the corresponding partition function. A detailed balance condition $P_{0,i}\Gamma_{i,j} = P_{0,j}\Gamma_{j,i}$ holds for every connected pair i, j.

Connectivity graphs are weighted directed graphs, each connection having a different weight according to the crossing direction. Simpler descriptions might nonetheless prove useful as well. By forgetting the dynamical weights on the connections one can define two matrices $\Gamma_{discrete}$ and $W_{discrete}$ that share the same relationship as Γ and W previously defined and such that the i, j−th element of $\Gamma_{discrete}$ is one if the two minima are connected and 0 otherwise. The discrete Laplacian matrix $L_{discrete} = -W_{discrete}$ is of fundamental interest because the power–law behavior of the low frequency part of its spectral density allows to define the spectral dimension of the graph, a generalization of the Euclidean dimension for graphs that are not defined on a regular lattice (see Sec. 6.2.2.1).

6.2.1. *Renormalization of the graph*

As stated above, the most dynamically sensible definition of connectivity graph is that of a graph whose nodes correspond to metastable states of the system. Thus the graphs where each node corresponds to a minimum of the potential energy can be considered a good approximation of the connectivity graph only in the zero temperature limit. Actually, as temperature increases, regions of phase space previously corresponding to different metastable states might fuse in a unique metastable state because the energy barrier that divides them has become dynamically negligible. At realistic temperatures real metastable states consist in conglomerates of basins of attraction of different minima and are separated from each other by saddles whose energy is still significantly higher than the temperature.

In [6] a coarse graining procedure has been put forward, aiming at defining a reasonable approximation of the connectivity graph at a given temperature. The procedure goes as follows. We first of all preliminarily sort all saddles in ascending order according to their minimal energy barrier, that is, the barrier experienced by crossing each saddle coming from the minimum of higher energy between the two minima it connects. Then, for each saddle of minimal energy barrier lower than a temperature dependent renormalization threshold, we perform the following operations:

- after identifying the lower in energy between the two minima it connects, we substitute its index for that of the higher one in the annotations of all the saddles (that is: if the saddle connects the i-th and the j-th minimum, and i is lower in than j, we rename i for j everywhere).
- we erase all self connecting saddles (that is, the saddles whose header states that they connect minimum k with minimum k).

After this procedure has been iteratively repeated for the saddles with the appropriate minimal energy barrier, all saddles are screened for multiple connections and, in case that more than one saddle is found between two minima, the lower energy one is chosen. The nodes that survive this procedure actually correspond to clusters of minima that approximate the metastable states of the system.

The rate of transition between such metastable states can be assigned by specifying a procedure for the updating of the weights of the graph connections during the renormalization procedure. We follow the prescription given in [6]: we first of all assign to each node of the zero–temperature graph a reference probability Π_i that corresponds to the stationary probability of the corresponding minimum of the potential energy, then, each time two different metastable states i and j coalesce to a new one i', we perform the following operations:

- the rate $\Gamma_{i',k}$ of the connection from i' to k is assigned as the weighted mean of the rates of the connections from i to k and from j to k, the weights being the reference probabilities of i and j respectively:

$$\Gamma_{i',k} = \frac{\Pi_i \Gamma_{i,k} + \Pi_j \Gamma_{j,k}}{\Pi_i + \Pi_j} \qquad (6.5)$$

- the rate of the connection from k to i' is assigned as the sum of the rates of the connections from k to i and from k to j

$$\Gamma_{k,i'} = \Gamma_{i,k} + \Gamma_{j,k} \qquad (6.6)$$

- the new reference probability of i' is assigned as the sum of the reference probabilities of i and j:

$$\Pi_{i'} = \Pi_i + \Pi_j \qquad (6.7)$$

It can be shown that the reference probability as defined by this renormalization prescription is still a stationary state of the Laplacian matrix of the renormalized graph. Moreover this probability distribution ensures that the renormalized transition rates satisfy the detailed balance.

We stress that connectivity graphs obtained by applying the renormalization protocol here described depend on temperature in a twofold manner: the obvious Arrhenius dependency of the jumping rates and the somewhat arbitrary renormalization threshold that dictates which connections to coalesce. In order to realistically reproduce the metastable states of the system the latter must be of the same order of magnitude of the temperature. It is however difficult to devise any theoretical argument suggesting what should be its precise numerical value. In [6] it has been shown that the large scale dynamical features of the system do not crucially depend on the choice of this parameter and that the renormalization procedure does not alter the long timescales of the dynamics on the graph.

6.2.2. *Two-dimensional models*

The idea of approximating the thermalized dynamics of a polypeptidic chain by a stochastic dynamics on a directed graph was tested using a very simple model, first introduced in [17]. This is a slight modification of the 2–d off-lattice HP model originally proposed by Stillinger *et al.* in [18]. It is defined by the Hamiltonian

$$H = \sum_{i=1}^{L} \frac{p_{x,i}^2 + p_{y,i}^2}{2} + \sum_{i=1}^{L-1} V_1(r_{i,i+1})$$

$$+ \sum_{i=2}^{L-1} V_2(\theta_i) + \sum_{i=1}^{L-2} \sum_{j=i+2}^{L} V_3(r_{ij}, \xi_i, \xi_j) \tag{6.8}$$

which contains the phenomenological potentials

$$V_1(r_{i,i+1}) = \alpha(r_{i,i+1} - r_0)^2 \,,$$

$$V_2(\theta_i) = \frac{1 - \cos\theta_i}{16} \,, \tag{6.9}$$

$$V_3(r_{i,j}) = \frac{1}{r_{i,j}^{12}} - \frac{c_{i,j}}{r_{i,j}^6} \,.$$

All the parameters are expressed in terms of adimensional arbitrary units: for instance, α and r_0 are fixed to the values 20 and 1, respectively. The model Hamilonian represents a one-dimensional chain of L point-like monomers corresponding to the residues of a real protein. Only two types of residues are considered: hydrophobic, H, and polar, P . Accordingly, a heteropolymer is identified by a sequence of discrete variables $\{\xi_i\}$ (with $i = 1, \ldots, L$) along the chain: $\xi_i = \pm 1$ indicates that the i-th residue is of type H or P, respecively. The intramolecular potential is composed of three terms: a stiff nearest-neighbour harmonic potential V_1, which keeps the bond distance almost constant, a three-body potential V_2, which measures the energetic cost of local bending, and a long–range Lennard-Jones potential V_3 acting between all pairs of monomers i and j such that $|i - j| > 1$. For the sake of simplicity, the monomers are assumed to have the same unitary mass. The space coordinates of the i-th monomer are (x_i, y_i) and their conjugated momenta are $(p_{x,i}, p_{y,i}) = (\dot{x}_i, \dot{y}_i)$. The variable $r_{i,j} = \sqrt{(x_i - x_j)^2 + (y_i - y_j)^2}$ is the distance between i-th and j-th monomer and θ_i is the bond angle at the i-th monomer. V_3 is the only contribution that depends on the nature of the monomers. The coefficients $c_{i,j} = \frac{1}{8}(1 + \xi_i + \xi_j + 5\xi_i\xi_j)$ are defined in such a way that the interaction is attractive if both residues are either hydrophobic or polar (with $c_{i,j} = 1$ and $1/2$, respectively), while it is repulsive if the residues belong to different species ($c_{ij} = -1/2$).

Here, we focus our investigation on three sequences of twenty monomers that represent the three classes of different folding behaviors observed in this model:

- [S0] a homo-polymer composed of 20 H residues;
- [S1]=[HHHP HHHP HHHP PHHP PHHH] a sequence that has been identified as a fast-folder in [19];
- [S2]=[PPPH HPHH HHHH HHHP HHPH] a randomly generated sequence, that has been identified as a slow-folder in [17].

The three characteristic temperatures T_θ (collapse temperature), T_f (folding temperature), T_g ("glassy" temperature) determined by means of Langevin simulations in [10] are reported in Table 6.1 for the three analyzed sequences.

Table 6.1.   Characteristic temperatures for the analyzed sequences.

	homopolymer S0	fast-folder S1	slow-folder S2
T_θ	0.16	0.11	0.13
T_f	0.044	0.061	0.044
T_g	0.022	0.048	0.025

A thorough discussion of numerical techniques that can be used to effectively reconstructing the connectivity graph for the models above and and to test the reliability of the graph in reproducing the observed equilibrium and nonequilibrium properties of the models can be found in [6].

6.2.2.1. *Topological properties of the renormalized connectivity graph*

We now consider the topological properties of the connectivity graph and in particular we focus on how the renormalization procedure affects the topology. We perform this analysis on all the three sequences previously introduced with the aim to describe how the topology of the renormalized connectivity graph influences the folding propensity of the system (see [6] for the details).

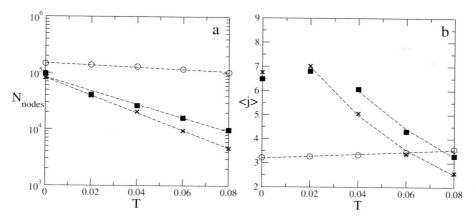

Fig. 6.1. Order (a) and average connectivity (b) of the renormalized connectivity graph as a function of temperature. Empty circles correspond to the homopolymer, filled squares and crosses to the slow–folding and fast–folding heteropolymers, respectively. Dashed lines represent least–square exponential fits to the data.

Preliminary to the study any topological characteristic we analyze how renormalization affects the size of the graphs under study. In Fig. 6.1(a) we show the order of the renormalized connectivity graph, i. e., the number of nodes it is composed of, for a temperatures that encompasses both the glassy and transitions $T = 0.00$, 0.02, 0.04, 0.06, 0.08 of all the sequences under study. The effect of renormalization on graph sizes are significantly weaker for the homopolymer than for the two heteropolymers. While in the first case the order of the connectivity graph reduces by a mere 30%, in the latter cases it decreases by an order of magnitude or more.

The different sensitivity of the sizes of connectivity graphs to temperature changes is an effect of the differences in the distribution of the energy barrier heights W. Figure 6.2 shows the distribution of W for the three analyzed sequences both at high and low energies (main panel and inset, respectively). A general shift toward high barriers can be observed in the case of the homopolymer. The consequent decrease in the fraction of low energy barriers explains the relative insensitivity of the homopolymer to renormalization: in this temperature range the same increase in temperature causes the coalescence of many less connections in the homopolymer than in the heteropolymer.

As far as topology is concerned, Fig. 6.1(b) shows the average node connectivities of the sequences under study. The homopolymer, that is characterized by the presence of many more metastable states at all temperatures, has instead a lower connectivity than heteropolymers, almost 3 against 6 at $T = 0$. This discrepancy however attenuates as temperature increases: while the homopolymer connectivity remains substantially unchanged with temperature, the connectivity of heteropolymers markedly decreases. This drop reflects a tendency of the connectivity distributions of the two heteropolymers to peak at low values.

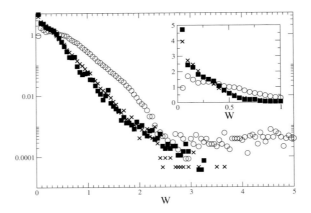

Fig. 6.2. Histogram of energy barrier heights for the three analyzed sequences. Empty circles correspond to the homopolymer, filled squares and crosses respectively to the slow–folding and fast–folding heteropolymers. In the inset the rightmost part of the histogram is reported in lin–lin scale.

The difference in the effects of renormalization on connectivities deserves to be investigated in deeper detail. Indeed, while the number of nodes clearly decreases with renormalization, it is much less obvious how should behave an intensive quantity as the local connectivity. Analyzing the details of the renormalization process one might notice that, at each renormalization step, the average number of connections might both increase or keep unchanged according to the amount of shared connections between the two coalescing nodes. If the coalescing nodes have k and k' connections of which they share m, the resulting renormalized node will have $k + k' - m - 2$, where the -2 term takes into account the disappearing connection between the two nodes. The two opposite cases that might take place are $m = 0$ (no shared connections) and $m = k - 1 = k' - 1$ (the two nodes share all connections) which lead to a final node with $k + k' - 2$ and $k - 1$ connections respectively. During renormalization on a realistic graph m will in principle take all possible values. Therefore, if m is small in average the connectivity of the graph will tend to increase, while it will decrease for large m values.

In order to confirm that the observed drop in the connectivity of heteropolymers actually depends on an high percentage of shared connections between coalescing nodes we analyze the relation between node connectivity and their size, that is the number of zero–temperature nodes that coalesce on a given node. In Fig. 6.3 we plot the degree of the nodes of the renormalized graphs of the three sequences under study at $T = 0.08$ versus their size. Data have been averaged at every node size for display purposes. A least–square fit with a power law (dashed lines in the figure) highlights a clear difference between homo and heteropolymers. While in the first case the best fitting exponent is 1, in the other two cases it is significantly smaller, indicating that renormalization in the heteropolymers causes the deletion of many connections.

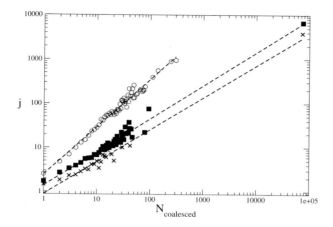

Fig. 6.3. Average degree of the nodes of the renormalized graph at $T = 0.08$ versus their size, measured in number of nodes of the un–renormalized graph that coalesced in each of them ($N_{\text{coalesced}}$). Data have been binned along the horizontal axis with step 1. Stars refer to the fast–folding heteropolymer, filled squares to the slow-folding heteropolymer and empty circles to the hydrophobic homopolymer. Dashed lines are power law fits of exponents 0.73, 0.75 and 1.1.

As previously anticipated, the connectivity of the lowest energy node of the graph deserves a separate consideration. First, it is important to stress that, following renormalization, this becomes by far the most connected node in the graph. This effect is definitely more pronounced in the case of the heteropolymers as one can see in Fig. 6.3 where the lowest energy nodes correspond to the points on the extreme right of the curves there displayed. We can conclude that, as far as connectivity is concerned, the lowest energy node behaves somehow opposite to the rest of the connectivity graph upon renormalization: it tends to acquire connections while the rest of the graph tends to lose it. The effect is strong for the heteropolymers and much weaker for the homopolymer.

One of the most informative topological quantity when studying diffusion over a graph is the *spectral dimension* of the graph itself. It has long been known that recurrence and first passage times in Markov chains on regular lattices diverge with the system size for lattice dimensions higher than 2. These results have been recently extended to general graphs by making use of the spectral dimension [20] that is the exponent describing how the spectral density of the graph vanishes at low eigenvalues. More specifically the spectral density of the discrete Laplacian matrix L_{discrete} of an infinite graph vanishes for low eigenvalues as $\omega^{\tilde{d}-1}$ where $\tilde{d}$ is defined as the spectral dimension. This quantity still controls the rate of divergence of first passage times but loses many relations to other topological quantities: for example, the dimensionality of a regular lattice is proportional to its average connectivity, the spectral dimension of an irregular graph is not.

Although $\tilde{d}$ is rigorously defined only for infinite graphs, a closely related quantity, $\bar{d}$, can be computed also for finite graphs by means of an asymptotic power–law

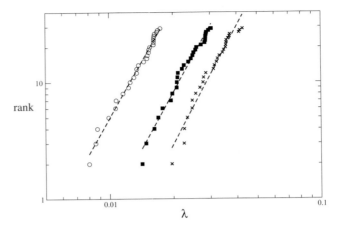

Fig. 6.4. Rank–to–eigenvalue plot for the discrete Laplacian matrix of the 0-temperature connectivity graph for the homopolymer (empty circles), the slow-folding heteropolymer (filled squares) and the fast-folding heteropolymer (crosses). The three dashed lines refer to power–law least square fits of exponents 3.2, 3.3 and 3.4 respectively. In order to ease the distinction between different curves the x-variable of the curve referring to the slow-folder has been multiplied by an arbitrary factor.

fit of the low frequency part of the spectral density. This last quantity approximates the spectral dimension in the sense that it converges to it as the graph size goes to infinity. It still however retain the same basic meaning also for finite system sizes. In fact, while in finite graphs, as those under study, the recurrence and first passage times are always finite, a lower $\bar{d}$ still implies a higher difficulty in exploring the graph. In order to better elucidate this concept we recall that the spectral dimension is computed by diagonalizing the discrete Laplacian matrix, that is a Laplacian matrix where off–diagonal entries do not correspond to the full transition rates but only to zeroes and ones depending on the existence of a connection between the corresponding nodes. Such a matrix retains the information about the graph topology but looses any information about the kinetics of the system. As a consequence its spectral properties have to be interpreted in terms of a discrete hopping process between nodes: the eigenvalues of the discrete Laplacian matrix don't relate to time scales but rather to path lengths. In this respect high spectral dimensions imply a higher density of long relaxation paths.

In Fig. 6.4 we plot the rank of the first 30 eigenvaluess of the discrete Laplacian matrix against their numerical value for each of the zero–temperature connectivity graphs associated with the three analyzed sequences. The resulting curves are proportional to the integral of the spectral density of the discrete Laplacian matrix, and can be thus least–square–fitted with a power–law in order to estimate the corresponding spectral dimension, which is equal to twice the fitted exponent. This procedure holds an essentially constant result, namely $\bar{d} = 6.4$ for the homopolymer, $\bar{d} = 6.5$ for the slow–folding heteropolymer and $\bar{d} = 6.8$ for the fast–folding one.

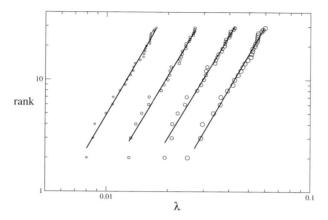

Fig. 6.5. Rank–to–eigenvalue plot for the discrete Laplacian matrix of the homopolymer for five different temperatures: $T = 0.00, 0.02, 0.04, 0.06, 0.08$. Increasing circle size corresponds to increasing temperature. Continuous lines refer to power–law least–square fits of exponents 3.2, 3.1, 3.05 and 3.1 respectively. In order to ease the distinction of different curves the x-variable of each curve has been multiplied by an arbitrary factor.

It is important to notice that the spectral dimension is a very robust quantity, being invariant for topological rescaling such as the renormalizing procedure we introduced, local link redirection, decimation or creation. This property, which is again rigorously true only in case of infinite graphs, still holds approximately for sufficiently large graphs. Such topological invariance is very useful when dealing with graphs that are a sample of other graphs, because it implies that the spectral dimension will not change dramatically if some connections are lost or wrongly assigned.

In order to study how $\bar{d}$ behaves with temperature we repeat the same evaluation procedure for the renormalized graph at the four usual temperatures: $T = 0.02, 0.04, 0.06, 0.08$. The resulting rank–to–eigenvalue plots are shown, together with the 0-temperature case, in Figs. 6.5, 6.6, 6.7. While Fig. 6.5 shows that, in the case of the homopolymer, the measured spectral dimension does not change with temperature, the two figures referring to the heteropolymers (6.7, 6.6) show a substantial decrease in this quantity for increasing temperatures. More precisely, while the spectral dimension of the zero–temperature connectivity graph of the two heteropolymers is around 6.5, at higher temperatures it stabilizes around 5, suggesting that the configuration space of heteropolymers appears more compact to a random exploration than that of a homopolymer.

In order to rule out that this change in the dimensionality of the system is a finite size effect we recall that between $T = 0.02$ and $T = 0.08$, a temperature range where spectral dimension is constant for all sequences, renormalization causes the number of nodes of the connectivity graph of the heteropolymers to decrease by an order of magnitude (see Fig. 6.1). Also the graph diameter, measured as the maximum

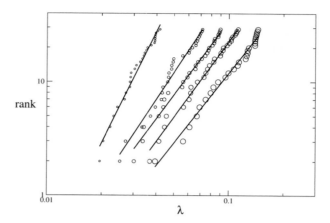

Fig. 6.6. Rank–to–eigenvalue plot for the discrete Laplacian matrix of the slow-folding heteropolymer for five different temperatures: $T = 0.00, 0.02, 0.04, 0.06, 0.08$. Increasing circle size corresponds to increasing temperature. Continuous lines refer to power–law least square fits of exponents 3.3, 2.4, 2.2, 2.2 and 2.05, respectively.

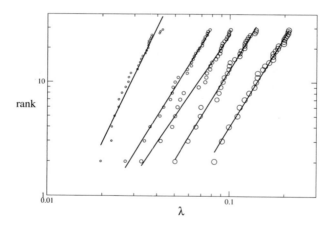

Fig. 6.7. Rank–to–eigenvalue plot for the discrete Laplacian matrix of the fast–folding heteropolymer for five different temperatures: $T = 0.00, 0.02, 0.04, 0.06, 0.08$. Increasing circle size corresponds to increasing temperature. Continuous lines refer to power–law least–square fits of exponents 3.4, 2.6, 2.4, 2.6 and 2.6, respectively.

shortest–path from from the minimal energy node (here not very rigorously, yet reasonably, considered as the center of the graph), decreases rather smoothly in the temperature range considered, further confirming that the observed effect depicts a genuine topological change and is not due to the limited system size.

The early drop in the spectral dimension that we observe implies that much of the dimensionality of the system lies in those subnetwork that collapse on a single node at very low temperatures. The corresponding nodes represent the minima of the potential energy separated by the smallest activation barriers. The energy landscape in heteropolymers is thus showing a different dimensionality when imaged at different energy resolutions. Diffusion on such a landscape will therefore be not only slower but also more involved at very low temperatures, requiring a lengthy wandering on an involved network of tiny wrinkles.

The scenario emerging from the analysis of the spectral properties of connectivity graphs of these very simplified models (which nonetheless exhibit the same qualitative kinds of behaviors of real proteins as the approach to the folded state is considered) is that the zero–temperature connectivity graph, which contains no information at all about saddle height, is not able to discern any detail of amino–acidic sequence of the system. As soon as some information about saddles is introduced in the form of a finite temperature connectivity graph, the distinction between homo and heteropolymers becomes evident and it is highlighted by a smaller spectral dimension for the connectivity graph of heteropolymers. In this model the higher frustration of homopolymers does indeed result in a more complicated energy landscape but only when looking at a coarse grained level the energy landscape: the fine details are frustration independent.

6.2.3. *Three-dimensional models*

Let us now briefly discuss some results recently obtained applying the techniques described above to a fully three-dimensional off-lattice coarse-grained model for short peptides that has been recently studied by Clementi and coworkers [7, 21]: further details can be found in [22].

The analysis of the latter model is confirming many of the features described in Sec. 6.2.2, while at the same time providing a glimpse on the phenomenological variability of these systems. A totally new phenomenon that is detected is a gradual transformation of the connectivity graph in a *star* as temperature approaches the T_θ from below. Star graphs are graphs characterized by the presence of a unique central node connected to many "leaves" which are themselves not linked to each other. The central node is actually the fastest growing one and corresponds to the minimal energy node.

Notably enough, in this model, the spectral dimension of the connectivity graph is basically independent on temperature, at least before leaf nodes appear and the graph starts its transformation in to a star. At variance with the two-dimensional case, it does show already at $T = 0$ an interesting correlation with the properties of the aminoacidic sequence. Actually for three analyzed sequences of the same length, a homopolyer and two heteropolymers characterized by different folding propensities, we notice that the spectral dimension of the bad folding heteropolymer is comparable to that of the homopolymer ($\bar{d} = 17 \pm 1$ in the first case and $\bar{d} = 16 \pm 1$ in the latter). On the contrary the good folder is characterized by a much smaller dimensionality: $\bar{d} = 10.3 \pm 0.6$. In this three-dimensional model the spectral dimension seems then to be an even more useful quantity than in the 2–d case to discriminate between good and bad folders.

6.3. Geometry of Energy Landscapes

The local strategy to analyze energy landscapes described in Sec. 6.2 requires a huge computational effort. So the following question arises: is there some *global* property of the energy landscape which can be easily computed numerically as an average along dynamical trajectories and which is able to identify polymers having a protein-like behavior? We shall show in the following that such a quantity indeed exists, at least for the minimalistic model we considered, and that it is of a geometric nature. In particular, the fluctuations of a suitably defined curvature of the energy landscape clearly mark the folding transition while do not show any remarkable feature when the polymer undergoes a hydrophobic collapse without a preferred native state. This is at variance with thermodynamic global observables, like the specific heat, which show a very similar behavior in the case of a folding transition and of a simple hydrophobic collapse.

The intuitive reason why geometric information on the landscape, and especially curvature, could be relevant to the problem of folding is that the dynamics on a landscape would be heavily affected by the local curvature: minima of the energy landscape are associated to positive curvatures and stable dynamics, while saddles involve negative curvatures, at least along some direction, thus implying some instability. One can reasonably expect that the arrangement and detailed properties of minima and saddles might reflect in some global feature of the distribution of curvatures of the landscape, when averaged along a typical trajectory.

The definition of the curvature of a manifold M depends on the choice of a metric g [23, 24]: once the couple (M, g) is given, a covariant derivative and a curvature tensor $R(e_i, e_j)$ can be defined; the latter measures the noncommutativity of the covariant derivatives in the coordinate directions e_i and e_j. A scalar measure of the curvature at any given point $P \in M$ is the the sectional curvature

$$K(e_i, e_j) = \langle R(e_i, e_j)e_j, e_i \rangle , \qquad (6.10)$$

where $\langle \cdot, \cdot \rangle$ stands for the scalar product. At any point of an N-dimensional manifold there are $N(N-1)$ sectional curvatures, whose knowledge determines the full curvature tensor at that point. One can however define some simpler curvatures (paying the price of losing some information): the Ricci curvature $K_R(e_i)$ is the sum of the K's over the $N-1$ directions orthogonal to e_i,

$$K_R(e_i) = \sum_{j=1}^{N} K(e_i, e_j) , \qquad (6.11)$$

and summing also on the N directions e_i one gets the scalar curvature

$$\mathcal{R} = \sum_{i=1}^{N} K_R(e_i) = \sum_{i,j=1}^{N} K(e_i, e_j) ; \qquad (6.12)$$

then, $\frac{K_R}{N-1}$ and $\frac{\mathcal{R}}{N(N-1)}$ can be considered as average curvatures at a given point.

Although one expects the association between minima and positive curvatures on the one side and negative curvatures along some directions and saddles on the other side to be essentially true for most choices of the metric, a particular choice of g among the many possible ones must be made in order to perform explicit calculations. The most immediate choice would probably be that of considering as our manifold M the N-dimensional surface $z = V(q_1, \ldots, q_N)$ itself, i.e., the graph of the potential energy V as a function of the N coordinates $q_1, \ldots, q_N$ of the configuration space, and to define g as the metric induced on that surface by its immersion in $\mathbb{R}^{N+1}$. Although perfectly reasonable, this choice has two drawbacks: (i) the explicit expressions for the curvatures in terms of derivatives of V are rather complicated and (ii) the link between the properties of the dynamics and the geometry is not very precise, i.e., one cannot prove that the geometry completely determines the dynamics and its stability. For these reasons in Refs. [25] a particular choice of (M, g), referred to as the Eisenhart metric [26], such that the link between geometry and dynamics is more clear was proposed. The properties of this metric have been previously thoroughly reviewed in [27, 28] so that we refer the reader to the latter references for the details. Let us only recall that given a potential energy $V(q_1, \ldots, q_N)$ this (pseudo-Riemannian) metric is defined on a the configuration space with two extra dimensions, $M \times \mathbb{R}^2$, with local coordinates $(q_0, q_1, \ldots, q_N, q_{N+1})$, and its arc-length is

$$ds^2 = \delta_{i,j} dq^i dq^j - 2V(q)(dq^0)^2 + 2dq^0 dq^{N+1} . \tag{6.13}$$

The metric tensor will be referred to as g_E and its components are

$$g_E = \begin{pmatrix} -2V(q) & 0 & \cdots & 0 & 1 \\ 0 & 1 & \cdots & 0 & 0 \\ \vdots & \vdots & \ddots & \vdots & \vdots \\ 0 & 0 & \cdots & 1 & 0 \\ 1 & 0 & \cdots & 0 & 0 \end{pmatrix} \tag{6.14}$$

as can be derived by Eq. (6.13).

The geodesics of this metric are the natural motions of a Hamiltonian system with standard kinetic energy and potential energy V (see [27]). The nonvanishing components of the curvature tensor are

$$R_{0i0j} = \partial_i \partial_j V ; \tag{6.15}$$

it can then be shown that the Ricci curvature (6.11) in the direction of motion, i.e., in the direction of the velocity vector v of the geodesic, is given by

$$K_R(v) = \triangle V , \tag{6.16}$$

where $\triangle V$ is the Laplacian of the potential V, and that the scalar curvature $\mathcal{R}$ identically vanishes. We note that $K_R(v)$ is nothing but a scalar measure of the average curvature "felt" by the system during its evolution; we will refer to it simply as K_R dropping the dependence on the direction. Another feature of K_R is its very simple analytical expression which simplifies both analytical calculation and numerical estimates. It is also worth noticing that expression (6.16) is a very natural and intuitive measure of the curvature of the energy landscape, as it can be seen as a naive generalization of the curvature $f''(x)$ of the graph of a one-variable function to the graph of the N-dimensional function $V(q_1, ..., q_N)$: the Laplacian of the function. However, the previous discussion shows that it is much more than a naive measure of curvature and that it contains information on the local neighborhood of the dynamical trajectories.

The Ricci curvature defined in Eq. (6.16) will be used to characterize the geometry of the energy landscape.

6.3.1. *Curvature of the energy landscape of simple model proteins*

Let us now describe the model whose energy landscape geometry was studied. We considered a simple model able to describe protein-like polymers as well as polymers with no tendency to fold; the different behaviors being selected upon the choice of the amino acidic sequence. The model we chose is a minimalistic model originally introduced by Thirumalai and coworkers [29]. In order to characterize its energy landscape geometry, we sampled the value of the Ricci curvature K_R defined in Eq. (6.16) along its dynamical trajectories using Langevin simulations [25].

The Thirumalai model is a three-dimensional off-lattice model of a polypeptide which has only three different kinds of amino acids: polar (P), hydrophobic (H) and neutral (N). The potential energy is

$$V(\vec{r}_1, \ldots, \vec{r}_N) = V_{\text{bond}}(|\vec{r}_i - \vec{r}_{i-1}|) + V_{\text{angular}}(|\vartheta_i - \vartheta_{i-1}|)$$
$$+ V_{\text{dihedral}}(\psi_i) + V_{\text{non-bonded}}(\vec{r}_1, \ldots, \vec{r}_N) \tag{6.17}$$

where

$$V_{\text{bond}} = \sum_{i=1}^{N-1} \frac{k_r}{2}(|\vec{r}_i - \vec{r}_{i-1}| - a)^2 \,; \tag{6.18}$$

$$V_{\text{angular}} = \sum_{i=1}^{N-2} \frac{k_\vartheta}{2}(|\vartheta_i - \vartheta_{i-1}| - \vartheta_0)^2 \,; \tag{6.19}$$

$$V_{\text{dihedral}} = \sum_{i=1}^{N-3} \{A_i[1 + \cos\psi_i] + B_i[1 + \cos(3\psi_i)]\} \,; \tag{6.20}$$

$$V_{\text{non-bonded}} = \sum_{i=1}^{N-3} \sum_{j=i+3}^{N} V_{ij}(|\vec{r}_{i,j}|) \,, \tag{6.21}$$

where $\vec{r}_i$ is the position vector of the i-th monomer, $\vec{r}_{i,j} = \vec{r}_i - \vec{r}_j$, ϑ_i is the i-th bond angle, i.e., the angle between $\vec{r}_{i+1}$ and $\vec{r}_i$, ψ_i the i-th dihedral angle, that is

Table 6.2.   The six sequences considered and their estimated characteristic temperatures T_θ and T_f. $(X)_y$ means that X is repeated y times.

name	sequence	T_θ	T_f
S_g^{22}	$PH_9(NP)_2NHPH_3PH$	0.65 ± 0.05	0.55 ± 0.1
S_b^{22}	$PHNPH_3NHNH_4(PH_2)_2PH$	0.55 ± 0.15	none
S_i^{22}	$P_4H_5NHN_2H_6P_3$	0.75 ± 0.1	0.7 ± 0.2
S_h^{22}	H_{22}	0.65 ± 0.05	none
S_g^{46}	$P(HP)_5N_3H_9N_3(HP)_4N_3H_9$	0.65 ± 0.03	0.65 ± 0.05
S_h^{46}	H_{46}	0.70 ± 0.03	none

the angle between the vectors $\hat{n}_i = \vec{r}_{i+1,i} \times \vec{r}_{i+1,i+2}$ and $\hat{n}_{i+1} = \vec{r}_{i+2,i+1} \times \vec{r}_{i+2,i+3}$, $k_r = 100$, $a = 1$, $k_\vartheta = 20$, $\vartheta_0 = 105°$, $A_i = 0$ and $B_i = 0.2$ if at least two among the residues $i, i+1, i+2, i+3$ are N, $A_i = B_i = 1.2$ otherwise. As to V_{ij}, we have

$$V_{ij} = \frac{8}{3}\left[\left(\frac{a}{r}\right)^{12} + \left(\frac{a}{r}\right)^6\right] \tag{6.22}$$

if $i, j = P, P$ or $i, j = P, H$,

$$V_{ij} = 4\left[\left(\frac{a}{r}\right)^{12} - \left(\frac{a}{r}\right)^6\right] \tag{6.23}$$

if $i, j = H, H$ and

$$V_{ij} = 4\left(\frac{a}{r}\right)^6 \tag{6.24}$$

if either i or j are N [29]. Such a model is clearly very similar to the two-dimensional one described in Sec. 6.2.

We considered six different sequences of "amino acids" H, P and N: four of 22 monomers, S_g^{22}, S_b^{22}, S_i^{22}, S_h^{22}, and two sequences of 46 monomers, S_h^{46}, S_g^{46}. The six sequences are listed in Table 6.2 together with their estimated collapse and folding temperatures; the latter has been defined using a different protocol with respect to the case of Sec. 6.2.2 (see [25]). Sequences S_g^{22} and S_g^{46} are good folders [25, 29, 30]: below a given temperature S_g^{22} (resp. S_g^{46}) always reach the same β-sheet-like structure (resp. β-barrel-like structure). Homopolymers S_h^{22} and S_h^{46}, on the other hand, show a hydrophobic collapse but no tendency to reach a particular configuration in the collapsed phase, as expected. Sequence S_b^{22} (which has the same overall composition of S_g^{22} rearranged in a different sequence) behaved as a bad folder and did not reach a unique native state, while S_i^{22} was constructed by us to show a somehow intermediate behavior between good and bad folders: it always forms the same structure involving the middle of the sequence, while the beginning and the end of the chain fluctuate also at low temperature.

L. Bongini & L. Casetti

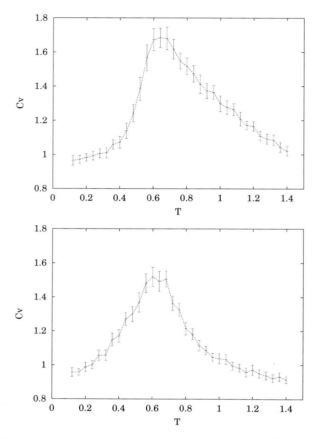

Fig. 6.8. Specific heat c_V vs. temperature T for the homopolymer S_{h}^{22} (left) and for the good folder S_{g}^{22} (right). The curves are a guide to the eye.

As to standard thermodynamic observables, all the sequences showed very similar behaviors: to give an example, in Fig. 6.8 we compare the specific heat c_V of the homopolymer S_{h}^{22} and of the good folder S_{g}^{22}: both exhibit a peak at the transition, and on the sole basis of this picture it would be hard to discriminate between a simple hydrophobic collapse and a folding. The same happens in the case of the longer homopolymer S_{h}^{46} and the good folder S_{g}^{46} (data not shown). On the other hand, a dramatic difference between the homopolymers and the good folders shows up if we consider the geometric properties of the landscape. In particular, a lot of information appears to be encoded in the fluctuations of the Ricci curvature K_R, i.e., of the Laplacian of the potential energy—see Eq. (6.16). We defined a relative adimensional curvature fluctuation σ as

$$\sigma = \frac{\sqrt{\frac{1}{N}(\langle K_R^2 \rangle_t - \langle K_R \rangle_t^2)}}{\frac{1}{N}\langle K_R \rangle_t} \tag{6.25}$$

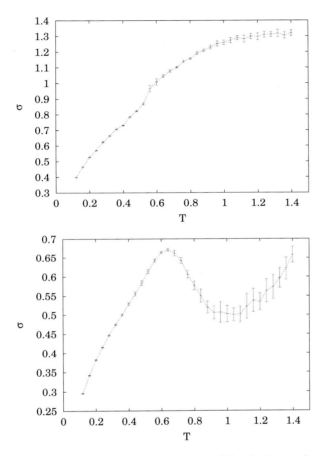

Fig. 6.9.   Relative curvature fluctuation σ vs. temperature T for the homopolymer $S_{\rm h}^{22}$ (left) and for the good folder $S_{\rm g}^{22}$ (right). The curves are a guide to the eye.

where $\langle \cdot \rangle_t$ stands for a time average: in Fig. 6.9 we plot σ as a function of the temperature T for the homopolymer $S_{\rm h}^{22}$ and for the good folder $S_{\rm g}^{22}$. A clear peak shows up in the case of the good folder, close to the folding temperature T_f below which the system is mostly in the native state, while no particular mark of the hydrophobic collapse can be seen in the case of the homopolymer. A similar situation happens for the longer sequences, the good folder $S_{\rm g}^{46}$ and the homopolymer $S_{\rm h}^{46}$ (Fig. 6.10).

The relative curvature fluctuation σ of the energy landscape appears then to be a good marker of the presence of a folding transition, at least in the simple model considered here. We stress that the comparison between sequences of length 22 and 46 clearly indicates that the peak in σ is really related to the folding and not to the hydrophobic collapse, because σ for the long homopolymer $S_{\rm h}^{46}$ is even smoother than in the case of the short homopolymer $S_{\rm h}^{22}$, at variance with the specific heat

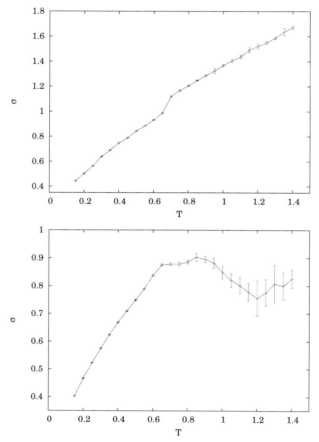

Fig. 6.10. As in Fig. 6.9 for the homopolymer S_h^{46} and the proteinlike sequence S_g^{46}.

which develops a sharper peak, consistently with the fact that the system is closer to the situation where a thermodynamic θ-transition exists.

As to the other sequences, for the bad folder S_b^{22}, $\sigma(T)$ is not as smooth as for the homopolymers, but only a very weak signal is found at $T \simeq 0.4$, lower than T_θ; below this temperature the systems seems to behave as a glass. For the "intermediate" sequence S_i^{22} a peak is present at the "quasi-folding" temperature, although considerably broader than in the case of S_g^{22} (see Fig. 6.11).

Where does the peak in $\sigma(T)$ near T_f come from? As shown in [25] one can argue, using an analysis of the behavior of the curvature along the dynamical trajectories, that it is due to the presence of two macroregions in the energy landscape, one corresponding to the native state and the other—charatcterized by a smaller average curvature—corresponding to the unfolded state. Then the dynamics is effectively two-state: close to T_f, the system often jumps between the two basins and this explains the growth of the fluctuations of K_R.

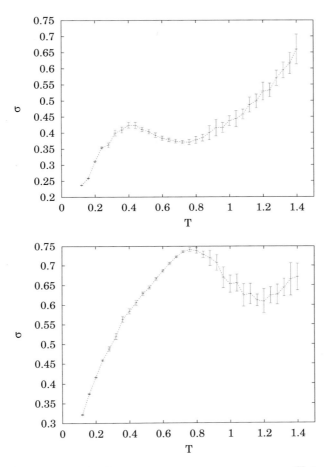

Fig. 6.11. $\sigma(T)$ for the sequence S_b^{22} (on the left) and for the sequence S_i^{22} (on the right). The curves are a guide to the eye.

We have shown that a geometric quantity which measures the amplitude of the curvature fluctuations of the energy landscape, σ, when plotted as a function of the temperature T shows a dramatically different behavior when the system undergoes a folding transition with respect to when only a hydrophobic collapse is present; $\sigma(T)$ can thus be used to mark the folding transition and to identify good folders, within the model considered here.

It must be stressed that no knowledge of the native state is necessary to define σ, and that it can be computed with the same computational effort needed to obtain the specific heat and other thermodynamic observables. This means that using e.g. reweighting histogram techniques one can reliably estimate the behavior of σ as a function of T using few simulation runs at properly chosen temperatures, if not a single one. Hence, if tested successfully on other, maybe more refined models of proteins, the calculation of the curvature fluctuations might prove a useful tool in the search of protein-like sequences. Preliminary results on more refined minimalistic models [31] completely confirm the scenario presented here.

6.4. Concluding Remarks

The scientific community studying protein folding has proposed a large number of competitive models many of which are capable of reproducing, albeit at a qualitative level, the main thermodynamic features of the process. A comparative study of the kinetic, geometric and topological properties of the energy landscape is therefore mandatory in order to spot which of them are weak and model dependent and which are instead generalized features of any systems characterized by a rough and funnelled energy landscape. Needless to say, the latter are likely to be shared by real proteins, too.

The two approaches described here appear as complementary. Topological and global geometrical features of the energy landscape seem to contain a lot of information on the folding propensity of given amino acidic sequences, at least as minimal models are considered. These approaches are however in their infancy so that further investigation along these lines will be necessary to devise effective methods able to elucidate the nature of the energy landscape of protein-like heteropolymers.

Acknowledgments

We would like to thank all the colleagues in collaboration with whom the results discussed here have been obtained and especially M. Baiesi, R. Franzosi, R. Livi, L. N. Mazzoni, A. Politi, L. Tattini and A. Torcini. We also acknowledge useful discussions with R. Burioni and D. Cassi. This work is part of the EC (FP6-NEST) project *Emergent organisation in complex biomolecular systems (EMBIO)* (EC contract n. 012835) whose financial support is acknowledged.

References

[1] C. Anfinsen, Science **181**, 223 (1973).
[2] A. Finkelstein and O. B. Ptitsyn, *Protein physics* (Academic Press, London, 2002).
[3] D. J. Wales, *Energy landscapes* (Cambridge University Press, Cambridge, 2004).
[4] P. G. Debenedetti and F. H. Stillinger, Nature **410**, 259 (2001).
[5] J. N. Onuchic, P. G. Wolynes, Z. Luthey-Schulten, and N. D. Socci, Proc. Natl. Acad. Sci. USA **92**, 3626 (1995).
[6] L. Bongini, L. Casetti, R. Livi, A. Politi, and A. Torcini, *Stochastic dynamics of model proteins on a directed graph*, Phys. Rev. E **79**, 061925 (2009).
[7] P. Das, S. Matysiak, and C. Clementi, Proc. Natl. Acad. Sci. USA **102**, 10141 (2005).
[8] C. Clementi, Curr. Op. Struct. Biol. **18**, 10 (2008).
[9] M. A. Miller and D. J. Wales, J. Chem. Phys. **111**, 6610 (1999).
[10] L. Bongini, R. Livi, A. Politi, and A. Torcini, Phys. Rev. E **68**, 061111 (2003).
[11] L. Bongini and A. Rampioni, Gene **347**, 231 (2005).

[12] J. Gil Kim and T. Keyes, J. Phys. Chem. B **111**, 2647 (2007).

[13] M. Nakagawa and M. Peyrard, Proc. Natl. Acad. Sci. USA **103**, 5279 (2006); Phys. Rev. E **74**, 041916 (2006).

[14] L. Bongini, R. Livi, A. Politi, and A. Torcini, Phys. Rev. E **72**, 051929 (2005).

[15] J. S. Langer, Ann. Phys. **54** 258, (1969).

[16] L. Bongini, R. Livi, A. Politi, and A. Torcini, Phys. Rev. E **72**, 051929 (2005).

[17] A. Torcini, R. Livi, and A. Politi, J. Biol. Phys. **27**, 181 (2001).

[18] F. H. Stillinger, T. Head-Gordon, and C. L. Hirshfeld, Phys. Rev. E **48**, 1469 (1993).

[19] A. Irbäck, C. Peterson, and F. Potthast, Phys. Rev. E **55**, 860 (1997).

[20] R. Burioni and D. Cassi, Phys. Rev. Lett. **76**, 1091 (1996).

[21] A. Mossa and C. Clementi, Phys. Rev. E **75**, 046707 (2007).

[22] M. Baiesi, L. Bongini, L. Casetti, and L. Tattini, *A graph theoretical analysis of the energy landscape of a model protein*, ArXiv:0812.0316, Phys. Rev. E (2009 to appear).

[23] M. Nakahara, *Geometry, Topology and Physics* (IOP Publishing, London, 2003).

[24] M. P. do Carmo, *Riemannian geometry* (Birkhauser, Boston, 1992).

[25] L. N. Mazzoni and L. Casetti, Phys. Rev. Lett. **97**, 218104 (2006); Phys. Rev. E **77**, 051917 (2008).

[26] L. P. Eisenhart, Ann. Math. **30**, 591 (1929).

[27] L. Casetti, M. Pettini, and E. G. D. Cohen, Phys. Rep. **337**, 237 (2000).

[28] M. Pettini, *Geometry and topology in Hamiltonian dynamics and statistical mechanics*, Interdisciplinary applied mathematics series (Springer-Verlag, New York, 2007).

[29] T. Veitshans, D. Klimov, and D. Thirumalai, Fold. Des. **2**, 1 (1997).

[30] J. D. Honeycutt and D. Thirumalai, Proc. Natl. Acad. Sci. USA **87**, 3527 (1990).

[31] L. N. Mazzoni, R. Franzosi, L. Casetti, and C. Clementi, in preparation.

Chapter 7

Elastic Network Models For Biomolecular Dynamics: Theory and Application to Membrane Proteins and Viruses

Timothy R. Lezon, Indira H. Shrivastava, Zheng Yang and Ivet Bahar[*]

Department of Computational Biology, School of Medicine, University of Pittsburgh, Suite 3064, Biomedical Science Tower 3, 3051 Fifth Ave., Pittsburgh, PA 15213

7.1. Introduction

Elastic network models (ENMs) have over the last decade enjoyed considerable success in the study of macromolecular dynamics. These models have been used to predict the global dynamics of a variety of proteins and protein complexes, ranging in size from single enzymes to macromolecular machines (Keskin *et al.*, 2002), ribosomes (Tama *et al.*, 2003; Wang *et al.*, 2004) and viral capsids (Tama & Brooks, 2002; Tama & Brooks, 2005; Rader *et al.*, 2005). They have provided insights into a wide range of protein behaviors, such as mechanisms of allosteric regulation (Ming & Wall, 2005; Bahar *et al.*, 2007; Chennubhotla *et al.*, 2008), protein-protein binding (Tobi & Bahar, 2005), anisotropic response to uniaxial tension and unfolding (Eyal & Bahar, 2008; Sulkowska *et al.*, 2008), co-localization of catalytic sites and key mechanical sites (e.g., hinges) (Yang & Bahar, 2005), interactions at the binding sites (Ming & Wall, 2006), and energetics (Miller *et al.*, 2008), to name a few. ENMs allow the global motions of a molecule to be quickly calculated, making them an ideal complement to conventional molecular dynamics (MD) simulations. Increasingly, variants of ENMs are being applied to non-equilibrium situations, such as the prediction of transition pathways between functional states separated by low energy barriers (Zheng *et al.*, 2007) or driving MD simulations (Isin *et al.*, 2008).

At its core, the ENM provides a simplified representation of the potential energy function of a system, in this case a macromolecule or macromolecular

---

[*]Corresponding Author: *bahar@ccbb.pitt.edu http://www.ccbb.pitt.edu*

assembly, near equilibrium. The nodes of the network are the building blocks, such as atoms, nucleotides or amino acids, from which the system is composed (Fig. 7.1). Each node is typically represented as a point particle in three dimensions (3D), and the edges of the network, or the springs joining the nodes, represent harmonic restraints on displacements from the equilibrium structure. Thus, the ENM provides an intuitive and quantitative description of behavior near equilibrium: The starting conformation resides at the bottom of a harmonic well, and any deviations from equilibrium will increase the energy and result in a linear net force directed toward restoring the system to its lowest energy state.

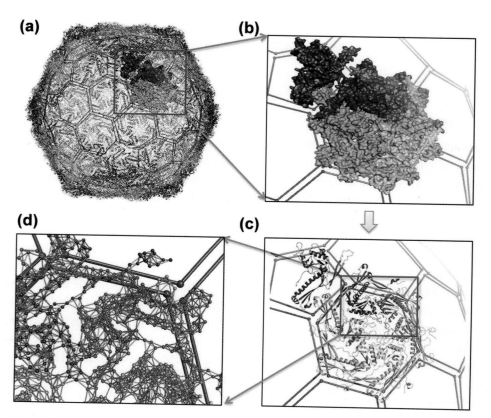

Fig. 7.1. *From protein assemblies to network models.* (a) External view of the intact viral capsid HK97 colored by chain, generated using the PDB file 2FT1 deposited by Johnson and coworkers (Gan *et al.*, 2006). The capsid consists of 420 identical proteins arranged into 12 pentamers and 60 hexamers. (b) One asymmetric unit from panel (a) is enlarged. Each chain is in a distinctive color, indicating a possible scheme for rigid building blocks. (c) One asymmetric unit shown as secondary structures, in the same viewpoint and color scheme of panel (b). (d) A cartoon of the ENM in which the nodes are C^α atoms (spherical dots) and the edges represent the springs (or elastic couplings) connecting pairs of nodes located within a distance of r_c.

The harmonic approximation is ubiquitous in physics, and similar models have been used for calculating elastic properties of bulk polymers and classical lattice vibrations in crystals (see, for example, Marder, 2000). Proteins, on the other hand, are small and somewhat flexible, and it was not until Tirion (1996) demonstrated that a harmonic potential faithfully captures the global dynamics of proteins that ENMs saw wide use in theoretical investigations of protein dynamics. Subsequent studies reinforced this observation, bestowing the ENM with particular relevance to proteins (Cui & Bahar, 2006). It is well established that a relationship exists between protein structure and protein function. Here we have a simple network model that enables the calculation of global dynamics from structure alone, suggesting that protein dynamics is intermediate to structure and function, and ENMs provide an easily employable conduit between structure and dynamics.

An attractive feature of ENMs that keeps them in continuous use is that they provide a wealth of information at low computational cost. Construction of the EN is a matter of straightforwardly defining and linking nodes provided that information on structure, or simply on inter-residue contact topology, is available. The standard technique for determining dynamics or statistical distributions from an ENM is to conduct a mode decomposition using spectral graph theory and methods (e.g., Gaussian Network Model (GNM) (Bahar *et al.*, 1997; Bahar *et al.*, 1998) inspired by polymer network theory (Flory, 1976)) or a normal mode analysis (NMA) with uniform harmonic potentials, both of which provide analytical solutions to the equations of motion, bypassing the need to sample conformation space. Although NMA can also be applied to potentials derived from more detailed force fields, such calculations either require an initial energy minimization that inevitably distorts the input conformation, or else they risk producing energetically unstable solutions. ENMs, on the other hand, can take any conformation as input and guarantee that it resides at a minimum of the potential and therefore has physical motions. Furthermore, the few parameters used in ENMs can be easily adjusted as seen fit, giving ENMs uncommon adaptability. There are, however, some limitations to ENMs as predictive tools. Although ENMs robustly predict collective global motions, they do not fare as well in providing reliable descriptions of local motions. Also, the harmonic approximation requires a potential minimum, limiting the utility of ENMs for modeling non-equilibrium dynamics. A corollary of this second point is that only motions in the neighborhood of the global energy minimum can be accurately predicted by the ENMs, and one must be careful when interpreting the results from the model not to exceed the limits of the model.

Here we will outline the theory behind the ENM and some of its extensions, and then we will present some recent applications. We will focus on two groups of proteins, membrane proteins and viral capsids. Membrane proteins form one of the most ubiquitous classes of proteins, accounting for more than 25% of the proteins in most genomes (Wallin & von Heijne, 1998). Their functions cover a wide range of spectrum from transport of metabolites in prokaryotes to regulating and maintaining intra-cellular communications in eukaryotes by transporting ions. In mammalians, these proteins are responsible for maintaining the electrochemical gradients across cell-membrane, which is vital for efficient functioning of the central nervous system. Malfunctioning of membrane proteins leads to potentially fatal diseases, such as Alzheimer's, multiple slereosis and arrhythmia (Ashcroft, 2000). Membrane proteins indeed constitute a large fraction of proteins currently targeted by approved drugs. Understanding the general principles of their structural dynamics and thereby mechanisms of function is thus essential in the rational design of therapeutics that target membrane proteins. By way of applications to a number of membrane proteins, we will illustrate in the present chapter how ENM approaches can provide insights into gating and/or signal transduction mechanisms.

Viruses constitute another group of proteins that are difficult to examine by all-atom simulations (due to their sizes of the order of Megadaltons), but are amenable to ENM analyses. The viral capsids in particular possess solid-like behavior, and can be well represented by elastic network models and their material properties. We will show how ENMs can greatly enhance our understanding of the complex dynamics of viral capsids, and open the way to simple descriptions in terms of measurable material properties. In summary, both groups of applications illustrate the utility of ENM approaches in providing simple descriptions of highly complex structures' dynamics, and gaining insights into potential mechanisms of biomolecular functions.

7.2. Theory and Assumptions

7.2.1. Statistical mechanical foundations

Potential energy. The elastic network model theory follows the same formalism that is commonly presented for studying small oscillations (Goldstein, 1953). Here the physical system is a molecule or molecular assembly consisting of N constituent particles, where each particle may be an atom, a residue, or some other structural element acting as a node in the network. The changes in generalized coordinates are defined by the vector $\mathbf{q} = (q_1, \ldots, q_n)^T$ of displacements from equilibrium. Typically the three Cartesian coordinates of each node are considered separately, giving $\mathbf{q}$ a total of $3N$ components (notable exceptions include the GNM, in which $\mathbf{q}$ has N components, and highly symmetric systems, such as viral capsids, for which symmetry can be exploited to reduce the dimensionality of $\mathbf{q}$). Near the equilibrium structure, the potential energy can be expanded as a power series in $\mathbf{q}$ as

$$V(\mathbf{q}) = V(0) + \sum_i \left(\frac{\partial V}{\partial q_i} \right)_0 q_i + \frac{1}{2} \sum_{ij} \left(\frac{\partial^2 V}{\partial q_i \partial q_j} \right)_0 q_i q_j + \cdots. \tag{7.1}$$

The first term of the above expression is a constant that may be set to zero, and the second term is identically zero at a potential minimum. To second order, the potential is a sum of pairwise potentials

$$V(\mathbf{q}) = \frac{1}{2} \sum_{ij} \left(\frac{\partial^2 V}{\partial q_i \partial q_j} \right)_0 q_i q_j \tag{7.2}$$

$$= \frac{1}{2} \sum_{ij} q_i U_{ij} q_j \tag{7.3}$$

$$= \frac{1}{2} \mathbf{q}^T \mathbf{U} \mathbf{q}, \tag{7.4}$$

where $\mathbf{U}$ is the matrix of second derivatives of the potential with respect to the generalized coordinates. It should be noted that $\mathbf{U}$ is symmetric and nonnegative definite.

Equation of motion. The kinetic energy can similarly be written in compact form as

$$T = \frac{1}{2} \dot{\mathbf{q}}^T \mathbf{M} \dot{\mathbf{q}}, \tag{7.5}$$

where the elements of the diagonal matrix $\mathbf{M}$ are the masses of the nodes and $\dot{\mathbf{q}}$ is the time derivative of $\mathbf{q}$. The equations of motion of the system are

$$\mathbf{M}\ddot{\mathbf{q}} + \mathbf{U}\mathbf{q} = 0. \tag{7.6}$$

Here the double dot denotes the second derivative with respect to time.

Analytical solution. We solve Eq. (7.6) by transforming to mass-weighted coordinates, $\mathbf{r} = \mathbf{M}^{1/2}\mathbf{q}$, $\mathbf{K} = \mathbf{M}^{-1/2} \mathbf{U} \mathbf{M}^{-1/2}$, which yield

$$\ddot{\mathbf{r}} = -\mathbf{K}\mathbf{r}, \tag{7.7}$$

Note that the potential energy (Eq. (7.4)) can be expressed in terms of the changes in mass-weighted coordinates and the mass-weighted stiffness matrix $\mathbf{K}$ as $V(\mathbf{r}) = \frac{1}{2}\mathbf{r}^T\mathbf{Kr}$. The solution to Eq. (7.7) is

$$\mathbf{r}(t) = \mathbf{a}e^{-i\omega t} . \tag{7.8}$$

From Eq. (7.7) and Eq. (7.8), we find that the coefficients $\mathbf{a}$ solve the eigenvalue equation $\mathbf{Ka} = \lambda\mathbf{a}$, where $\mathbf{a}$ is a vector of displacements along a normal mode of vibration, and the eigenvalue, λ, is the square of the normal mode frequency ω.

In most cases, $\mathbf{K}$ is not invertible, but has a well-defined number of eigenvalues that are identically zero. This occurs because the potential energy only depends on internal degrees of freedom and places no energetic restrictions on rigid-body rotations and translations. With this in mind, the inverse of $\mathbf{K}$, when required, is replaced by the pseudo-inverse, defined as

$$\mathbf{K}^{-1} = \sum\nolimits_{\lambda_k \neq 0} \frac{\mathbf{v}_k \mathbf{v}_k^T}{\lambda_k} , \tag{7.9}$$

where λ_k are the nonzero eigenvalues of $\mathbf{K}$, and $\mathbf{v}_k$ are their associated eigenvectors.

ENM partition function. The system's partition function can be calculated by integrating the potential over all possible changes in structure:

$$Z = \int d^n r \exp\left\{ -\frac{1}{2k_B T}\mathbf{r}^T\mathbf{Kr} \right\} \tag{7.10}$$

$$= (2\pi k_B T)^{n/2} [\det(\mathbf{K}^{-1})]^{1/2} , \tag{7.11}$$

where k_B is the Boltzmann constant and T is the absolute temperature. As $\det(\mathbf{K}^{-1})$ is simply the product of the reciprocal nonzero eigenvalues of $\mathbf{K}$, the lowest frequency modes contribute most to the partition function. These modes are also of highest interest when seeking to determine the most probable *global* fluctuations of a molecule. Indeed, the low-frequency, or 'slow', modes of an ENM are robust to variations in network topology, the level of resolution adopted in describing the network (see for example Doruker *et al.*, 2002) and the force field adopted in NMA, and they reflect the intrinsically accessible motions that are endowed upon the molecule by its structure.

Mean-square fluctuations and cross-correlations. Expectation values for dynamical variables predicted by the ENM can be directly compared to experimental measurements. X-ray temperature factors (B-factors) provide a measure of the mean-square fluctuations of individual atoms. Similarly, an ensemble of NMR structures can be used to calculate the correlations between the displacements of atoms. The correlations between node fluctuations in the ENM are given by

$$\langle r_i r_j \rangle = \frac{1}{Z} \int d^n r \exp\left\{-\frac{1}{2k_B T} \mathbf{r}^T \mathbf{K} \mathbf{r}\right\} r_i r_j \qquad (7.12)$$

$$= k_B T (\mathbf{K}^{-1})_{ij}, \qquad (7.13)$$

and from Eq. (7.9) it is clear that the most significant contributions to the inter-residue correlations are also from the slowest modes.

7.2.2. *Anisotropic network models*

The most common ENMs are anisotropic network models (ANM) (Doruker *et al.*, 2000; Atilgan *et al.*, 2001; Tama & Sanejouand, 2001) that use the $3N$ mass-weighted coordinates of the nodes as generalized coordinates: $\mathbf{r} = (\Delta x_1, \Delta y_1, \Delta z_1, \ldots, \Delta x_N, \Delta y_N, \Delta z_N)^T$, where $\Delta x_i = x_i - x_i^0$ is the x-component of the displacement of node i from its equilibrium position, $\mathbf{r}_i^0$. In this case the interaction matrix $\mathbf{K}$ is the $3N \times 3N$ Hessian matrix, $\mathcal{H}$, of mixed second derivatives of the potential with respect to the coordinates of the residues. The Hessian might be thought of as an $N \times N$ matrix of 3×3 submatrices, each of which describes the energetic contribution from the interaction of two nodes. The elements of $\mathcal{H}$ can be calculated from the potential energy,

$$V = \frac{1}{2}\sum_{ij}\gamma_{ij}(R_{ij} - R_{ij}^0)^2, \qquad (7.14)$$

where γ_{ij} is the spring constant between nodes i and j, R_{ij} is their distance, and R_{ij}^0 is their equilibrium distance. The second derivatives of the potential function at equilibrium have the general form

$$\frac{\partial^2 V}{\partial x_i \partial y_j} = -\frac{\gamma_{ij}(x_j - x_i)(y_j - y_i)}{R_{ij}^2}, \qquad (7.15)$$

where x_i and y_j are the x- and y-coordinates of nodes i and j, respectively. Using the notation $x_{ij} = (x_j - x_i)$, and similarly for y_{ij} and z_{ij}, the off-diagonal super-elements of $\mathcal{H}$ are

$$\mathcal{H}_{ij} = -\frac{\gamma_{ij}}{R_{ij}^2}\begin{bmatrix} x_{ij}^2 & x_{ij}y_{ij} & x_{ij}z_{ij} \\ x_{ij}y_{ij} & y_{ij}^2 & y_{ij}z_{ij} \\ x_{ij}z_{ij} & y_{ij}z_{ij} & z_{ij}^2 \end{bmatrix}, \qquad (7.16)$$

where the components of the matrix refer to the equilibrium distance vectors, and the diagonal super-elements satisfy

$$\mathcal{H}_{ii} = -\sum\nolimits_{j; j \neq i} \mathcal{H}_{ij}. \tag{7.17}$$

Diagonalization of $\mathcal{H}$ yields $3N - 6$ normal modes, each of which has a 3-vector component for every node. The remaining 6 modes have zero eigenvalue and correspond to rigid-body rotations and translations of the system.

The spring constants γ_{ij} are the only adjustable parameters in this model, and a variety of methods are used to select their values. Pairwise interactions are predominantly local, and a common practice is to assign a uniform spring constant, $\gamma_{ij} = \gamma$, to all pairs of nodes separated by less than some cutoff distance, and $\gamma_{ij} = 0$ for all others. It has been found empirically (Eyal *et al.*, 2006) that when the nodes are taken to be the α-carbons of a protein, a cutoff distance of about 15Å results in residue mean-square fluctuations that correlate well with experimental B-factors. An alternative approach (Hinsen, 1998) that agrees comparably with experiments is to assign spring constants that decay with distance. Recent studies (Kondrashov *et al.*, 2006) show that the adoption of stiffer force constants for the springs that connect first neighbors along the sequence further enhances the correlation with B-factors.

7.2.3. *Gaussian network model*

A simplification of the above-described model is the Gaussian network model (GNM) (Bahar *et al.*, 1997). This model uses the assumptions that node fluctuations are isotropic and Gaussian to reduce the interaction matrix from a $3N \times 3N$ Hessian to an $N \times N$ Kirchhoff matrix. Interestingly, this model often agrees better with experimental data than does its anisotropic counterpart, because the underlying potential penalizes the vectorial changes $\Delta \mathbf{R}_{ij} = \mathbf{R}_{ij} - \mathbf{R}_{ij}^0$ in internode distances (as opposed to penalizing the changes in the magnitudes $|\mathbf{R}_{ij}| - |\mathbf{R}_{ij}^0|$ only, as in the ANM; see Eq. (7.14)).

The inherent assumption of vibrational isotropy allows GNM to predict the size of motions and their cross-correlations, but not their directions. It also predicts the displacements along normal coordinates (e.g., slow modes) and permits us to define the domains engaged in concerted motions in the global modes, but not their mechanism/direction of concerted rearrangements. Note that the GNM uses only a single parameter, the elastic constant γ that defines the interactions between nodes that are separated by a distance less than a cutoff distance, r_c. When applied to proteins at the residue level, a value of r_c between

6.5Å and 7.5Å, corresponding to the radius of the first coordination shell, is typically selected for use in GNM.

The potential in Eq. (7.14) is a sum over pairwise potentials, each of which depends on the difference between the instantaneous distance between two nodes and their equilibrium separation. By assuming isotropic fluctuations, we can separate the spatial components of each node's motion, resulting in the GNM potential

$$V_{GNM} = \frac{1}{2}\sum_{ij}\gamma_{ij}[(R_{ij} - R_{ij}^0)\cdot(R_{ij} - R_{ij}^0)] \tag{7.18}$$

$$= \frac{1}{2}\sum_{ij}\gamma_{ij}\left[(x_{ij} - x_{ij}^0)^2 + (y_{ij} - y_{ij}^0)^2 + (z_{ij} - z_{ij}^0)^2\right] \tag{7.19}$$

$$= \frac{\gamma}{2}\sum_{ij}\Gamma_{ij}(\Delta x_{ij}^2 + \Delta y_{ij}^2 + \Delta z_{ij}^2) \tag{7.20}$$

$$= \frac{\gamma}{2}(\Delta\mathbf{x}^T\Gamma\Delta\mathbf{x} + \Delta\mathbf{y}^T\Gamma\Delta\mathbf{y} + \Delta\mathbf{z}^T\Gamma\Delta\mathbf{z}) \tag{7.21}$$

In the preceding lines we use the notation $\Delta x_{ij} = x_{ij} - x_{ij}^0$, and similarly for Δy_{ij} and Δz_{ij}. Likewise, we used the notation $\Delta\mathbf{x} = (\Delta x_1, \Delta x_2, \Delta x_3, ..., \Delta x_N)^T$ and similar expressions for $\Delta\mathbf{y}$ and $\Delta\mathbf{z}$, where Δx_i is the x-component of the vector Δr_i describing the fluctuation in the position of node i. The Kirchhoff adjacency matrix, Γ, has elements

$$\Gamma_{ij} = \begin{cases} -1 \text{ if } R_{ij} \leq r_c \text{ and } i \neq j \\ 0 \text{ if } R_{ij} > r_c \text{ and } i \neq j \\ -\sum_{k;k\neq i}\Gamma_{ik} \text{ if } i = j \end{cases} \tag{7.22}$$

The equations of motion separate into three identical equations, one for each spatial coordinate, and the normal modes are obtained by diagonalizing Γ. The assumption of isotropy essentially reduces the system to one dimension, and Γ has $N-1$ non-zero eigenvalues. Correlations between nodes can be found as before with

$$\langle \Delta\mathbf{r}_i \cdot \Delta\mathbf{r}_j \rangle = 3k_BT(\Gamma^{-1})_{ij}, \tag{7.23}$$

which is identical to Eq. (7.13) with Γ taking the place of $\mathbf{K}$ and a factor of three from the summation over spatial coordinates.

7.2.4. *Rigid block models*

One of the advantages of ENMs over more detailed force fields and simulations is their ability to produce analytical results with a coarse-grained model. One can quickly calculate the dynamics of a moderately large macromolecule simply by building and diagonalizing its Hessian matrix, an $O(N^3)$ computation. Occasionally, though, the memory requirements of the calculation overwhelm contemporary computers, and further coarse-graining is necessary. For example, if one wishes to consider the motion of side chains, a Hessian based on α-carbon coordinates alone is insufficient, and an ENM using all atoms must be used instead. Such refinement increases the system size approximately tenfold, possibly beyond the tolerance of available computational resources. Similarly, very large molecular assemblies such as viral capsids can contain upwards of 10^5 residues, and their Hessian matrices cannot be easily handled by conventional computing resources.

One way to overcome the problem of excessive system size is to bundle several elements of the physical system into a single node. This method faithfully reproduces the global dynamics of the system (Doruker *et al.*, 2000), but does not produce detailed motions for all of the original nodes. Mixed models (Kurkcuoglu *et al.*, 2003) that rely on multiple levels of coarse-graining can provide detailed results only for specific regions of interest. A good method for surmounting the problem of finding the normal modes of very large systems is the rotations and translations of blocks (RTB) (Tama *et al.*, 2000), also called block normal mode (BNM) (Li & Cui, 2002) method. This method assumes that the system is constructed of n_b rigid blocks, and that the normal modes of the system can be expressed as rigid body rotations and translations of its constituent blocks. Each block has 6 degrees of freedom, and the approximation reduces the size of the system from $3N$ to $6n_b$.

Consider a system of N nodes that can be collected into $n_b < N$ rigid blocks connected by elastic springs. As before, the generalized mass-weighted coordinates form the $3N$-vector $\mathbf{r}$, and the system's Hessian, $\mathcal{H}$, is calculated from the topology of the elastic network. We define the $3N \times 6n_b$ projection matrix, $\mathbf{P}$, from the $3N$-dimensional space of all nodes into the $6n_b$-dimensional space of rotations and translations of the rigid blocks. The relationship between linear motion of a rigid body and the motion of its constituent components is captured by the conservation of linear momentum in mass-weighted coordinates:

$$\sqrt{M}\,\dot{\mathbf{r}}_{CM} = \sum_k \sqrt{m_k}\,\dot{\mathbf{r}}_k\,, \tag{7.24}$$

where the summation is over all nodes in the rigid block, m_k and $\mathbf{r}_k$ are the mass and position of node k, and $M = \Sigma_k\, m_k$ and $\mathbf{r}_{CM}$ are the mass of the block and position of its center of mass. Note that Eq. (7.24) holds not only for velocities, but also for positions and higher time derivatives. The matrix elements projecting from the full $3N$-dimensional space to translations in the block space are

$$\frac{\partial(\dot{\mathbf{r}}_{CM})_\mu}{\partial(\dot{\mathbf{r}}_k)_v} = \sqrt{\frac{m_k}{M}}\,\delta_{\mu v}, \qquad (7.25)$$

where μ and v indicate x-, y- or z-components of the vector (or tensor) enclosed in parentheses, and $\delta_{\mu v}$ is the Kronecker delta function. Similarly, angular momentum conservation gives

$$\mathbf{I}^{1/2}\dot{\theta} = \sum_k \sqrt{m_k}\,(\mathbf{r}_k \times \dot{\mathbf{r}}_k), \qquad (7.26)$$

where $\mathbf{I}$ and θ are the block's moment of inertia tensor (the elements of which are given by $I_{\mu\alpha} \equiv \Sigma_k\, m_k\, (r_k^2\delta_{\mu\alpha} - (\mathbf{r}_k)_\mu\,(\mathbf{r}_k)_\alpha)$) and angular displacement vector. The components of $\dot{\theta}$ are

$$\dot{\theta}_\mu = \sum_k \sqrt{m_k} \sum_\alpha (\mathbf{I}^{-1/2})_{\mu\alpha}(\mathbf{r}_k \times \dot{\mathbf{r}}_k)_\alpha \qquad (7.27)$$

$$= \sum_k \sqrt{m_k} \sum_\alpha (\mathbf{I}^{-1/2})_{\mu\alpha} \sum_{\beta v}(\mathbf{r}_k)_\beta(\dot{\mathbf{r}}_k)_v\varepsilon_{\alpha\beta v}, \qquad (7.28)$$

where $\varepsilon_{\alpha\beta v}$ is the permutation symbol ($\varepsilon_{123} = \varepsilon_{231} = \varepsilon_{312} = 1$; $\varepsilon_{213} = \varepsilon_{321} = \varepsilon_{132} = -1$; otherwise $\varepsilon_{\alpha\beta v} = 0$), also known as Levi-Civita symbol. Differentiating the components of $\dot{\theta}$ with respect to the components of $\dot{\mathbf{r}}_k$, the matrix elements projecting from the full $3N$-dimensional space to rotations in the block space are

$$\frac{\partial\theta_\mu}{\partial(\dot{\mathbf{r}}_k)_v} = \sum_{\alpha\beta} \sqrt{m_k}\,(\mathbf{I}^{-1/2})_{\mu\alpha}(\mathbf{r}_k)_\beta\varepsilon_{\alpha\beta v}. \qquad (7.29)$$

Using Eqs. (25) and (29) and the notation of Li and Cui (2002), the elements of the $3N \times 6n_b$ projection matrix $\mathbf{P}$ are

$$P_{J,\,jv}^\mu = \begin{cases} \sqrt{m_j/M_J}\,\delta_{\mu v} & \text{for } \mu = 1, 2, 3 \\[2mm] \sum_{\alpha\beta}(\mathbf{I}^{-1/2})_{(\mu-3),\alpha}\sqrt{m_j}\,(r_j - r_J^{\,0})_\beta\varepsilon_{\alpha\beta v} & \text{for } \mu = 4, 5, 6 \end{cases} \qquad (7.30)$$

where J is the rigid block index, $\mathbf{r}_J^{\,0}$ is the position vector of the mass center of block J, j is the node index, μ designates the rigid block translation ($1 \leq \mu \leq 3$) or rotation ($4 \leq \mu \leq 6$), and v is the node displacement component ($1 \leq v \leq 3$). The

Hessian is projected into the space of rigid blocks with the transformation $\mathcal{H}_{BLK} = \mathbf{P}^T \mathcal{H} \mathbf{P}$, $\mathcal{H}_{BLK}$ is diagonalized with $\mathbf{V}_{BLK}^T \mathcal{H} \mathbf{V}_{BLK} = \Lambda_{BLK}$, and the resulting eigenvectors are projected back into the full $3N$-dimensional space with the inverse projection $\mathbf{V} = \mathbf{P}^T \mathbf{V}_{BLK}$.

An example application of the RTB/BNM formalism is to virus maturation. Tama and Brooks (2005) modeled viral capsids of different sizes and symmetries using an ENM in which each capsomer was taken to be a rigid block. Their analysis showed that the conformational changes that occur during viral maturation can be largely accounted for with only a few icosahedrally symmetric slow modes. Another example application is to molecular motors. Li and Cui (2004) found that the conformational changes that occur in myosin and the calcium transporter Ca^{2+}-ATPase are dominated by a small number of slow modes, suggesting that Brownian motions have an essential role in the function of molecular motors.

7.2.5. *Treatment of perturbations*

It is often interesting to compare the global dynamics of a system in the absence and presence of some environmental perturbation applied at a given position, such as ligand binding. In such cases the perturbation takes the form of additional nodes, and the Hessian is calculated for the extended system, including these additional nodes. Comparing the normal modes of the original system to those of the perturbed system is not straightforward: Including the perturbation provides additional degrees of freedom, so the normal modes of the perturbed system are not necessarily orthogonal when projected into the space of the unperturbed system. It is useful to have an effective Hessian that will account for the influence of the perturbation without modifying the size of the system. This can be calculated as follows.

The state vector for an N node system is $\mathbf{r} = (r_1, ..., r_{3N})^T$, and the state vector for the same system in the presence of a perturbation, $\mathbf{e} = (e_1, ..., e_{3n})^T$, by a system of n nodes is $\mathbf{r}' = (s_1, ..., s_{3N}, e_1, ..., e_{3n})^T = (\mathbf{r}^T \ \mathbf{e}^T)^T$, where the first $3N$ components refer to the original system, and the last $3n$ to the environment or perturbing molecule. As demonstrated by Ming and Wall (2005), and by Zheng and Brooks (2005), the Hessian of a molecule within a specific environment can be decomposed as follows:

$$\mathcal{H} = \begin{pmatrix} \mathbf{H}_{ss} & \mathbf{H}_{se} \\ \mathbf{H}_{se}^T & \mathbf{H}_{ee} \end{pmatrix}, \tag{7.31}$$

where $\mathbf{H}_{ss}$ contains contributions from interactions of the original system with itself, $\mathbf{H}_{ee}$ accounts for interactions of the environment with itself, and $\mathbf{H}_{se}$ contains interactions between the system and its environment. Note that $\mathbf{H}_{ss}$ is not simply the unperturbed Hessian, but has different diagonal super-elements due to environmental contributions. The potential energy can be written as

$$V = \frac{1}{2}(\mathbf{r}')^T \mathcal{H}\mathbf{r}' \tag{7.32}$$

$$= \frac{1}{2}\mathbf{r}^T \mathbf{H}_{ss}\mathbf{r} + \mathbf{r}^T \mathbf{H}_{se}\mathbf{e} + \frac{1}{2}\mathbf{e}^T \mathbf{H}_{ss}\mathbf{e}, \tag{7.33}$$

using $\mathbf{e}^T \mathbf{H}_{se}{}^T \mathbf{r} = \mathbf{r}^T \mathbf{H}_{se} \mathbf{e}$. At equilibrium, $\partial V / \partial e_i = 0$ for all environmental nodes, giving

$$0 = \mathbf{H}_{es}\mathbf{r} + \mathbf{H}_{es}\mathbf{e}, \tag{7.34}$$

which yields

$$\mathbf{e} = -\mathbf{H}_{ee}^{-1}\mathbf{H}_{es}\mathbf{r}. \tag{7.35}$$

Substitution of this expression into Eq. (7.33) permits us to write the potential energy in terms of the $3N$ components of $\mathbf{r}$ as

$$V = \frac{1}{2}\mathbf{r}^T \overline{\mathbf{H}}\mathbf{r} \tag{7.36}$$

where $\overline{\mathbf{H}}$ is a pseudo-Hessian with the same dimensionality as the unperturbed Hessian, but which includes the environmental effects:

$$\overline{\mathbf{H}} = \mathbf{H}_{ss} - \mathbf{H}_{se}\mathbf{H}_{ee}^{-1}\mathbf{H}_{se}^T. \tag{7.37}$$

Diagonalizing $\overline{\mathbf{H}}$ leads to the normal modes in the presence of the perturbation, and these can be directly compared to the modes of unperturbed system. This approach has been used to examine conformational changes in myosin and kinesin nucleotide-binding pockets. Zheng and Brooks (2005) employed a model in which the binding pockets of motor proteins constituted the system, and the remainder of the protein made up its environment. They showed that the dynamics relevant to the myosin binding pocket are coupled to its global modes, in agreement with hypothesized pathways between actin binding and force generation. Ming and Wall (2006) used this method to demonstrate that substrate allosteric proteins usually bind their substrates at sites that induce significant perturbation in the collective dynamics.

7.2.6. Langevin dynamics

The equations of motion that are most commonly adopted in ENM studies do not generally account for frictional forces (see Eq. (7.6)). Nonetheless, biomolecules exist in viscous environment, and viscous drag may alter their normal modes of motion. It is therefore useful to have a technique for calculating normal modes of motion in the presence of damping forces. Perhaps the simplest way to introduce viscous drag is through the Langevin equation:

$$\mathbf{M\ddot{q}} = -\mathbf{Uq} - \zeta\dot{\mathbf{q}} + \boldsymbol{\xi}(t).$$ (7.38)

Here ζ is a velocity-dependent damping term and $\xi(t)$ is a time-dependent vector of random forces, also called white noise, which satisfies the conditions

$$\langle \xi_i(t) \rangle = 0$$ (7.39)

$$\langle \xi_i(t)\xi_j(t') \rangle = 2\zeta_{ij}\delta(t-t')k_BT ,$$ (7.40)

In mass-weighted coordinates, Eq. (7.38) becomes

$$\ddot{\mathbf{r}} = -\mathbf{Kr} - \mathbf{F\dot{r}} + \mathbf{R}(t),$$ (7.41)

with $\mathbf{K}$ as defined earlier, $\mathbf{F} = \mathbf{M}^{-1/2}\zeta\,\mathbf{M}^{-1/2}$ is the mass-weighted friction matrix, and $\mathbf{R} = \mathbf{M}^{-1/2}\boldsymbol{\xi}$.

Defining the $6N \times 6N$ matrix (Miller *et al.* 2008)

$$\mathbf{A} = \begin{pmatrix} 0 & I \\ -\mathbf{K} & -\mathbf{F} \end{pmatrix},$$ (7.42)

in which $\mathbf{I}$ is the $3N \times 3N$ identity matrix, Eq. (7.41) may be re-written as

$$\begin{pmatrix} \dot{\mathbf{r}} \\ \ddot{\mathbf{r}} \end{pmatrix} = \mathbf{A}\begin{pmatrix} \mathbf{r} \\ \dot{\mathbf{r}} \end{pmatrix} + \begin{pmatrix} 0 \\ \mathbf{R}(t) \end{pmatrix},$$ (7.43)

and the normal modes can be solved analytically by diagonalizing $\mathbf{A}$. The first $3N$ components of the eigenvectors of $\mathbf{A}$ provide the displacements along the normal modes, and the last $3N$ components correspond to the mode velocities. The eigenvalues of $\mathbf{A}$ are complex; their imaginary parts are the oscillatory frequencies of the modes, and their real parts are the exponential decay constants of their amplitudes. This approach has been used by Miller *et al.* (2008) to estimate the fractional free energy loss in the myosin power stroke.

In the limit of strong friction, all of the modes are over-damped and the system obeys Brownian dynamics. Hinsen *et al.* (2000) demonstrated that the normal modes in this limit are found by diagonalizing the friction-weighted force constant matrix,

$$\hat{\mathbf{U}} = \zeta^{-1/2}\mathbf{U}\zeta^{-1/2}. \tag{7.44}$$

In such a case, the system does not oscillate, but displacements from the equilibrium conformation will relax along the eigenvectors of $\hat{\mathbf{U}}$ with relaxation constants given by the corresponding eigenvalues. This technique has been used to calculate scattering functions of proteins, and to investigate the sources behind damping in global protein motions (Hinsen *et al.*, 2000).

7.3. Applications

7.3.1. *Membrane proteins*

Membrane proteins are typically composed of three domains: an extracellular (EC) domain exposed to the periplasm, an intracellular/cytoplasmic (IC or CP) domain buried in the cytoplasm, and a transmembrane (TM) domain embedded in the lipid bilayer. Some membrane proteins, known as receptors, are involved in signal transmission via recognition and binding of substrate/ligand to the EC domain, which triggers conformational changes in the CP domain. The allosteric coupling between different domains or the concerted motions, permit the protein to recognize, bind or translocate substrates. Other membrane proteins serve as ion channels or substrate transport. Permeations of ions and/or substrates thus require collective relaxation mechanism or cooperative motions, which are usually amenable to ENMs. Here we focus on recent progress made in delineating the dynamics of four groups of membrane proteins, potassium channels, acetylcholine receptors, rhodopsin and mechanosensitive channels using ENM-based methods.

Potassium channels: Common gating mechanism observed in different potassium channels. The TM domain of K$^+$ channels is composed of a bundle of eight α–helices contributed by four identical monomers (Fig. 7.2). At the center of these helices is a narrow *selectivity filter* (towards the EC region), followed by a large cavity in the middle, and a long gating region, also called *pore*, that connects to the CP region (Fig. 7.3a). MD studies have provided us with insights in to the mechanism of function at the selectivity filter, including the preferential selectivity of potassium over sodium (Shrivastava & Sansom, 2000; Shrivastava *et al.*, 2002; Bernèche & Roux, 2000) and the free energy

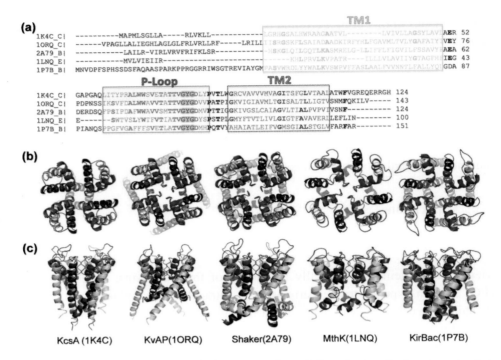

Fig. 7.2. **Sequence and structure of the pore region of five structurally known potassium channels.** (a) Alignment of the pore region sequences. The regions corresponding to the helices TM1 and TM2 and the selectivity filter are indicated by yellow, blue and red blocks, respectively. The alignment was performed using ClustalW (Thompson *et al.*, 1994). Fully or highly conserved regions are shown in bold. The signature motif GYG at the selectivity filter, is highlighted and shaded in red (b) Structural comparison of the pore forming regions of the K⁺ channels aligned in panel a. These are all tetrameric structures, the monomers of which contain either 2 TM helices (KcsA, MthK and KirBac) colored yellow (TM1) and blue (TM2), or 6 TM helices (KvAP and Shaker) denoted as S1-S6. Only the pore forming helices S5 and S6, equivalent to TM1 and TM2, are displayed here, along with the selectivity filter, (which is colored red in all the structures).

profile along the selectivity filter (Bernèche & Roux, 2003). Yet, a fundamental question that remained unanswered until recently has been the mechanism of *gating*, i.e., the conformational events that allow for the transfer of ions from the central cavity to the CP region through the channel-pore. In the X-ray structures, the radius of the pore is too small to let the ions through. The question was: how does the narrow pore open up to permit the permeation of potassium ions?

Toward gaining an understanding of the potential mechanism of pore opening, we recently examined five K⁺ channels, KcsA, KirBac, MthK, KvAP and Shaker (Shrivastava & Bahar, 2006). Figure 7.2a shows a comparison of the

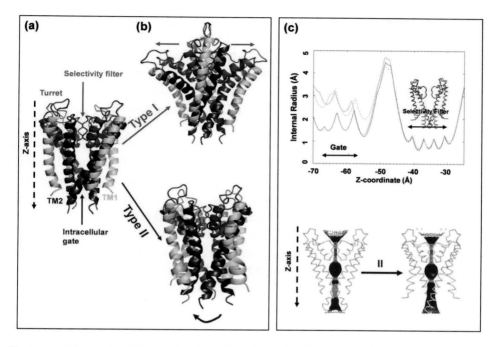

Fig. 7.3. *Mobilities in global modes shown by color-coded diagrams for KcsA.* (a) The different regions of the protein are indicated, namely, the inner and outer helix (TM1 and TM2 respectively), the putative intracellular gate, the selectivity filter and the turret region. (b) The slowest global mode (mode type I), which is two fold degenerate (top), is an opening/stretching motion, with the turrets opening and closing above the selectivity filter. The second lowest frequency mode (mode type II) (bottom), corresponds to a twisting/torsion motion, leading to a widening of the pore, shown in the next panel. (c) Pore-radius profiles (top panel) as a function of the position along the cylindrical (Z-) axis. Bottom panel shows a solid-sphere representation of the inner surface of the channel at the pore region for the crystal structure (left), for the model of the open form (right). The color code for the solid-sphere representation of the pore region is: red, pore radius < 1.15 Å; green, 1.15 Å < radius < 2.30 Å; and blue, radius > 2.30 Å. The pore radius profiles were generated using HOLE (Smart *et al.*, 1993). In the inset of panel (c) is the backbone of the crystal structure (blue) superimposed onto the model of the open form (red). Two monomers have been deleted for clarity.

sequences and pore region structures of these five channels. The observed structural similarities in the pore-forming region suggested a common gating mechanism, which was indeed verified by GNM/ANM calculations. The equilibrium dynamics of these five channels were found to obey similar patterns on a global scale. Mainly, two types of highly cooperative motions were identified at the low frequency end of the mode spectrum, shared by all five structures: The first (referred to as type I) is an alternating expansion/contraction of the EC and/or CP via anti-correlated fluctuations of oppositely located pairs of monomers (Fig. 7.3a). The second (type II) is global torsion of the helical bundle similar to a cork-screw mechanism, with the net result of inducing an enlargement of the pore region (Figs. 7.3b and 7.3c). The change in the relative spacing of the TM2 helices (Shrivastava & Bahar, 2006) was observed to be in accord with the models based on site-directed spin labeling and EPR spectroscopy data (Perozo *et al.*, 1999) and experimental structures of the open form.

This study reinforces the observation that proteins have an inherent ability to undergo conformational changes required for their biological function (Ma, 2005; Bahar *et al.*, 2007; Tama & Sanejouand, 2001; Xu *et al.*, 2003). Computational studies performed by Sansom and coworkers for investigating the dynamics of inward rectifying potassium channels (Kirs) (Haider *et al.*, 2005) further indicated a good agreement (correlation coefficient of 0.87) between the mean-square fluctuations of Kir3.1 residues obtained from molecular dynamics (MD) simulations and those predicted by ANM. The lowest frequency mode from ANM indicated an asymmetric dimer-of-dimers motion, which is also in agreement with that inferred from MD simulations, suggesting that this mechanism of motion is a robust property of the structure. (Sansom *et al.*, 2005). A good correlation was also reported more recently between ENM results and site-directed mutagenesis experiments for KcsA and Mthk (Haliloglu & Ben-Tal, 2008).

nAcetyl Choline Receptor (nAChR): Gating via global twist of the quaternary structure. As a member of the receptor family of membrane proteins, this homo- or hetero-pentamer switches between ion-permeable and –impermeable conformations upon binding or releasing its neurotransmitter substrate, acetylcholine (ACh). The ACh binding site is located at the boundary between the subunits in the EC region. Binding of ACh promotes a transient opening of the channel. Several models have been proposed for the structural transition mediating the signal transmission (Changeux & Edelstein, 1998; Taly *et al.*, 2005; Liu *et al.*, 2008; Szarecka *et al.*, 2007). Normal mode analysis on the complete structure revealed a concerted symmetric quaternary twist motion (Taly *et al.*, 2005), with the EC and IC domains rotating in opposite directions resulting in a wide opening of the pore, compatible with experimental observations. GNM/ANM analyses (Szarecka *et al.*, 2007) of structural models based on cryo-electron microscopy data (Unwin, 2005) also revealed two types of quasi-symmetric twisting motions: Type I inducing a twisting of ligand binding domain (LBD) in opposite direction to that of TM and IC domains; and Type II where the central TM domain undergoes counter-rotation with respect to the EC and CP domains (Figs. 7.4a and 7.4c). Both motions induce an increase in the pore diameter (Figs. 7.4b and 7.4d).

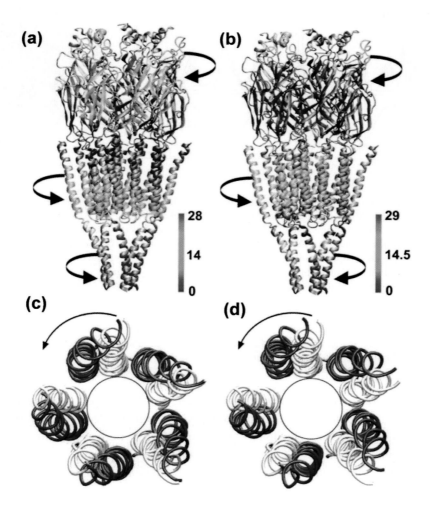

Fig. 7.4. ***Quasisymmetric twisting motion of heteropentameric nAChR.*** GNM mean-square (MS) fluctuations of the global mode types I and II are shown, mapped onto a ribbon diagram of the protein. The fluctuation values are color-coded on the nACHr structure (red:high, green:moderate, blue:low). The black arrows indicate the phases of twisting directions from the correlated ANM modes. The TM2 domains from the models of open-pore structures (magenta) calculated from ANM eigenvectors and eigenvalues are compared with the starting close-pore structure (gray) for type I (c) and type II (d) twists. Note the widening of the pore as a result of the twists (*Figure adapted from **Szarecka et al., 2007***).

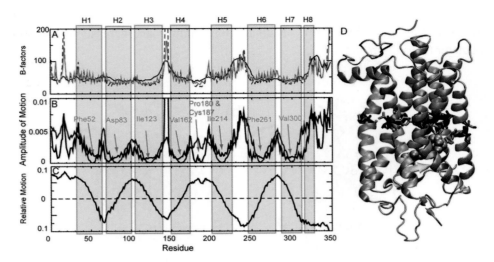

Fig. 7.5. **Dynamics of rhodopsin predicted by the GNM and ANM.**   (a) Experimental(black), GNM (red) and ANM (dashed blue) predicted thermal B-factors. (b) Distribution of square displacements of residues predicted by GNM (blue) and ANM (black). The non-TM regions exhibit higher mobilities, in general. ANM yields two additional minima: Pro180 and Cys187 near the EC entrance to the chromophore binding pocket. (c) The global mode eigenvector calculated with the GNM, indicating relative motions of different regions of the proteins along the principal mode coordinate. Positive and negative regions delineate structural blocks subject to concerted motions. The locations of the helices (1-8) are indicated on the upper abscissa and distinguished by gray bands. (d) Ribbon diagram of rhodopsin color-coded according to the relative motions in (b) in order of increasing mobilities: blue (lowest mobility), cyan, green, yellow, orange, red (highest mobility). Side chains are shown for the seven GNM hinges labeled in (b) and 11-*cis*-retinal is shown in light blue space-filling representation.

Rhodopsin: An activation mechanism coupling the EC and CP domains. As the only structurally determined member of the G-protein coupled receptors (GPCR) family, rhodopsin has been widely studied by both experimental and computational techniques. Capture of the substrate G-protein triggers a highly cooperative conformational change in rhodopsin accompanied by the isomerization of its chromophore (11-*cis*-retinal) at the TM region. The chromophore binding pocket is a highly packed region. The perturbation of the structure at this region drives the propagation of the conformational change via cooperative rearrangements of TM helices to the cytoplasmic (CP) domain, to induce the active state of rhodopsin, metarhodopsin II (Meta-II) (Isin *et al.*, 2006). The application of GNM to two dark state structures of rhodopsin 1L9H (Okada *et al.*, 2002) and 1U19A (Okada *et al.*, 2004) yielded the B-factor profile (Fig. 7.5A) in close agreement with the experimental data (correlation coefficient of 0.837). The loops between helices 3 and 4 and the C-terminus are distinguished by their enhanced mobilities.

A clearer picture of the relative mobilities of the different structural components in the long-time regime is obtained upon examination of the slowest mode profile obtained by GNM in Fig. 7.5 (panels B and C). The slowest mode divides the structure into two regions subject to anticorrelated fluctuations. Mainly, the positive and negative portions of the mode 1 eigenvector plotted in Fig. 7.5C define the two anticorrelated regions. The cross-over regions between them form the minima in the square displacements profile shown in Fig. 7.5B. The corresponding residues occupy central positions in the TM helices (Fig. 7.5D). Since they also lie at the interface between the two anticorrelated regions of the molecule, these residues play an important role in transmitting conformational perturbations. Many residues lying in this critical region (e.g., D83, V162, F261) participate in the retinal binding pocket, and efficiently propagate local conformational changes between the CP and extracellular (EC) ends of the molecule (Isin *et al.*, 2006). ANM analysis shows that this mode essentially drives a global twisting of the TM helices, which results in an overall expansion at the two ends. These conformational changes agree well with experimental data (Isin *et al.*, 2006). In particular, the mobility of spin-labeled side chains at the buried surfaces of TM helices 1, 2, 3, 6 and 7 were found to increase upon isomerization, indicating a reduced packing consistent with the expansion of the pore, in accord with ANM results.

Mechanosensitive Channel (MscL): Channel widening upon global twisting and torsion. These proteins act as a "safety-valve" in *E. Coli*: they open up when the osmotic pressure is beyond a certain threshold (Hamill & Martinac, 2001; Anishkin & Kung, 2005), thus preventing membrane breakdown and cell lysis. The diameter of the gate region, as inferred from the X-ray structure of the closed form (Chang *et al.*, 1998), is ~2 Å, whereas in the open form it is ~30-35 Å (Sukharev *et al.*, 1999), suggesting a significant conformational change. ENM studies have elucidated the dynamics of this mechanosensitive protein (Valadie *et al.*, 2003; Haliloglu & Ben-Tal, 2008). Two major kinds of motions were identified: Type I (Fig. 7.6), a symmetrical motion that corresponds to an overall iris-like opening, exhibited by the non-degenerate modes; and Type II, which resulted in a global bending/tilting. Notably, three non-degenerate modes (modes 11, 31 and 64) (Valadie *et al.*, 2003) accessible to the closed state can alone account for 65% of the conformational change observed between the closed and open states, while the first 100 modes describe 76% of the transition. As to the opposite change, five non-degenerate modes recover 65% of the conformational change.   Interestingly, in this protein as well, the twisting and tilting motions result in widening up of the channel pore.

Although more than 25% of the human genome is made up of membrane proteins (Bond *et al.*, 2007), only ~100 X-ray structures of membrane proteins are known to date (White, 2004). Moreover, most of these structures are from prokaryotic organisms. Thus insights on mechanics of biological functions of these proteins that are made from these structures are vital in understanding the functioning of the respective human homologues. In particular, the 'twist-to-open' mechanism instrumental in the gating function of most of the membrane proteins discussed here, suggests a common mechanism of pore-opening when the pore architecture exhibits a cylindrical symmetry with funnel-like organization of a bundle of helices.

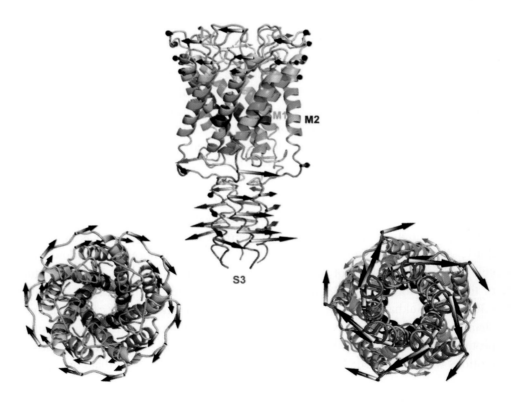

Fig. 7.6. *Twisting and torsion MscL in the slow-frequency twisting/torsion mode.* The top panel illustrates the side view with the protein with a vector representation of the amplitude and direction of motion predicted by the ANM (see also Valadie *et al.*, 2003). The lengths of the arrows scale with the amplitude of the motion. The bottom panel shows the motion of the protein as viewed from top (left) and bottom (right).

(a)   **(b)**

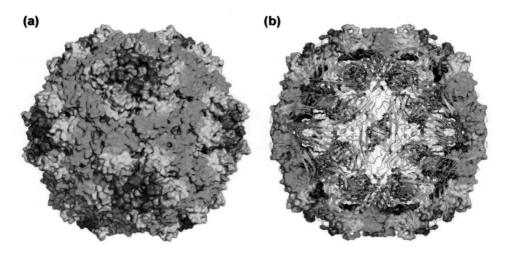

Fig. 7.7. *The structure of STMV.* (a) The structure of STMV represented by its capsid, colored by geometric positions to illustrate the icosahedral symmetry. (b) A cross section view of the STMV, highlighting the capsid and enclosed genetic material (RNA). The RNA, shown as cartoon representation in yellow, on which spherical dots represents the positions of atoms P, C2 and C4'. These atoms are used to build up the elastic network in ANM.

7.3.2. *Viruses*

Viral capsids are some of the largest systems to have been studied with ENMs. Their large size – on the order of a million atoms – places many viruses beyond the reach of MD. Even NMA of viral ENMs is computationally cumbersome, necessitating the use of various techniques to further simplify normal mode calculation. The RTB method was used (Tama & Brooks, 2002; Tama & Brooks, 2005) to investigate swelling of viral capsids, leading to the observation that capsid maturation in several viruses can be largely accounted for with only a few non-degenerate (icoasahedrally symmetric) slow modes. The symmetry of viral capsids was exploited (Kim *et al.*, 2003; Kim *et al.*, 2004) to construct simplified ENMs for studies of capsid maturation, while the maturation of the HK97 bacteriophage was explored using the GNM (Rader *et al.*, 2005). The dynamics of sufficiently small viruses can thus be analyzed through direct application of an ENM without further simplification, as exemplified below.

The satellite tobacco mosaic virus (STMV) is one of the smallest viruses known, consisting of 60 identical protein subunits arranged in an icosahedral shell about a single-stranded RNA with 1058 nucleotide bases (Figs. 7.7a and 7.7b) (Dodds, 1998; Day *et al.*, 2001). Recent studies of STMV include the structural analysis of its RNA, MD simulations of the intact capsid (Freddolino *et al.*, 2006) and direct measurements of its elastic properties. We analyzed the normal modes of STMV using the ANM (Doruker *et al.*, 2000; Atilgan *et al.*, 2001). Starting from the PDB (Berman *et al.* 2000) structure 1A34 (Larson *et al.*, 1998), we built an elastic network using the atoms C^{α} on the capsid proteins, and the atoms P, C2 and C4' on the RNA nucleotides as the network nodes. A cutoff distance of 15 Å was used to define connections between nodes, and the force constant was set to 1 N/m (note the conversion factor 1 N/m = 1.44 kcal/mol/ $Å^2$).

Results from ANM analysis of STMV. The group of rotations that preserve icosahedral symmetry has finitely many irreducible representations: 1, 3, 4 and 5 dimensional representations (Tinkam, 1964; Widom *et al.*, 2007). Since the dimensionality of the irreducible representations determines the degeneracy of each normal mode, the allowed degeneracies of vibrational frequency are simply 1 (nondegenerate), 3, 4 and 5. No other degeneracies may occur. Figure 7.8 shows the eigenvalues calculated using the ANM in the cases of STMV with RNA and the STMV protein coat alone.

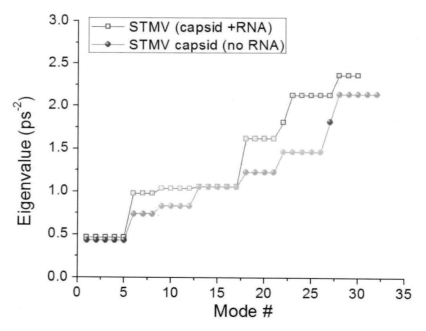

Fig. 7.8. *Dispersion of mode frequencies.* The curves display the eigenvalues calculated using the ANM for STMV with RNA (empty dots) and STMV protein coat alone (solid spheres). They are colored in groups. Each distinctive color represents a type of motion, with certain degeneracy (1, 3, 4 or 5). Higher resolution figure can be provided upon request.

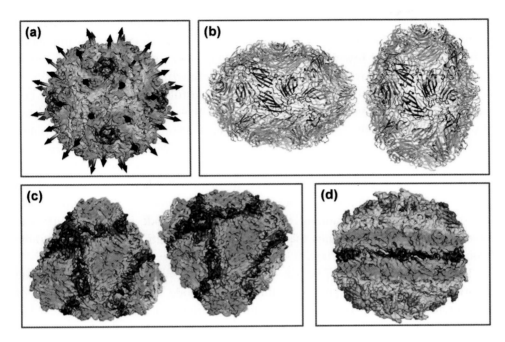

Fig. 7.9. ***Global mode shapes predicted for the intact STMV.*** *Results are based on ANM* (a) The $l = 0$ breathing mode, mode 22 in Fig. 7.8. (b) The $l = 2$ squeezing mode, 5-fold degenerate modes 1-5. (c) The $l = 3$ squeezing mode, modes 6-9. (d) The torsional mode (13–17). All panels are colored according to their mobilities from small (blue) to large (red).

The collective eigenmodes can be divided into two distinctive types (Coccia *et al.*, 1998). The first type is torsional modes, in which the deformations have no radial component. An example of this type of motion is modes 13-17, shown in Fig. 7.9d. The other type is spheroidal modes (e.g. modes 1-5, Fig. 7.9b), in which the eigenvectors contain both tangential and radial components. Each mode can be described by a wave number, l, that corresponds to the degree of the spherical harmonic that best aligns with the mode. Qualitatively, the wave number may be thought of as the degree of symmetry of a mode: $l = 0$ indicates spherical symmetry, $l = 2$ indicates deformation along a single axis, and so on. Figures 7.9a–c illustrates the spheroidal modes with wave numbers 0, 2 and 3, respectively.

Figure 7.9a displays a typical $l = 0$ spheroidal mode, corresponding to mode 22 of STMV (capsid + RNA) in the mode frequency distribution shown in Fig. 7.8. It is non-degenerate because it preserves icosahedral symmetry. Such modes correspond to purely radial motions — shrinking or swelling (*breathing*) of the entire structure. This type of deformation occurs in response to strong internal pressure, such as that exerted by the genome encapsulation, or external pressure, such as osmotic pressure. It is notable that the eigenvalue of this mode is higher than other low frequency modes (Fig. 7.8), implying that the capsid exhibits a relatively stronger resistance against deformations of this type.

Figure 7.9b shows the slowest mode with five-fold degeneracy ($l = 2$), i.e., modes 1-5 in Fig. 7.8. The motion induced in these modes can be visualized physically as the result of *squeezing* a sphere radially inwards at the poles, allowing it to bulge outwards at the equator, and vice versa. Such a deformation occurs when a molecule is probed with an atomic force microscope. Our recent calculations (Yang *et al.*, 2008) showed that hollow spheres are quite soft in response to this mechanism of deformation (Michel *et al.*, 2006; Kol *et al.*, 2006). This is the top ranking (lowest frequency) mode in both STMV and STMV+RNA, which indicates that this kind of deformation is highly favorable from an energetic point of view.

The $l = 3$ mode (Fig. 7.9c) is similar to the $l = 2$ mode above, but the deformation direction is split into three. This mode involves two groups of degenerate modes, 6-8 and 9-12. The torsional mode illustrated in Fig. 7.9d is a fivefold degenerate mode with $l = 2$. This mode can be visualized as the result of twisting the upper and lower hemispheres in opposite directions.

Effect of RNA. There is no difference between the first 21 modes of the STMV capsid (alone) and those of the capsid with RNA in so far as the mechanism of motion is concerned, but their orders (or relative frequencies) exhibit slight changes. The eigenvalues of the capsid with RNA are slightly higher than their counterparts for the protein coat only (Fig. 7.8), but this is essentially due to the larger number of nodes and higher mass of the capsid with RNA. The eigenvalue of the $l = 2$ spheroidal squeezing modes (modes 1-5) increases by 8% in the presence of the genome, whereas that of the $l = 2$ torsional twisting modes (modes 13-17) increases by only 0.4%. Differences appear after the 21st mode; for example, the breathing mode appears earlier in the case of the capsid with RNA.

7.4. Conclusion

Elastic network models lead to a unique analytical solution and provide a thorough sampling of the energy landscape near the energy minimum. Advances in accurate representation of systems and in validation procedures have highlighted coarse-grained approaches as valuable tools for analysis, allowing a direct comparison with experimental results on macromolecular dynamics.

Based on a simplified harmonic potential, ENMs have been enhanced to account for a variety of effects that influence the dynamics of biological molecules. An advantage of ENMs is that they usually provide an accurate description of the mechanisms of motions, although no absolute time scale and size of motions can be predicted. Another advantage is to simplify our understanding of the complex and diverse interactions in biological systems with the help of a simple model, and a small number of parameters. It should be noted, however, that in a strict sense they are applicable to the close neighborhood of the native (or equilibrium) state. They essentially inform us on the intrinsic dynamic preferences of biomolecular systems, which are verified in numerous applications to be functionally relevant (Bahar *et al.*, 2007). ENMs also provide us with physical insights: The overall topology of the protein plays a major role in the mechanical behavior of the protein, implying that proteins related by evolution are expected to show similar quantitative behavior, as seen in the case of potassium channels, discussed above. Basic research in biology and biochemistry along with statistical mechanical and analytical methods will thus lead to improved transferability and predictability of such approaches.

References

Ashcroft, F.M. (2000) Ion Channels and Disease. Academic Press, San Diego, p. 472.

Atilgan, A.R., Durell, S.R., Jernigan, R.L., Demirel, M.C., Keskin, O. and Bahar, I. (2001) Anisotropy of fluctuation dynamics of proteins with an elastic network model. *Biophys. J.* 80: 505-515.

Bahar, I., Atilgan, A.R., and Erman, B. (1997) Direct evaluation of thermal fluctuations in proteins using a single-parameter harmonic potential. *Folding & Design* 2: 173-181.

Bahar, I., Atilgan, A.R., Demirel, M.C., and Erman, B. (1998) Vibrational dynamcis of folded proteins: significance of slow and fast motions in relation to function and stability. *Phys. Rev. Lett.* 80: 2733-2736.

Bahar, I., Chennubhotla, C., and Tobi, D. (2007) Intrinsic dynamics of enzymes in the unbound state and relation to allosteric regulation. *Cur. Op. Struct. Biol.* 17: 633-640.

Bahar, I., and Rader, A.J. (2005) Coarse-grained normal mode analysis in structural biology. *Curr Opin Struct Biol.* 15:1-7.

Bernèche, S. and Roux, B.(2003) A microscopic view of ion conduction through the K^+ channel. Proc. *Natl. Acad. Sci. USA.* 100:8644-8648.

Berman, H.M., Westbrook, J., Feng, Z., Gilliland, G., Bhat, T.N.,Weissig, H., Shindyalov, I.N., Bourne, P.E. (2000) The Protein Data Bank. Nucl Acids Res 28: 235-242.

Coccia, E., Fafone, V., Frossati, G., Lobo, J.A., Ortega, J.A. (1998) Hollow sphere as a detector of gravitational radiation. *Phys. Rev. D* 57: 2051-2060.

Changeux, J.P. and Edelstein, S.J. (1998) Allosteric receptors after 30 years. *Neuron.* 21:959-980.

Chennubhotla, C., Yang, Z., and Bahar, I. (2008) Coupling between global dynamics and signal transduction pathways: a mechanism of allostery for chaperonin GroEL. *Mol. Biosystems* 4: 287-292.

Cui, Q., and Bahar, I. (2006) Normal Mode Analysis: Theory and Applications to Biological and *Chemical Systems.* Boca Raton: Chapman & Hall/CRC.

Day, J., Kuznetsov, Y.G., Larson, S.B., Greenwood, A., McPherson, A. (2001) Biophysical Studies of the RNA Cores of Satellite Tobacco Mosaic Virus. *Biophys J* 80: 2364-2371.

De Pablo, P.J., Schaap, I.A.T., MacKintosh, F.C., Schmidt, C.F. (2003) Deformation and Collapse of Microtubules on the Nanometer Scale. *Phys. Rev. Lett.* 91: 098101.

Dodds, J.A .(1998) Satellite Tobacco Mosaic Virus. *Ann. Rev. Pathol.* 36: 295-310.

Doruker, P., Atilgan, A.R., and Bahar, I. (2000) Dynamics of proteins predicted by molecular dynamics simulations and analytical approaches: application to alpha-amylase inhibitor. *Proteins* 40: 512-524.

Doruker, P., Jernigan, R.L., and Bahar, I. (2002) Dynamics of large proteins through hierarchical levels of coarse-grained structures. *J. Comp. Chem.* 23: 119-127.

Eyal, E., and Bahar, I. (2008) Toward a molecular understanding of the anisotropic response of proteins to external forces: insights from elastic network models. *Biophys. J.* 94: 3424-3435.

Eyal, E., Yang, L.-W., and Bahar, I. (2006) Anisotropic network model: systematic evaluation and a new web interface. *Bioinformatics* 22: 2619-2627.

Flory, P. (1976) Statistical thermodynamics of random networks. *Proc. R. Soc. Lond. A.* 351: 351-380.

Freddolino, P.L., Arkhipov, A.S., Larson, S.B., McPherson, A., Schulten, K. (2006) Molecular Dynamics Simulations of the Complete Satellite Tobacco Mosaic Virus. *Structure* 14: 437-449

Gan, L., Speir, J.A., Conway, J.F., Lander, G., Cheng, N., Firek, B.A., Hendrix, R.W., Duda, R.L., Liljas, L., Johnson, J.E. (2006) Capsid Conformational Sampling in HK97 Maturation Visualized by X-Ray Crystallography and Cryo-EM. *Structure* 14: 1655-1665

Goldstein, H. (1953) *Classical Mechanics.* Cambridge: Addison-Wesley.

Haider, S., Grottesi, A., Hall, B.A., Ashrcoft, F.M. and Sansom, M.S.P. (2005) Conformational dynamics of the ligand-binding domain of inward rectifier K channels are revealed by molecular dynamics simulations: toward an understanding of Kir-channel gating. *Biophys. J.* 88: 3310-3320.

Haliloglu, T. and Ben-Tal, N. (2008) Cooperative transition between open and closed conformations in potassium channels. *PLoS Comp. Biol.* 4: Pcbi.1000164.

Hinsen, K. (1998) Analysis of domain motions by approximate normal mode calculations. *Proteins* 33: 417-429.

Hinsen, K, Petrescu, A.-J., Dellerue, S., Bellissent-Funel, M.-C., and Kneller, G.R.. (2000) Harmonicity in slow protein dynamics. *Chem. Phys.* 261: 25-37.

Isin, B., Rader, A.J., Dhiman, H.K., Klein-Seetharaman, J. and Bahar, I. (2006) Predisposition of the dark state of rhodopsin to functional changes in structure. *Proteins: Struct. Func. Bioinf.* 65: 970-983.

Isin, B., Schulten, K., Tajkhorshid, E. and Bahar, I. (2008) Mechanism of signal propagation upon retinal isomerization: insights from molecular dynamics simulations of rhodopsin restrained by normal modes. *Biophys. J.* 95: 789-803.

Keskin, O., Bahar, I., Flatow, D., Covell, D.G., and Jernigan, R.L. (2002) Molecular mechanisms of chaperonin GroEL-GroES function. *Biochemistry* 4: 491-501.

Kim, M.K., and Jernigan, R.L. (2005) Rigid-cluster models of conformational transitions in macromolecular machines and assemblies. *Biophys. J.* 89: 43-55.

Kim, M.K., Jernigan, R.L. and Chirikjian, G.S. (2003) An elastic network model of HK97 capsid maturation. *J. Struct. Biol.* 143: 107-117.

Kol, N., Gladnikoff, M, Barlam, D., Shneck, R.Z., Rein, A., Rousso, I. (2006) Mechanical Properties of Murine Leukemia Virus Particles: Effect of Maturation. *Biophys J* 91: 767-774.

Kondrashov, D.A., Cui,Q., and Phillips, G.N. (2006) Optimization and evaluation of a coarse-grained model of protein motion using x-ray crystal data. *Biophys. J.* 91: 2760-2767.

Kurkcuoglu, O., Jernigan, R.L., and Doruker, P. (2003) Mixed levels of coarse-graining of large proteins using elastic network model succeeds in extracting the slowest motions. *Polymer* 45: 649-657.

Kurkcuoglu, O., Jernigan, R.L. and Doruker, P. (2005) Collective dynamics of large proteins from mixed coarse-grained elastic network model. *QSAR Comb. Sci.* 24: 443-448.

Larson, S.B., Day, J., Greenwood, A., and McPherson, A. (1998) Refined structure of satellite tobacco mosaic virus at 1.8 Å resolution. *J. Mol. Biol.* 277: 37-59.

Li, G., and Cui, Q. (2002) A coarse-grained normal mode approach for macromolecules: an efficient implementation and application to CA2+-ATPase. *Biophys. J.* 83: 2457-2474.

Li, G., and Cui, Q. (2004) Analysis of functional motions in brownian molecular machines with an efficient block normal mode approach: myosin-II and Ca2+-ATPase. *Biophys. J.* 86: 743-763.

Liu, X., Xu ,Y., Li, H., Jiang, H., and Barrantes, F.J. (2008) Mechanics of Channel Gating of the Nicotinic Acetylcholine Receptor. *PLoS. Comp. Biol.* 4:pcbi0040019

Ma, J. (2005) Usefulness and limitations of normal mode analysis in modeling dynamics of biomolecular complexes. *Structure* 13:373-380.

Marder, M.P. (2000) *Condensed Matter Physics.* New York : John Wiley & Sons.

Michel, J.P., Ivanovska, I.L., Gibbons, M.M., Klug, W.S., Knobler, C.M., Wuite, G.J.L.,

Schmidt, C.F. (2006) Nanoindentation studies of full and empty viral capsids and the effects of capsid protein mutations on elasticity and strength. *Proc. Natl. Acad. Sci.* USA 103: 6184-6189.

Miller, B.T., Zheng, W., Venable, R.M., Pastor, R.W., and Brooks, B.R. (2008) Langevin network model of myosin. *J. Phys. Chem. B* 112: 62746281.

Ming, D., and Wall, M.E. (2005) Allostery in a coarse-grained model of protein dynamics. *Phys. Rev. Lett.* 95: 198103.

Ming, D., and Wall, M.E. (2006) Interactions in native binding sites cause a large change in protein dynamics. *J. Mol. Biol.* 358: 213-223.

Okada, T., Fujiyoshi, Y., Silow, M., Navarro, J. Landau, E. M. and Shichida, Y. (2002) Functional role of internal water molecules in rhodopsin revealed by x-ray crystallography. *Proc. Nat. Acad. Sci. USA.* 99: 5982-5987.

Okada, T., Sugihara, M., Bondar, A.N., Elstner, M., Entel, P., and Byss, V. (2004) The retinal conformation and its environment in rhodopsin in light of a new 2.2 Å crystal structure. J. Mol. Biol. 342: 571-583.

Perozo, E., Cortes, D.M. and Cuello, L.G. (1999) Structural rearrangements underlying K⁺ channel activation. *Science* 285: 73-78.

Rader, A.J., Vlad, D.H., and Bahar, I. (2005) Maturation dynamics of bacteriophage HK97 capsid. *Structure* 13: 413-421.

Roux, B., Bernèche, S. and Im,W. (2000) Ion Channels, Permeation, and Electrostatics: Insight into the Function of KcsA. *Biochemistry,* 39 (44), 13295 -13306.

Shrivastava, I.H., and Sansom, M.S.P. (2000) Simulations of Ion Permeation through a Potassium Channel: Molecular Dynamics of KcsA in a Phospholipid Bilayer. *Biophys J.* 78: 557-570

Shrivastava, I.H., Tieleman, D.P., Biggin, P.C., and Sansom, M.S.P.(2002) K⁺ versus Na⁺ Ions in a K Channel Selectivity Filter: A Simulation Study. *Biophys J,* 83: 633-645.

Shrivastava, I.H., and Bahar, I. (2006) Common mechanism of pore opening shared by five different potassium channels. *Biophys J* 90: 3929-3940.

Sulkowska, J.I., Kloczkowski, A., Sen, T.Z., Cieplak, M., and Jernigan, R.L. (2008) Predicting the order in which contacts are broken during single molecule protein stretching experiments. *Proteins* 71: 45-60.

Szarecka, A., Xu, Y., and Tang, P. (2007) Dynamics of heteropentameric nictotinic acetylecholine receptor:Implications of the gating mechanism. *Proteins* 68: 948-960.

Taly, A., Delarue, M., Grutter, T., Nilges, M., Le Novère, N., Corringer, P.J., and Changeux, J.P. (2005) Normal mode analysis suggests a quartenary twist model for the Nicotine Receptor gating mechanism. *Biophys J.* 88: 3954-3965.

Tama, F., and Brooks III, C.L. (2002) The mechanism and pathway of pH induced swelling in Cowpea Chlorotic mottle virus. *J. Mol. Biol.* 318: 733-737.

Tama, F., and Brooks, III, C.L. (2005) Diversity and identity of mechanical properties of icosahedral viral capsids studied with elastic network normal mode analysis. *J. Mol. Biol.* 345: 299-314.

Tama, F., and Sanejouand, Y.-H. (2001) Conformational change of proteins arising from normal mode calculations. *Protein Eng.* 14: 1-6.

Tama, F., Gadea, F.X., Marques,O., and Sanejouand, Y.-H. (2000) Building-block approach for determining low-frequency normal modes of macromolecules. *Proteins* 41: 1-7.

Tama, F., Valle, M., Frank, J., and Brooks, C.L. (2003) Dynamic reorganization of the functionally active ribosome explored by normal mode analysis and cryo-electron microscopy. *Proc. Natl. Acad. Sci. USA* 100: 9319-9323.

Tinkham, M. (1964) Group Theory and Quantum Mechanics: McGraw Hill.

Tirion, M. M. (1996) Large amplitude elastic motions in proteins from a single-parameter, atomic analysis. *Phys. Rev. Lett.* 77: 1905-1908.

Tobi, D., and Bahar, I. (2005) Structural changes involved in protein binding correlate with intrinsic motions of proteins in the unbound state. *Proc. Natl. Acad. Sci. USA* 102: 18908-18913.

Unwin, N. (2005) Refined structure of nicotinic acetylcholine receptor at 4Å resolution. *J. Mol. Biol.* 346: 967-989.

Valadie, H., Lacapere, J.J., Sanejouand, Y.-H., and Etchebest, C. (2003) Dynamics properties of MScL of Escherirchia coli: A normal mode analysis *J. Mol Biol.* 332, 657-674.

Wallin, E. and von Heijne, G.(1998) Genome-wide analysis of integral membrane proteins from eubacterial, archeana nd eukaryotic organisms. *Protein Sci.* 7: 1029-1038.

Wang, Y.M., Rader, A.J., Bahar, I., and Jernigan, R.L. (2004) Global ribosome motions revealed with elastic network model. *J. Struct. Biol.* 147: 302-314.

Widom, M., Lidmar, J., and David, R.N. (2007) Soft modes near the buckling transition of icosahedral shells. *Phys. Rev. E.* 76: 031911.

White, S.H. (2004) The progress of membrane protein structure determination. *Protein Sci.* 13: 1948-1949.

Xu, C., Tobi, D., and Bahar, I. (2003) Allosteric Changes in Protein Structure Computed by a Simple Mechanical Model: Hemoglobin T --> R2 Transition. *J. Mol. Biol.* 333:153-168

Yang, L.-W., and Bahar, I. (2005) Coupling between catalytic site and collective dynamics: a requirement for mechanochemical activity of enzymes. *Structure* 13: 893-904.

Zheng, W., and Brooks, B.R. (2005) Probing the local dynamics of nucleotide-binding pocket coupled to the global dynamics: myosin versus kinesin. *Biophys. J.* 89: 167-178.

Zheng, W., Brooks, B.R., and Hummer. G. (2007) Protein conformational transitions explored by mixed elastic network models. *Proteins* 69: 43-57.

Chapter 8

Metabolic Networks

Maria Concetta Palumbo[1], Lorenzo Farina[2], Alfredo Colosimo[1] and
Alessandro Giuliani[3*]

[1] *Department of Physiology and Pharmacology,*
University of Rome 'La Sapienza', P.le Aldo Moro 10,
00182, Rome, Italy

[2] *Department of Computer and Systems Science "A. Ruberti",*
University of Rome, 'La Sapienza', Via Ariosto 25,
00185, Rome, Italy

[3] *Department of Environment and Health,*
Istituto Superiore di Sanità, Viale Regina Elena 299,
00161, Rome, Italy

The use of the term 'network' is more and more widespread in all fields of biology. It evokes a systemic approach to biological problems able to overcome the evident limitations of the strict reductionism of the past twenty years. The expectations produced by taking into considerations not only the single elements but even the intermingled 'web' of links connecting different parts of biological entities, are huge. Nevertheless, we believe that the lack of consciousness that networks, beside their biological 'likelihood', are modeling tools and not real entities, could be detrimental to the exploitation of the full potential of this paradigm.

Like any modeling tool the network paradigm has a range of application going from situations in which it is particularly fit to situations in which its application can be largely misleading. In this chapter we deal with an aspect of biological entities that is particularly fit for the network approach: the intermediate metabolism. This fit derives both from the existence of a privileged formalization in which the relative role of nodes (metabolites) and arches (enzymes) is immediately suggested by the system architecture. Here we will discuss some applications of both graph theory based analysis and multidimensional statistics method to metabolic network studies with the emphasis on the derivation of biologically meaningful information.

---

[*]Corresponding author. E-mail: alessandro.giuliani@iss.it

8.1. Introduction

The network paradigm is the prevailing metaphor in nowadays biology. We can read about gene networks, protein networks, metabolic networks, ecological networks, protein folding networks, as well as signaling networks. The network paradigm is an horizontal construct, basically different from the classical top-down molecular biology view, dominant until not so many years ago, in which there was a privileged flux of information from DNA down to RNA and proteins. The shift from a top-down approach to the network paradigm stems from both the dramatic failure in terms of biotechnological applications and the decline of the molecular biology central dogma after the discovery of alternative and in many cases much more efficient fluxes of information with respect to the classical DNA-RNA-protein cascade, RNA editing, epigenetics heritage, post-translational modification.

This sudden crisis of molecular biology fundamentals just after the completion of the genome sequencing enterprise stimulated a resurgence of interest of a system-based biology in a style very similar to the cybernetic wave of the fifties with two basic differences with respect to Norbert Wiener and von Bertalanffy Post II World War speculations on general systems theory: the massive use of computers and the emphasis on molecular data with respect to physiological signals.

The general concept of a network as a collection of elements (nodes) and the relationships among those (arcs), cannot be separated by the definition of a "system" in dynamical systems theory, where the basic elements (nodes) are time varying functions and relationships are differential or difference equations. In this respect, purely topological approaches in which all the nodes and edges are considered as equivalent, and dynamical approaches in which the relations take the form of differential equations are two sides of the same coin, from a purely graphical point of view, being the separation of the purely topological and dynamical approaches based only on practical circumstances coming from the difficulty to get reliable kinetic data from biological experimentation.

Thus when we talk about topological properties of biological networks and discuss how different kinds of wiring architectures like random or small-world network, essentially we are asking ourselves the basic question 'how far we can go knowing only the wiring diagram of a biological system?', one hundreds year ago in life sciences the same question should be formulated as 'what we could infer from the pure anatomical perspective without any information about physiology?'.

Here we will describe metabolic networks along the perspective of the comparison between static (anatomical, purely topological) and dynamic (physiological, kinetic) information, highlighting the particular relations holding between these two approaches in the biochemical regulation of cells.

a) What is a metabolic network?

Biological systems, by the exploitation of suitable energy sources, achieve spontaneous self-organization (order) allowing them to reach high levels of diversity and complexity by means of adaptive processes. From the thermodynamic point of view, the actual decrease of entropy of the system, relative to its organization, is balanced by the entropy increase of the surrounding environment.

The whole level emergent properties of which the most basic of all being the fact an organism can perform a metabolism sufficient to sustain its life by the utilization of the energy embedded in the chemical bonds of lipids, carbohydrates and proteins coming from food, impose the constraints to the molecular organization, but these constraints can be managed in a relatively flexible way by the microscopic level atomisms due to their extreme redundancy and richness of interaction patterns. This allows for the display of a huge repertoire of possible solutions that appear as equivalent in terms of the perceived result (the organisms can live in a myriad of different environments by the use of very different energy sources and passing through many diverse intermediate states).

The presence of multiple solutions to the same problem (and thus the basic degeneracy of the structure/function problem in biological settings) arises very early in biological organization: a single protein (an object in the twilight zone between chemistry and biology) presents a multiplicity of almost equally energetically available configurations, and this multiplicity of possible states allows the protein to display a rich dynamics that is necessary for playing its physiological role, moreover the same basic 'average structure' can be obtained by completely different sequences or different 'structures' generated on demand by the same sequence. Moreover, there are many evidences of proteins that 'moonlight', or have more than one role in an organism. The same degeneracy holds at all the levels of biological organization from genetic regulatory networks up to ecological communities. This is particularly evident in the case of metabolism, looking at the myriads of possible chemical reactions transforming a small organic molecule (metabolites) into another by one or more enzymes that, drastically lowering the activation barrier between a reactant and a product molecule makes the reaction kinetically feasible.

Figure 8.1 reports a relatively small portion of yeast metabolic network in which the nodes are the metabolites and the edges correspond to the presence of one (or more) enzymes able to catalyse the chemical transformation of metabolite *i* at an extreme into metabolite *j* at the opposite end. A well known metabolic pathway (fatty acid biosynthesis) is reported in both a biochemical meaningful appearance and in a synthetic graph formalization. The correspondence between the two representations can be inferred from Table 8.1.

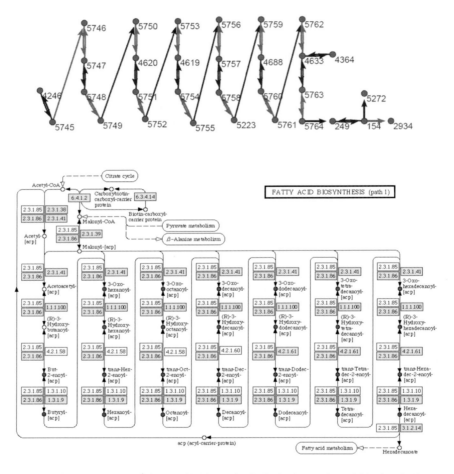

Fig. 8.1. The figure reports the fatty acid biosynthesis in both graph and biochemical ways of representations (upper and lower panel). The first one is obtained using Pajek software on Ma and Zeng database, the second one comes from KEGG database. It is worth noting that only green labelled enzymes are present in yeast.

Table 8.1

Node number	Metabolite name
4246	**But-2-enoyl-[acp]**
5745	**Butyryl-[acp]**
5746	**3-Oxohexanoyl-[acp]**
5747	**(R)-3-Hydroxyhexanoyl-[acp]**
5748	**trans-Hex-2-enoyl-[acp]**
5749	**Hexanoyl-[acp]**
5750	**3-Oxooctanoyl-[acp]**
4620	**(3R)-3-Hydroxyoctanoyl-[acyl-carrier protein]**
5751	**trans-Oct-2-enoyl-[acp]**
5752	**Octanoyl-[acp]**
5753	**3-Oxodecanoyl-[acp]**
4619	**(3R)-3-Hydroxydecanoyl-[acyl-carrier protein]**
5754	**trans-Dec-2-enoyl-[acp]**
5755	**Decanoyl-[acp]**
5756	**3-Oxododecanoyl-[acp]**
5757	**(R)-3-Hydroxydodecanoyl-[acp]**
5758	**trans-Dodec-2-enoyl-[acp]**
5223	**Dodecanoyl-[acyl-carrier protein]**
5759	**3-Oxotetradecanoyl-[acp]**
4688	**(3R)-3-Hydroxytetradecanoyl-[acyl-carrier protein]**
5760	**trans-Tetradec-2-enoyl-[acp]**
5761	**Tetradecanoyl-[acp]**
5762	**3-Oxohexadecanoyl-[acp]**
4633	**(3R)-3-Hydroxypalmitoyl-[acyl-carrier protein]**
4364	2-Hexadecenoyl-[acyl-carrier protein]
5763	**trans-Hexadec-2-enoyl-[acp]**
5764	**Hexadecanoyl-[acp]**
249	Palmitate
154	Palmitoyl-CoA
5272	trans-Hexadec-2-enoyl-CoA
2934	3-Dehydrosphinganine

The presence of multiple pathways to perform the same general chemical transformations both in the sense of derivation of energy from the breaking of food molecules chemical bonds (catabolism) and in the opposite direction of the actual construction by energy expenditure of complex organic molecules relevant for the cell life (anabolism) are consistent with the existence of many possible synthetic 'solutions' with equivalent energetic costs.

This implies that simple energetic considerations are not endowed with a sufficiently discriminant power to guide our research toward a unique and satisfying solution allowing us to predict the behaviour of the network when presented with a given input. It is important to stress that this inability does not come from a lack of knowledge of the metabolic network wiring that, at odds with other biological networks as protein interaction networks, is very well known, but from the often encountered difficulties in finding biologically meaningful optimization principles for the system in a natural environment.

In order to understand the organization of a biological system we need both classical energy constraints and topological (energetically neutral) invariants emerging as bottom up (not necessarily induced by superimposed energy minimization principles) organizational principles. Even without advocating a vitalistic principle that we consider outside the range of science, we must in any case think of still neglected dimensions of optimisation that could be "energetic" but outside the reach of what we nowadays call energy balances.

Nevertheless, as we said above, even if we cannot predict which of the multiple possible routes a given metabolic process will take, we have a very detailed and reliable picture of 'what can in principle happen' as reported in the metabolism charts. The situation is analogue to have a map reporting all the roads but with no indication of the relative traffic on them, we will discover in the following that this sole information is embedded with many important information about the whole system behaviour at the biochemical level.

b) How a metabolic network is built?

Notwithstanding the wide spectrum of diversity existing in nature, metabolic networks of different organism show some very important invariant features.

The presence of some 'universal' molecules like ADP, ATP, CO2, H2O, NADPH... intervening in the great majority of ongoing reactions and thus constituting a sort of 'continuous phase' in which the network is embedded. The ubiquitary character of these molecules implies the impossibility to consider them as 'nodes' like the other metabolites and obliges the modeller to eliminate such molecules from the picture in order not to distort the network architecture. This choice is not new in network modelling: exactly the same happens with hydrogen atoms in graph modelling of organic molecules.

A metabolic network is composed of four distinct classes of nodes (Fig. 8.2). The first one is the strongly connected component (SCC), made of nodes each other linked by direct paths. This is the 'central portion' of the network where the multiplicity of possible paths giving rise to the same general behaviour in terms of molecular profile (concentrations of different molecular species) is maximal.

The 'reactant subset' consists of metabolites entering the system as reactants from the surrounding environment. They are called 'sources' because they can reach the SCC, but cannot be reached from it. The third class is represented by the 'product subset', made of metabolites that are accessible from SCC, but don't have any connections to it. Such metabolites are positioned at the end of a given pathway and thus exiting the system as products (sinks). The metabolites in the last class, called 'isolated subset', are inserted in autonomous pathways with respect to SCCs but nevertheless they have the same kind of links with both sources and sinks than SCCs.

There are two kind of redundancies in the metabolic network wiring: the first kind deals with the multiplicity of edges (enzymes) connecting the same two nodes i and j. This multiplicity stems from the existence of the so called isozymes, distinct enzymes able to catalyse the same chemical reaction. Even if the members of an isozyme family can differ in terms of kinetic parameters and thus of the efficiency of the reaction, nevertheless each member can guarantee the transformation of the metabolite i into metabolite j. The second kind of redundancies deals with reactions: same enzyme can catalyse different reactions. This implies the same molecular entity (a protein) acts as the arc connecting metabolites i and j in one portion of the network and as the arc connecting metabolites k and d in another portion.

The main differences between the metabolic networks of different species are in the reactants and product subsets, while the SCCs and isolated subsets share an elevated degree of similarity.

In the following section we will take into consideration the results of some experiments conducted on metabolic networks so to elucidate some emergent features of this particular kind of biological network system.

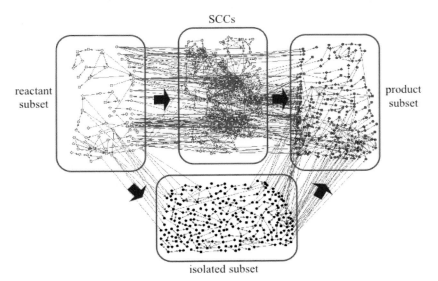

Fig. 8.2. General representation of a metabolic network. On the top, the SCC module corresponds to strongly connected components; on the left there is the 'reactant' subset (sources); on the right, the 'product' subset (sinks); the 'isolated' subset is on the bottom.

8.2. Relevant Biological Feature from Network Wiring

8.2.1. *Network topology and effect of mutations*

From a purely topological point of view, each node of a network is uniquely defined by its position in the graph. Obviously, when dealing with experimentally derived and not abstract networks, each node has a name (a particular gene, protein, metabolite) and the same is true for the edges. However, if we are interested in discovering what can be inferred solely from topological information, we should try and predict some relevant features of the studied organism without relying on the particular 'nature' of nodes and edges, but only taking into consideration their connectivity pattern. In other terms, all the properties relative to each node (edge) must be derived only by its pattern of relations. We checked for the possibility to derive, from purely topological information on the metabolic network of *Saccharomyces cerevisiae*, the lethal character of genetic mutations.

The metabolic network of microorganisms, at odds with different biological networks like protein-protein interaction network or genetic regulation network, is very well understood and characterized. It corresponds to those Boheringer's 'Charts of Metabolism', pinned on the walls of almost every biochemistry laboratory in the world, having enzymatic reactions as edges and metabolites as nodes. Since an enzymatic reaction is catalyzed by one or more enzymes, an arc can also represent the enzyme(s) involved in the reaction. This opens the way to a straightforward analysis of the possibility to derive biologically meaningful features from network topology: the elimination of an enzyme by means of a knock-out experiment implies the elimination from the network of the edge (or edges since, as reported above, the same enzyme can catalyze several biochemical reactions) corresponding to that particular enzyme.

If it is possible to pick up a connectivity descriptor able to unequivocally define essential enzymes (those enzymes whose lack provokes the yeast death) we can safely assume the biological relevance of the metabolism 'wiring structure', irrespective of the knowledge of the specific nature of the involved enzymes. This should correspond to the possibility to define the essentiality of single enzymes for the yeast life in terms of metabolic network topology. This also implies that the relevance of a specific enzyme derives from its role in a global architecture that is the 'effective system' that carries out the metabolic work in the cell. In other words, the possibility to predict the essential character of a specific enzyme on the sole basis of its pattern of connections in the

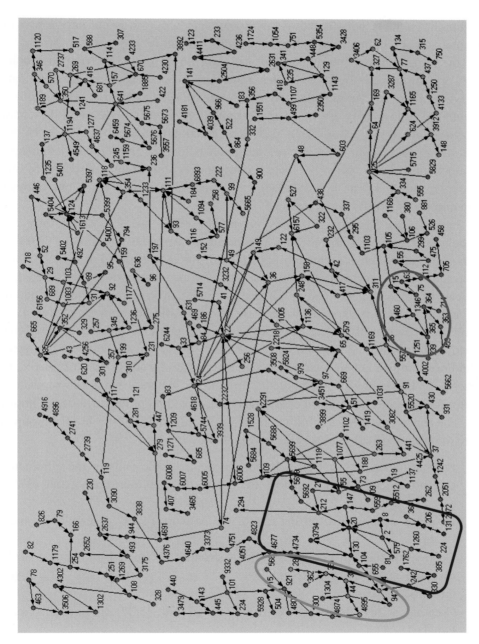

Fig. 8.3. Nodes coloured in blue, light blue, violet and green belong to the SCCs while red nodes pertain to 'reactant', 'product' and 'isolated' subsets. Yellow bars show the location of essential enzymes.

metabolic network tells us the metabolic network wiring it is not only a way of formalization but the image in light of a biological entity.

In the graphic representation of the *Saccharomyces cerevisiae* metabolic network shown in Fig. 8.3, nodes represent metabolites, and arcs indicate enzymes involved in the reaction necessary to synthesize a product from a reactant. Yellow bars show where essential enzymes are located. Essential enzymes are those whose lack provokes lethal effects for the yeast. Nodes colored in blue, light blue, violet and green belong to the strongly connected components of the network. These components constitute the cores of the network and all of the nodes belonging to one of them can reach one another because they are fully connected.

In order to base our analysis on reliable data we considered only the knockouts in which there was a substantial agreement between different experimental evidences, namely a specific enzyme is considered as essential, and consequently the correspondent arcs in the network representation are deleted, if at least the 75% of genes coding for that particular enzyme were demonstrated to be lethal in the knocked out organism. If a specific reaction is catalysed by multiple enzymes, the corresponding arc in the graph is deleted if the above condition is fulfilled for all the enzymes involved. These very stringent conditions allowed us to select only 37 lethal mutations out of the 412 relative to enzymes involved in metabolism, reported in the Stanford repository (http://www-equence.stanford.edu/group/yeast_deletion_project/deletions3.html).

The main effects of inhibitions of essential enzymes can be briefly grouped into the following categories (see also Fig. 8.4):

- Preventing the connection between the nodes at the ends of the eliminated arc: no other path is available to restore the previous connection (37/37 essential enzymes).
- Isolating a pathway from the rest of the network (29/37 essential enzymes, Panel a).
- Destroying the full connectivity in a strongly connected component (7/37 essential enzymes, corresponding to 4/37 in the giant set and 3/37 in the minor SCCs, Panel b).
- Creating a new smaller SCC from another one (1/37 essential enzymes, Panel c).
- Disrupting a cluster of nodes in the IS compartment (16/37 essential enzymes, Panel d).

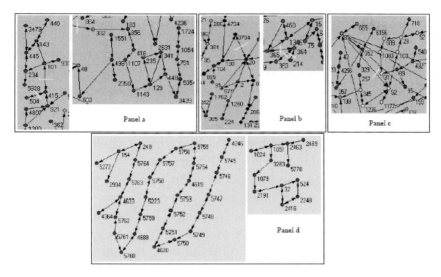

Fig. 8.4. Examples reporting the main effects of enzyme inhibitions. Panel a: isolating and separating a pathway from the rest of the network. Panel b: interrupting the full connectivity in a strongly connected component. Panel c: creating a new smaller SCC from another one. Panel d: disrupting a cluster of nodes in the IS compartment.

The analysis of the metabolic network wiring after the deletions corresponding to the knocking out of an essential enzyme, allowed us to discover that all of the enzymes corresponding to lethal mutations, when deleted, prevent the connections between the separated nodes. No other path is available to restore the broken arc and the involved metabolites (nodes) are no more connected by alternative pathways. The strict relation between lack of alternative pathways reaching a given metabolite and enzyme essentiality implies that the crucial elements of the network correspond to periphery edges (more central locations being more easily reachable by alternative paths along the graph): this is in line with the recently discovered importance of the so called non-hub connectors (elements linking different modules of the network).

The a-la-Barabasi approach to the study of scale-free networks is based on the effects produced by the elimination of nodes. In this perspective, essential nodes in the protein interaction network are often hubs (nodes at the center of the connection pattern). By contrast, the relevance of periphery in metabolic networks is revealed by considering the elimination of edges which corresponds to the elimination of the associated enzyme(s).

The factors influencing the relevance of the elimination of a node or an edge on the overall network are exactly the opposite. In fact, the entity of the "damage" produced by the elimination of a node is proportional to the number of relationships that the node was implied into, whereas in the case of the arc, we directly perturb only one relation (the one corresponding to the arc) so that the entity of the damage is proportional to the number of elements related to that arc. This is pretty intuitive if we think of the consequences on the traffic of the block of a 'peripheral' highway connecting two cities and the consequences of a block of a central road, which keeps many of alternative pathways open.

This last example allows us to comment another important biological feature emerging from our purely topological analysis of metabolic network wiring patterns: the lack of lethal 'purely kinetical' mutations.

We could imagine that cars and trucks could escape the block of the above mentioned highway by using more narrow alternative roads (kinetically non optimal solutions). Soon or later this solution will not be efficient at all, the narrow roads cannot sustain the traffic load of the highway and the flux will be again blocked due to the 'kinetic insufficiency' of the alternative paths. This situation never happens in our data, if an alternative pathway does exist, this will invariably sustain the yeast growth so pointing to a huge plasticity of the kinetic constants governing the entire metabolism.

This effect was still more clear in another experiment in which, instead of studying the location of essential mutations in the metabolic network, we concentrated on the 'double essential mutations' i.e. those mutation couples (two genes are knocked out simultaneously) whose each member is *per se* non lethal but the knocked out couple becomes lethal. In this case the essential character is emerging from the particular coupling and is not peculiar of the specific enzyme. In this case too only double mutations for which the continuity of the network is interrupted are lethal, whenever an alternative pathway, albeit non optimal in kinetic terms is present, the double mutation is not essential.

In a quite schematic form the "upgraded" relationship between biological and topological "missing alternative" property can be summarized by the following points:

- multiple genes synthesizing the same enzymatic function,
- supplementary enzymes participating in the same metabolic reaction,
- availability of other pathways in the graph connecting the separated nodes after the knockouts.

A minimal model has been used in order to build metabolic network: every gene involved in a reaction is considered as an arc; the network has 'multiple edges' between nodes if more than one gene participates in the same reaction (or reactions). A single knockout can disconnect two nodes in the network if the deleted gene (arc) is the only responsible for the catalysis of a reaction, no alternative enzyme can restore the missing link between the separated nodes, and supplementary pathways don't exist; otherwise the connection 'survives'. As for the experimental side, a double mutation can be classified as 'essential' if the double deletion causes the yeast death, or 'non essential' if the yeast survives.

Whereas single deletion knockouts data for *S. cerevisiae* are available from SGDP, the experimental analysis of double mutations by means of an exhaustive gene couples deletion is by now still largely incomplete. Thus, we made use of two approaches in order to derive such information: the Segrè group theoretical approach in which the essentiality of double mutations is obtained from the application of a flux balance analysis model, and a full text mining meta-analysis approach applied on the available literature on double mutants experiments.

Synthetic data from the Segrè group are available in many simulated different growth conditions and their synthetic-lethality range goes from 0 (essential couple) to 1 (non essential couple). Since many genes are relevant only under particular growth conditions, literature studies have been analysed in which many conditions have been tested. As an alternative, the lethality results of all the experimental works available have been compared.

The agreement between Segrè group simulated results, experimental data and our topological essentiality features is checked. This ended up with 25 couples of genes singularly non essential.

Two different properties have been taken into consideration in order to classify these couples: experimental essentiality/non essentiality (coded as ES/NES) and presence/missing of alternatives (A/MA) to maintain network connectivity in our topological analysis. Such alternatives can be, as already stated in section 3.4:

i)   other pathways,
ii)   multiple genes synthesizing the same enzyme,
iii) supplementary enzymes participating in the reaction.

As clearly shown by Table 8.2, our conjecture (namely that essentiality arises from the lack of alternative ways to link two or more nodes) is consistent with the obtained results, given that only one couple out of 25 appears into the 'forbidden class' (namely essential even if in presence of alternatives). Moreover, such inconsistency seems to be due to a lack of information. It is important to keep in mind that the MA condition must be considered as a necessary but not sufficient prerequisite for essentiality, since the corresponding missing metabolite may be included in the culture medium, or its role may be played by another one.

Table 8.2. Couples are classified by rows as 'essential' (ES) or 'not essential' (NES) if experimental data proof the yeast survival or death, and by column if they have topological 'alternatives' (A) or not (MA) to arcs deletions. These alternatives can be: other pathways (P), multiple genes synthesizing the same enzymatic function (G) or supplementary enzymes involved in the reaction (E).

	A	MA
ES	TDH2 (YJR009C) - TDH3 (YGR192C) (E,G)	ADE16 (YLR028C) - ADE17 (YMR120C)
		ARO3 (YDR035W) - ARO4 (YBR249C)
		PGM1 (YKL127W) - PGM2 (YMR105C)
		URA5 (YML106W) - URA10 (YMR271C)
		LYS20 (YDL182W) - LYS21 (YDL131W)
		GSY1 (YFR015C) - GSY2 (YLR258W)
		SER3 (YER081W) - SER33 (YIL074C)
		TKL1 (YPR074C) - TKL2 (YBR117C)
		ASN1 (YPR145W) - ASN2 (YGR124W)
NES	DAL7 (YIR031C) - MLS1 (YNL117W) (G,P)	FUR1 (YHR128W) - URA5 (YML106W)
	ADH1 (YOL086C) - ADH2 (YMR303C) (G)	ALD2 (YMR170C) - ALD3 (YMR169C)
	GDH1 (YOR375C) - GDH3 (YAL062W) (E)	APT1 (YML022W) - APT2 (YDR441C)
	GLT1 (YDL171C) - GDH3 (YAL062W) (E)	AAH1 (YNL141W) - APT2 (YDR441C)
	GLT1 (YDL171C) - GDH1 (YOR375C) (E)	SHM1 (YBR263W) - SHM2 (YLR058C)
	TDH1 (YJL052W) - TDH2 (YJR009C) (E,G)	SHM1 (YBR263W) - GCV1 (YDR019C)
	TDH1 (YJL052W) - TDH3 (YGR192C) (E,G)	GCV1 (YDR019C) - SHM2 (YLR058C)
	MIS1 (YBR084W) - SHM1 (YBR263W) (E,G)	

In some cases predictions of double essentiality deduced by the Segrè group do not agree with both the topological analysis and the experimental data. As an example, FBA predicts yeast survival after the deletion of both URA5 and URA10, while both the graph analysis and De Montigny experimental study say the opposite.

The same considerations apply when two more couples have been taken into account: SER3-SER33 (isozymes, whose essentiality is found from Albers), and LYS20-LYS21, whose lethality estimated by Segrè is 0.96 (thus considered non essential), but whose essentiality is experimentally proved by Feller and colleagues. This result demonstrates a slight superiority of our very simple method with respect to the most sophisticated FBA in predicting the biological consequences of double mutations.

Looking at Table 8.2 and focussing on the main effects of inhibitions of essential enzymes listed before, one can argue the existence of a close relationship between essentiality/non essentiality and the presence of topological connections. The causes and evolution of gene dispensability has remained a controversial issue. It is also intuitive that the existence of many alternatives is synonymous of dispensability and robustness: duplicated genes as well as more than one pathway are the main mechanisms of metabolic flexibility conferring robustness to a biological system.

The essential character of a given mutation as provided by network topological invariants, although derived from laboratory experiments and not in the natural environment where the competitive pressure is active, opens the possibility to consider the metabolic network wiring structure as a biological entity that, even if immaterial, deeply influences the behavior of the organism.

This considerations are reinforced by the possibility to derive phylogenetic trees by the comparison of metabolic network wiring patterns of different organisms so demonstrating that metabolic network topology is influenced by evolution process. In other words, metabolic network wiring is a phenotype as a whole, sensing the adaptive history of the organism.

8.2.2. *Evolution of network topologies*

In the process of adaptation of an organism to its particular environment, during evolution, what is actively exposed to selective forces is the so called 'phenotype' i.e. the entire set of behavioural, physiological, structural, biochemical features of the organism.

On the contrary the 'genotype' is considered as the hereditary information being transmitted thru different generations. If, and only if, a given phenotypical feature offering a specific advantage is determined by a genotypic change, it will be passed across generations and an evolutive step will be achieved.

The somewhat revolutionary acquisitions that followed the completion of the so called genome project abruptly changed our notion of what a 'genotype' is in phenotypic terms. Until twenty years ago everyone could answer 'genotype is the DNA sequence of a given organism'. As for now, the discovery of RNA editing, epigenetic heredity in terms of DNA methylation patterns, post translational modifications and so forth do not allow us to give a satisfactory answer to the question.

Notwithstanding that, we can safely assign, to different biological entities their relative distance from the 'pure phenotypic' (from Greek 'what appears, what is observable') or 'pure genotypic' (from Greek 'what generates', 'what produces') poles. Thus metabolic network wiring is a classical phenotype, given this wiring is the image in light of the repertoire of chemical reactions that are specifically catalysed in a cell producing its chemical make-up with wich the cell responds to environmental stimuli.

On the contrary, the aminoacid sequence of a specific enzyme of this network, even being – strictly speaking – a phenotype (depending on the sequence of the correspondent gene on DNA) it is much closer to the genotype pole (it is produced by the simple application of genetic code to the DNA sequence, setting aside for a moment all the complications of RNA editing, splicing and so forth.).

The classification of organisms on the basis of their phenotypic similarities is thought to be related with the classification based on corresponding genotypic make-up. When we discover a genotype/phenotype relation in terms of consistency of two classifications respectively based on a phenotypic and a genotypic character, we have the proof of a functional relation between the two features that somehow ‘sensed’ the evolutive history in the same way giving rise to congruent similarity spaces for the studied organisms.

In this application, we investigated the classification of organisms based on similarities as for metabolic wiring (phenotypic feature) in comparison with the classification of the same organisms based on the sequence similarities of the shared metabolic enzymes. The finding of a strong correlation between the two allowed us to both give a strong proof of ‘metabolic wiring’ as a specific phenotypic character under evolution constraints and to confirm, by means of a completely independent analysis, the crucial role exerted by non-hub connectors in metabolic networks.

The first computational problem to solve was finding a useful metrics for network comparison so to have a reliable measure of how much two wiring architectures differ from each other. A network of n nodes can be represented as a binary square adjacency matrix nxn where the entry at (i, j) position is 1/0 if there is/there is not an edge from vertex i to vertex j (see Fig. 8.5).

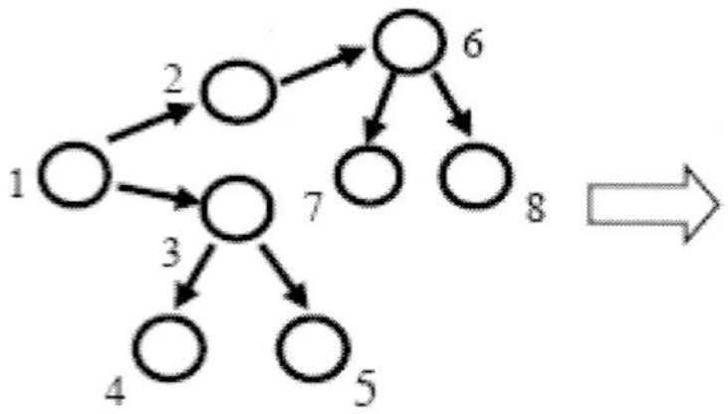

$$A = \begin{bmatrix} 0 & 1 & 1 & 0 & 0 & 0 & 0 & 0 \\ 0 & 0 & 0 & 0 & 0 & 1 & 0 & 0 \\ 0 & 0 & 0 & 1 & 1 & 0 & 0 & 0 \\ 0 & 0 & 0 & 0 & 0 & 0 & 0 & 0 \\ 0 & 0 & 0 & 0 & 0 & 0 & 0 & 0 \\ 0 & 0 & 0 & 0 & 0 & 0 & 1 & 1 \\ 0 & 0 & 0 & 0 & 0 & 0 & 0 & 0 \\ 0 & 0 & 0 & 0 & 0 & 0 & 0 & 0 \end{bmatrix}$$

Fig. 8.5. The network on the left is represented as an adjacency matrix, in which an element in i,j position is 1/0 if an arc exists/doesn't exist connecting node i to j.

In the case of metabolic networks, an arc can be imagined as an enzyme catalyzing a chemical reaction transforming a metabolite into the other. The adjacency matrix formalization was very useful in describing a lot of network structures, and the literature adopting this notation is particularly rich. The adjacency matrix allows to the development a straightforward metrics to compare different metabolic networks. The same network module (like glycolysis, purine metabolism, aminoacid biosynthesis...) gives rise to a peculiar adjacency matrix for each organism.

All these matrices have the same set of rows and columns corresponding to the maximal coverage of the whole set of intervening metabolites (it is enough that a given metabolite is present in a single network to allow inclusion). The distance between each pair of networks will be simply set to the Hamming distance between the two networks, i.e. to the number of discrepancies (1vs.0 or 0vs.1) scored in the corresponding elements of the two networks.

In order to make the metrics independent of the number of analyzed variables (metabolites), we divide the sum of the discrepancies by the total number of variables (maximal attainable distance) and multiply the ratio by 100. Thus, we obtain a 'percentage of dissimilarity' ranging from 0 (complete equivalence of the two networks) to 100.

It is worth noting that the application of Hamming distance is made possible by the peculiar character of metabolic network in which an edge has in any instance the same meaning, namely that 'metabolite i can be transformed into metabolite j', self-catalytic cycles do not exist, and the great majority of metabolites is shared among the different organisms.

This kind of metrics should be unfeasible for other more complex networks as gene regulation or protein-protein interaction networks. The above metrics, when applied to a set of n different networks will end into a symmetric $n \times n$ dissimilarity matrix conveying all the information linked to the pairwise similarities between the corresponding organisms in terms of the metabolic module analyzed.

Being the dissimilarity matrix fully quantitative, it can be analyzed by means of the whole range of multidimensional statistical techniques (multidimensional scaling, principal component analysis) as well as to be the basis for the construction of similarity trees. These can be considered as 'classification' trees analog to those based upon anyother biological character amenable of a given metrics. The idea of metabolic pathway comparisons was exploited by many authors starting from the pioneering work of Dandekar and colleagues to other recent publications. The method we present here is simpler than the above mentioned ones, thus allowing for a wider application range. We analyzed the glycolysis/gluconeogenesis module, and the general scheme of this metabolic network is reported in Fig. 8.6.

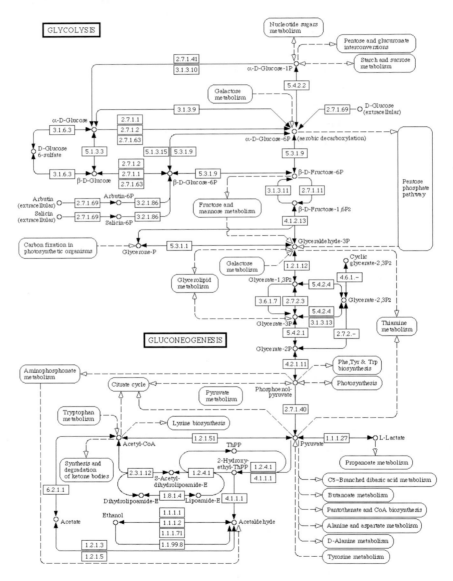

Fig. 8.6. Glycolysis/gluconeogenesis model pathway. The figure is taken from the KEGG database with the indication of the involved enzymes.

We re-iterated the same procedure using different modules (aminoacid biosynthesis, lipid metabolism...) as well as the entire full-range metabolic network of the different organisms obtaining exactly the same results, so pointing to a strong invariance of the metabolic characterization of the different organisms, irrespective of the particular point of view. This is another, albeit indirect, proof-of-concept of the fact that metabolic wiring is a unitary phenotype of the cell.

The inter-organisms network dissimilarity matrix was first computed on 25 microorganisms, the clustering tree relative to this dissimilarity matrix is depicted in Fig. 8.7, where a clear separation of the different species is evident. To test the robustness of the obtained result, an addition of other 18 microorganisms (archaea) to a total of 43 units was made. The general tree is reported in Fig. 8.8.

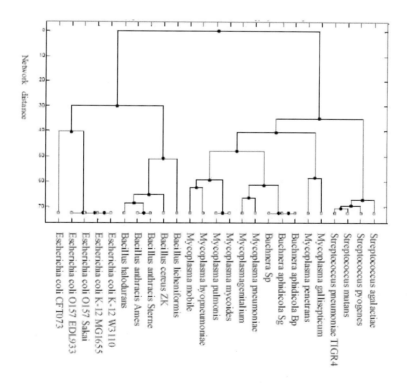

Fig. 8.7. Glycolysis/gluconeogenesis phenotypic tree in the case of the classification of 25 organisms.

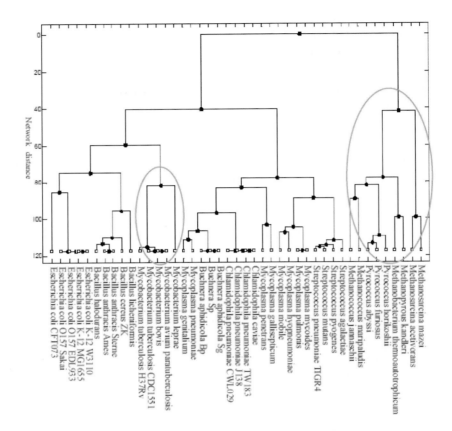

Fig. 8.8. Phylogenetic tree of metabolome BLAST distance data of Glycolysis/gluconeogenesis module from 43 organisms. Other 18 microrganisms shown in the ellipses (9 archaea and 9 bacteria (*Mycobacteria* and *Clamidophylae*)) were added to test the robustness of the obtained result.

A simple inspection of Fig. 8.8 allows for the immediate recognition of a two classes scheme separating archaea species from the others, while the initial 25 organisms classification was kept unchanged.

The classification of microorganisms based on network topological wiring not only maintains species specificity but has a remarkable stability so indicating network topology as a relevant phenotypic character for establishing biologically meaningful classifications.

The same procedure was operated with a much wider spectrum of species (including mammals, reptiles and fishes) and obtaining classification trees almost coincident with the usual Linnean classification of species.

The 25 analyzed organisms share 11 common enzymes of the glycolysis/gluconeogenesis pathway (EC 1.2.1.12, EC 1.2.4.1, EC 1.8.1.4, EC 2.7.1.11, EC

2.7.1.40, EC 2.7.2.3, EC 4.1.2.13, EC 4.2.1.11, EC 5.3.1.1, EC 5.3.1.9, EC 5.4.2.1) thus, each of these enzymes allows for a peculiar sequence space to be constructed and compared with the metabolic network space.

It is well known that the correlation of two protein distance spaces is a marker of some form of interaction between the two molecules, and this feature is routinely adopted for inferring protein-protein interactions. The need to support a viable interaction between the two proteins imposes a mutual constraint (resulting into a covariation) to the random mutation drift of the two systems.

This covariation results into correlated dissimilarity matrices (trees) relative to the two proteins as for a suitable set of organisms. This is exactly what we observed for the 11 common glycolysis/gluconeogenesis sequence based dissimilarity matrices that were significantly correlated to each other. The same was true when we correlated between the organisms dissimilarity matrix based on metabolic networks with the single enzymes sequence spaces. The metabolic network space scored the maximal correlation with the E.C. 4.2.1.11 (phosphopyruvate hydratase) sequence space correspondent to a Pearson $r = 0.94$, $p < 0.001$.

This implies an almost perfect coincidence of the two spaces. Phosphopyruvate hydratase is the enzyme at the interface between the glycolysis and gluconeogenesis modules. Moreover, it represents the link of the module with other biochemical processes like photosynthesis, aminoacid biosynthesis and cytrate cycle. This 'frontier' position is registered by the loading of its dissimilarity space on components other than the common variation axis.

On the basis of such considerations, we may hypothesize that it represents a crucial point of the module and a sort of 'summary' of the entire network behavior. This is probably the reason of the almost perfect linear superposition between metabolic and sequence spaces.

It is worth considering the genesis of the resemblance between metabolic-wiring and phosphopyruvate hydratase sequence based classifications. In the metabolic-wiring space the difference between two organisms A and B is based on the fact a given reaction can take place into organism A but not in the organism B or viceversa. This in general comes from the fact a specific enzyme able to catalyse that reaction is present only in one of the two organisms.

The great majority of the enzymes making the between organisms differences has to do with the processing of the 'reactant subset'. This is in line with the presence of specific 'ecological niches' for the analysed microrganims that allow certain species to grow in a given environment by the utilization of a specific carbon source.

On the contrary, by definition, all the considered microorganisms have the phosphopyruvate hydratase enzyme that in turn makes part of SCC. Nevertheless the inter-organisms difference in their carbon source utilizations (metabolic network distances) is mirrored by differences in the aminoacid sequence of a shared enzyme, metabolically 'very distant' from the reactant subset.

This result forces us to consider the metabolism of an organism as a strongly unitary system in which apparently 'distant' portions of the system, nevertheless share a 'family air' we are still far to rationalize with the current biological theories. Going down to a more usual biological 'reasoning' we can state that the attaining of a strong species specificity (much higher than the one attained by nucleic acid comparisons) and the stability of classifications indicates that pure topological wiring allows for a meaningful picture of the studied organisms.

The striking similarity between phosphopyruvate hydratase sequence space and the general metabolic network space is consistent with the recently discovered relevance of the so called non-hub-connectors described in the previous paragraph: this enzyme represents the major non-hub-connector of the studied system and the sequence/network mapping is in line with the crucial role these network elements are supposed to play.

In their 'cartographic' representation of the metabolic networks of twelve organisms, Guimera and colleagues in define different 'roles' for the nodes involved in a metabolic network based on its within-module degree and its participation coefficient, which define how the node is positioned in its own module with respect to the other modules.

The authors assess the plausible hypothesis that nodes with different roles are under different evolutionary constraints and pressure. In fact, they found that nodes called 'non-hub connectors', that is nodes with many links to other modules, are well conserved in the networks analyzed. This means that nodes connecting different modules, when deleted, have a larger impact on the global structure of fluxes in the network than nodes with many connections within a module. This result is in line with the above discussed 'essentiality-from-topology' problem: the crucial role played by the enzymes at the borders of metabolic modules is probably at the basis of their co-evolution with the metabolic wiring pattern.

Many applications can be imagined for this kind of comparative network studies, extending from the correlation between metabolic network shapes and the pattern of sensitivity to specific antibiotics to large scale environmental studies of ecological communities so as to correlate metabolic similarities to trophic networks. Clearly, the proposed metrics for comparing different networks is oversimplified not taking into account flux consideration as well as other elements like feedbacks or resonances. But this is the same for sequence with respect to structure comparisons of different proteins as well as for graph representations of organic molecules and we know how important both these research avenues were for the respective fields of investigation. Nevertheless the problem of the possible link between network topology and its dynamical properties at large is worth of consideration.

8.3. Network Dynamics

The mathematical description of a dynamical system is a collection of state variables changing with time according to a set of rules which determine the future state basing on a given present state. The finding of such rules for various natural systems is a central problem in science.

Once the dynamics is given, one of the tasks of mathematical dynamical systems theory is to investigate the patterns of how states change in the long run. Dynamics is often described by means of differential or difference equations, which are widely used (and sometimes misused) in virtually all fields of the applied sciences whenever dynamical aspects, *i.e.* time dependency, come into action. This is mainly due to its spectacular success in physics that has led many researchers to apply this mathematical tool to almost everything: economy, social sciences, analytical chemistry and so on.

The "amount" of differential equations used by a scientific discipline is often considered as a measure of how "mature" it is. This is a very dangerous attitude, since the use of differential equations, as any other computational tool, may be justified by very different paradigms, more or less convincing or appropriate. For example, in papers dealing with gene expression modeling, it is common to read that "the underlying molecular machinery is governed by mass action laws" so that differential equations are necessarily relevant. This argument may be very misleading, since it assumes a high degree of reductionism at many different levels, which is not necessarily needed in the context of complex systems, where the "language" of an emergent level may completely different from the subsiding level. Actually, some like to define a "complex system" as those systems whose descriptive items are fundamentally different at different levels.

The opposite approach relies on the fact that since differential equations are ubiquitous in science, than we can use it "everywhere" even for describing "love affairs"!. It is clear that such an approach is nothing more than "wishful thinking". This preliminary discussion is important to understand the impact of the differential equations approach to metabolic networks.

8.3.1. *Representing dynamical schemes*

The basic elements of a metabolic network are the different metabolite concentrations and the reactions, often catalyzed by enzymes. The change of concentration in time can be described using differential equations by considering m metabolites and r reactions, so that we have

$$\frac{dS_i}{dt} = \sum_{j=1}^{r} n_{ij} v_j \quad i = 1,...,m$$

where S_i is the concentration of metabolite i, v_j is the reaction rate and n_{ij} is the stoichiometric coefficient of metabolite i in reaction j. By rearranging the coefficients n_{ij} in a matrix, we obtain the stoichiometric matrix N. In the case of the network in Fig. 8.9, the stoichiometric matrix has four rows corresponding to A, B, C, D metabolites and seven columns, corresponding to the number of internal metabolites and reactions and is the following

$$N = \begin{pmatrix} 1 & 0 & 0 & 0 & -1 & -1 & 0 \\ 0 & 1 & 0 & -1 & 1 & 0 & -1 \\ 0 & 0 & 0 & 0 & 0 & 1 & 1 \\ 0 & 0 & -1 & 1 & 0 & 0 & 0 \end{pmatrix}.$$

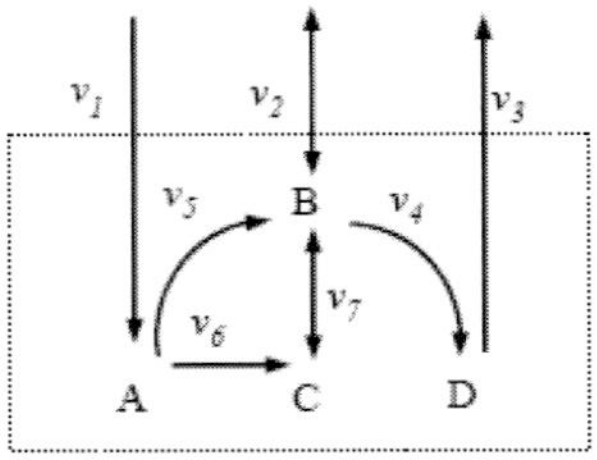

Fig. 8.9.   A simple network. A, B, C and D are internal metabolites; v_4, v_5, v_6, v_7 are internal reactions and v_1, v_2, v_3, are external fluxes. The area delimited by the dotted line denotes the internal metabolism.

Since all reactions may be reversible, in order to determine the signs of the coefficients n_{ij}, the directions of the arrows for reversible reactions are considered positive if the incoming arc is directed 'from left to right' and 'from top to bottom' by convention. For example, let us consider the second row (metabolite B) of matrix N: the value in the second column is positive because the reversible flux entering B is directed 'from top to bottom', while the reversible reaction v_7 entering B has the opposite direction (from bottom to top) and so it has to be considered as negative.

If irreversible reactions are taken into account, the signs of the coefficients are positive for the incoming arcs, or negative in the opposite case. As a consequence, in the second row of N there is a 1 in the fifth column because v_5 enters B and there is a -1 in the fourth column because v_4 is leading away from B. At steady state we have

$$\frac{dS}{dt} = Nv = 0$$

and the allowed fluxes are those for which $Nv = 0$ holds. The problem of finding solutions to this kind of equations is well known in linear algebra where it is called as the *kernel* (or null) of N and it is known that any possible solution can be found as linear combinations of special solutions called *basis vectors*, corresponding to basis stationary flows in the network. The question whether such flows are biologically feasible must be addressed separately, using experimental evidences. However, such decomposition of steady state flows in a minimal number of elementary flows, form the basic tool for the dynamic analysis of metabolic networks. When considering irreversible reactions, the stoichiometric matrix remains the same while constraints on the signs of some basis vector elements arise.

8.3.2.   *Elementary flux modes*

The stoichiometry of a metabolic network can be used to determine the so-called *flux modes*, which are sets of flow vectors v that lead from one external metabolite to another external metabolite. A flux mode is an *elementary flux mode* (EFM) if it uses a minimal number of reactions and cannot be further decomposed.

An elementary flux mode fulfills the following three conditions: 1) steady state condition; 2) feasibility (irreversible reactions have to proceed in the 'right' direction, meaning that all vectors are nonnegative); 3) non-decomposability, implying that the participating enzymes in one EFM are not a subset of the enzymes from another EFM.

The number of elementary flux modes is no less than the number of basis vectors in the null space of N. An *extreme pathway* (EP) is a particular EFM that satisfies two additional conditions: 4) each reaction must be classified either as an exchange flux, which allow a metabolite to enter or to exit the system, or as an internal reaction. All reversible internal reactions must be decomposed as two separate irreversible reactions, thus implying that no internal reaction can have a negative value. Exchanges fluxes can be reversible but the usefulness of this distinction is debatable. 5) Extreme pathways are the minimal set of EFMs and represent a convex basis.

In Fig. 8.11 and Fig. 8.12 the elementary modes for the network in Fig. 8.10 are shown. In Fig. 8.12 the internal reversible reaction v_7 is broken down into its forward and backward directions, giving rise to two reactions: v_7 and v_8. As a consequence the new stoichiometric matrix $\bar{N}$ is

$$\bar{N} = \begin{pmatrix} 1 & 0 & 0 & 0 & -1 & -1 & 0 & 0 \\ 0 & 1 & 0 & -1 & 1 & 0 & -1 & 1 \\ 0 & 0 & 0 & 0 & 0 & 1 & 1 & -1 \\ 0 & 0 & 1 & 1 & 0 & 0 & 0 & 0 \end{pmatrix}$$

and, with respect to Fig. 8.10, a new elementary mode is added, namely the one depicted in the last box.

For each reversible internal reaction, the set of EFMs and EPs in the modified network includes at least a 'cycle'. However, these particular fluxes are not considered, since EFMs and EPs are important to understand the exchange of fluxes between external and internal metabolites. In Table 8.3, the relations between EFMs and EPs in Fig. 8.10 are shown.

Nonnegative combinations of EPs in the third column are shown to determine EFM'1 and EFM'5.

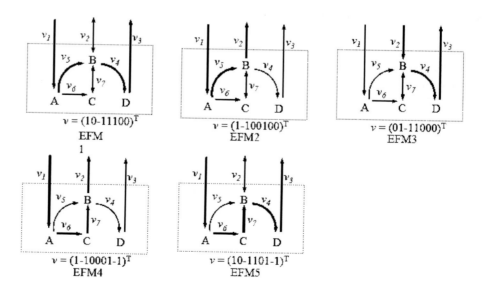

Fig. 8.10. Elementary flux modes for the network in Fig. 8.9. The corresponding flux vector in the null of the matrix N are shown at the bottom of each box.

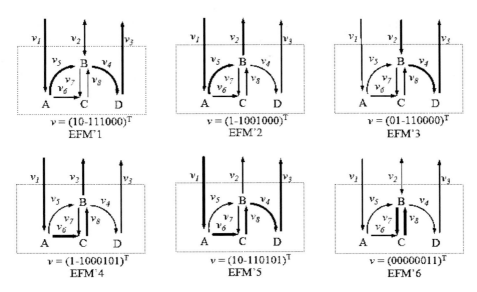

Fig. 8.11. Elementary flux modes for the network in Fig. 8.9 in which internal reversible reactions are split. The corresponding flux vector in the null of the matrix $\overline{N}$ are shown at the bottom of each box. Reversible reaction v_7 is broken down in its forward and backward directions, giving rise to reactions v_7 and v_8. As a consequence, a new elementary mode is added, depicted in the last box.

Table 8.3. Relations between EFMs and EPs. Nonnegative combinations of EPs in the third column are shown to determine EFM'1 and EFM'5.

EFMs	EPs	Sum of Eps
EFM'1		EP1 + EP2; EP1 + EP2 + EP4
EFM'2	EP1= EFM'2	
EFM'3	EP2= EFM'3	
EFM'4	EP3= EFM'4	
EFM'5		EP2 + EP3; EP2 + EP3 + EP4
EFM'6	EP4= EFM'6	

8.3.3. *Flux balance analysis*

The *flux balance analysis* (FBA) takes advantage of the mode decomposition of metabolic processes provided by the stoichiometric matrix to obtain information on actual fluxes operating during specific biological process characterized by optimizing some objective function (e.g. by maximizing ATP consumption in the glycolysis pathway). This kind of study does not need a deep knowledge of the detailed kinetics of individual processes and makes use of the fact that under steady-state conditions, the sum of fluxes producing or degrading any internal metabolite must be zero. The analysis consists of determining extreme pathways by intersecting the null subspace of N and the positive flux orthant, thus obtaining a polyhedral cone and by imposing constraints on the magnitude of individual metabolic fluxes (Fig. 8.12).

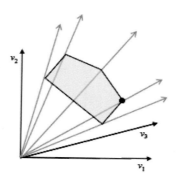

Fig. 8.12. In a three-dimensional flux space grey arrows correspond to the extreme pathways. They determine a convex cone inside the positive orthant. Capacity constraints on each flux limits the cone region to the polytope determined by the grey arrows and the light blue area. All possible flux distributions lie within such polytope. When the cost function is a linear combination of fluxes, then the optimal solution (black dot) usually lies in a corner of the polytope or along a whole edge.

The determination of a particular metabolic flux distribution can be obtained by maximizing (or minimizing) a linear or quadratic objective function using a linear programming approach. Examples of such objective functions are minimization of nutrient uptake, maximal yield of a desired product, maximal growth rate, flux minimization and so on.

8.3.4. *MOMA, PhPP and ROOM analyses*

Schuster and colleagues stated that although the assumption of optimality for a wild-type bacterium is justifiable, the same argument may not be valid for genetically engineered knockouts or other bacterial strains that were not exposed to long-term evolutionary pressure. So they address this point by introducing the method of *minimization of metabolic adjustment* (MOMA), whereby they test the hypothesis that knockout metabolic fluxes undergo a minimal redistribution with respect to the flux configuration of the wild type. MOMA employs quadratic programming to identify a point in flux space.

Phenotypic phase plane (PhPP) analysis is a constraint-based method used to obtain a global perspective of genotype-phenotype relationships in genome-scale metabolic networks. In PhPP analysis, flux balance analysis and linear programming are used to map all of the cellular growth conditions represented by two environmental variables onto a two-dimensional plane and identify phases with distinct metabolic pathway utilization patterns. Some applications of PhPP analysis include the study of optimal growth rates, adaptability of microorganisms, metabolic network functions and capacities, and the impact of gene regulations. Thus, PhPP analysis provides a way to guide experiments and analyze phenotypic functions based on genome-scale metabolic networks.

Regulatory on/off minimization (ROOM) is a constraint-based algorithm for predicting the metabolic steady state fluxes after gene knockout. This method is able to correctly identify short alternative pathways used for rerouting metabolic fluxes in response to gene deletions.

8.3.5. *Metabolic control analysis*

Metabolic control analysis (MCA) is a quantitative and qualitative method for studying how steady state properties of the network are interrelated with those of individual reactions method for analyzing how the control of fluxes and intermediate concentrations in a metabolic pathway is distributed among the different enzymes that constitute the pathway.

One of the most important applications of MCA is in biotechnological production processes, since it is important to reveal which enzyme(s) should be activated in order to increase the synthesis rate of some given metabolite.

The relationship between stationary flux distribution and kinetic parameters is highly nonlinear and to date, no general method to predict the effects of large parameter variations is known or proved to be effective. For this reason, it is convenient to consider small variations, i.e. assuming a linear relationship, so that precise mathematical expression can be derived and metabolic network behaviour quantitatively predicted.

In MCA, one studies the relative control exerted by each step (enzyme) on the system's variables (fluxes and metabolite concentrations). This control is measured by applying a perturbation to the step being studied and measuring the effect on the variable of interest after the system has settled to a new steady state. The basic mathematical tool is the sensitivity coefficient defined, for a generic quantity x depending on a parameter p, as follows:

$$c_p^x = \left(\frac{p}{x} \frac{\Delta x}{\Delta p} \right)_{\Delta x \to 0} = \frac{p}{x} \frac{\partial x}{\partial p} = \frac{\partial \ln x}{\partial \ln p}$$

$$c_p^x = \left(\frac{p}{x} \frac{\Delta x}{\Delta p} \right)_{\Delta x \to 0} = \frac{p}{x} \frac{\partial x}{\partial p} = \frac{\partial \ln x}{\partial \ln p}$$

An elasticity coefficient quantifies the sensitivity of a reaction rate with respect to a change of concentration or a kinetic parameter, while control coefficients refer to the change of steady state flux and concentration distributions due to a change of individual reactions rate. Elasticity coefficients are local properties regarding individual reactions and can be calculated for any given state while control coefficients are global properties related to the new steady state reached after a perturbation.

MCA was used in to model the human erythrocyte and in the specialized literature, there are plenty of applications.

8.4. Conclusion

It is important to realize how powerful the network approach can be and, at the same time, to understand that the search for 'the definitive network' of any organism is devoid of any sense: the actual shape of metabolism (as well as of gene regulation or protein interaction pattern) changes in time and with respect to the environmental conditions.

The possibility to design (and even to profitably study) topological Charts-of-Metabolism must not make scientists forget those charts report 'what in principle could happen' and not what necessarily happens in any situation. The same kind of relation holds between the entire genome sequence and the effective activation of certain genes. Nevertheless, the purely topological approach for metabolic networks has proved to be fruitful and rich of biological consequences.

This takes us to the basic question: how and when the network approach gives the scientist some doubt free advantages with respect to other methods?

We think the best criterion is embedded in the idea of what a network is: a bunch of nuclei (where matter as well as activities are highly concentrated) linked to each other by arcs (streets, mechanical junctions, power cables) passing through a much less dense environment.

In this view, the highways connecting different big cities constitute a network, because there is a sharp difference between the activity density (crowding, population density, energy expenditure) of the nodes (cities) with respect to the country along which the highway passes by. On the other hand, the intermingled texture of internal city streets is much less naturally formalized as a network, with the nodes/arcs distinction much less clear. This is the reason why the network paradigm was very useful to study metabolism where organic molecules are immediately evident 'crucial points' of the entire 'game', while it is much less natural to consider different RNAs transcription levels (like in microarray experiments) as the nodes of the differential gene expression.

References

Albert, R. and Othmer, H. G. (2003). The topology of the regulatory interactions predict the expression pattern of the segment polarity genes in *Drosophila melanogaster*, *J. Theor. Biol.*, 223, pp. 1-18.

Albers, E., Laizé, V., Blomberg, A., Hohmann, S. and Gustafsson, L. (2003). Ser3p (Yer081wp) and Ser33p (Yil074cp) are phosphoglycerate dehydrogenases in *Saccharomyces cerevisiae*, *J. Biol. Chem.*, 278, pp. 10264–10272.

Alfonzo, J. D., Croter, T. R., Guetsova, M. L., Daignan-Fornier, B. and Taylor, M. W. (1999). APT1 but Not APT2, codes for a functional adenine phosphoribosyltransferase in *Saccharomyces cerevisiae*, *J. Bacteriol.*, 181, pp. 347–352.

Artzy-Randrup, Y., Fleishman, S. J., Ben-Tal, N. and Stone, L. (2004). Comment on "Network Motifs: Simple Building Blocks of Complex Networks" and "Superfamilies of Evolved and Designed Networks", *Science*, 305, 1107c.

Avendano, A., DeLuna, A., Olivera, H., Valenzuela, L. and Gonzales, A. (1997). GDH3 encodes a glutamate dehydrogenase isozyme, a previously unrecognized route for glutamate biosynthesis in *Saccharomyces cerevisiae*, *J. Bacteriol.*, 179, pp. 5594–5597.

Bhalla, U. S. and Iyengar, R. (1999). Emergent properties of networks of biological signalling pathways, *Nature*, 283, pp. 381-387.

Blank, L. M., Kuepfer, L. and Sauer, U. (2005). Large-scale ^{13}C-flux analysis reveals mechanistic principles of metabolic network robustness to null mutations in yeast, *Genome Biology*, 6, pp. 1-15.

Boles, E., Liebetrau, W., Hofmann, M. and Zimmermann, F. K. (1994). A family of hexosephosphate mutases in *Saccharomyces cerevisiae*, *Eur. J. Biochem.*, 220, pp. 83-96.

Bork, P., Jensen, L. J., Von Mering, C., Ramani, A. K., Lee, I. and Marcotte, E. M. (2004). Protein interaction networks from yeast to human, *Curr. Opin. Struct. Biol.*, 14, pp. 292-299.

Branden, C. and Tooze, J. (1991). *Introduction to protein structure*, Garland Publishing Company, New York, NY.

Chua, G., Robinson, M. D., Morris, Q. and Hughes, T. R. (2004). Transcriptional networks: reverse-engineering gene regulation on global scale, *Curr. Opin. Microbiol.*, 7, pp. 638-646.

Crampin, E. J., Schnell, S. and McSharry, P. E. (2004). Mathematical and computational techniques to deduce complex biochemical reaction mechanism, *Prog. Biophys. Mol. Biol.*, 86, pp. 77-112.

Dang, V.-D., Valens, M., Bolotin-Fukuhara, M. and Daignan-Fornier, B. (1996). Cloning of the ASN1 and ASN2 genes encoding asparagines synthetases in *Saccharomyces cerevisiae*: differential regulation by the CCAAT-box-binding factor, *Mol. Microbiol.*, 22, pp. 681-692.

Dandekar, T., Schuster, S., Snel, B., Huynen, M. and Bork, P, (1999). Pathway alignment: application to the comparative analysis of glycolytic enzymes, *Biochem. J.*, 343, pp. 115-124.

De Jong, H. (2002). Modeling and simulation of genetic regulatory systems: A literature review, *J. Comput. Biol.*, 9, pp. 67-103.

De Montigny, J., Kern, L., Hubert, J.-C. and Lacroute, F. (1990). Cloning and sequencing of *Ura10*, a second gene encoding orotate phosphoribosyl transferase in *Saccharomyces cerevisiae*, *Curr. Genet.*, 17, pp. 105-111.

Delgado, M. L., O'Connor, J. E., Azorìn, I., Renau-Piqueras, J., Gil, M. L. and Gozalbo, D. (2001). The glyceraldehyde-3-phosphate dehydrogenase polypeptides encoded by the *Saccharomyces cerevisiae* TDH1, TDH2 and TDH3 genes are also cell wall proteins, *Microbiology*, 147, pp. 411-417.

DeLuna, A., Avendano, A., Riego, L. and Gonzalez, A. (2001) NADP-glutamate dehydrogenase isoenzymes of *Saccharomyces cerevisiae*. Purification, kinetic properties, and physiological roles, *J. Biol. Chem.*, 276, pp. 43775–43783.

Duarte, N. C., Palsson, B. O. and Fu, P. (2004). Integrated analysis of metabolic phenotypes in *Saccharomyces cerevisiae*, *BMC Genomics*, 5, pp. 63.

Dunker *et al.* (2002). Another window into disordered protein function, *Structure*, 15, pp.1026-1028.

Edwards, J. S. and Palsson, B. O. (2000). The *Escherichia coli* MG1655 *in silico* metabolic genotype: Its definition, characteristics, and capabilities, *Proc Natl Acad Sci USA*, 97, pp. 5528-5533.

Edwards, J. S., Ibarra, R. U. and Palsson, B. O. (2001). In silico prediction of Escherichia Coli metabolic capabilities are consistent with experimental data, *Nat. Biotechol.*, 19, pp. 125-30.

Farkas, I., Hardy, T. A., Goebl, M. G. and Roach P. J. (1991). Two glycogen synthase isoforms in *Saccharomyces cerevisiae* are coded by distinct genes that are differentially controlled, *J. Biol. Chem.*, 266, pp. 15602-15607.

Fell, D. A. (1992). Metabolic Control Analysis: a survey of its theoretical and experimental development, *Biochem. J.*, 286, pp. 313-30.

Fell, D. A. (1997). *Understanding the control of metabolism*, Portland Press, London, UK.

Feller, A., Ramos, F., Piérard, A. and Dubois, E. (1999). In *Saccharomyces cerevisiae*, feedback inhibition of homocitrate synthase isoenzymes by lysine modulates the activation of LYS gene expression by Lys14p, *Eur. J. Biochem.*, 261, pp. 163-170.

Fiehn, O. and Weckwerth, W. (2003). Deciphering metabolic networks, *Eur. J. Biochem.*, 270, pp. 579-588.

Finkelstein, A. V. and Galzitskaya, O. V. (2004). Physics of protein folding, *Phys. Life Rev.*, 1, pp. 23-56.

Fong, S. S., Marciniak, J. Y. and Palsson, B. O. (2003). Description and interpretation of adaptive evolution of Escherichia coli K-12 MG1655 using a genome-scale in silico metabolic model, *J. Bacteriol.*, 185, pp. 6400-6408.

Friedman, N. (2004). Inferring cellular networks using probabilistic graphical models, *Science*, 303, pp. 799-805.

Gardner, T. S. and Faith, J. J. (2005). Reverse-engineering transcriptional control networks, *Phys. Life Rev.*, 2, pp. 65-88.

Gardner, T. S., di Bernardo, D., Lorenz, D. and Collins, J. J. (2003). Inferring genetic networks and identifying compound mode of action via expression profiling, *Science*, 301, pp. 102-5.

Giersch, C. (2000). Matematical modelling of metabolism, *Curr. Opin. Plant. Biol.*, 3, pp. 249-253.

Goh, K. I., Kahng, B. and Kim, D. (2001). Spectra and eigenvectors of scale-free networks, *Phys. Rev. E.*, 64, 051903 1-5.

Gomperts, B. D., Mramer, I. M. and Tatham, P. E. R. (2002). *Signal transduction*, Academic Press, New York, NY.

Gu, X. (2003). Evolution of duplicate genes versus genetic robustness against null mutations, *Trends Genet.*, 19, pp. 354-356.

Gu, Z., Steinmetz, L. M., Gu, X., Scharfe, C., Davis, R. W. and Li, W. H. (2003). Role of duplicate genes in genetic robustness against null mutations, *Nature*, 421, pp. 31-32.

Guimera, R. and Nunes Amaral, L. A. (2005). Functional cartography of complex metabolic networks, *Nature*, 433, pp. 895-900.

Harbison, C. T. et al (2004). Transcriptional regulatory code of a eukaryotic genome, *Nature*, 431, pp. 99-104.

Hartemink, A. J., Gifford, D. K., Jaakkola, T. S. and Young, R. A. (2001). Using graphical models and genomic expression data to statistically validate models of genetic regulatory networks, *Pac. Symp. Biocomput.*, pp. 441-422.

Hartig, A., Simon, M. M., Schuster, T., Daugherty, J. R., Yoo H. S. and Cooper T. G. (1992). Differentially regulated malate synthase genes participate in carbon and nitrogen metabolism of *S.cerevisiae, Nucleic Acids Res.*, 20, pp. 5677-5686.

Hartman, J. L., Garvick, B., Hartwell, L. (2001). Principles for the buffering of genetic variation, *Science*, 291, pp. 1001-1004.

Heinrich, R. and Rapoport, T. A. (1974). A linear steady-state treatment of enzymatic chains. General properties, control and effector strength, *Eur. J. Biochem.*, 42, pp. 89-95.

Helmstaedt, K., Strittmatter, A., Lipscomb, W. N. and Braus, G. H. (2005). Evolution of 3-deoxy-D-arabino-heptulosonate-7-phosphate synthase-encoding genes in the yeast *Saccharomyces cerevisiae, Proc. Nat. Acad. Sci. USA*, 102, pp. 9784–9789.

Hirsh, M. W. (1984). The dynamical systems approach to differential equations, *Bullet. Am. Mathemat. Soc.*, 11, pp. 1-64.

Holtzhutter, H.-G. (2004). Analysis of complex metabolic networks on the basis of optimization principles, *Proc. Conference Complexity in the living*, pp. 122-38.

Husmeier, D. (2003). Sensivity and specificity of inferring genetic regulatory interactions from microarray experiments with dynamic Bayesian networks, *Bionformatics*, 19, pp. 2271-2282.

Ibarra, R. U., Edwards, J. S. and Palsson, B. O. (2002). Escherichia coli K-12 undergoes adaptive evolution to achieve in silico predicted optimal growth, *Nature*, 420, pp. 186-189.

Ideker, T. *et al.* (2001). Integrating genomic and proteomic analyses of a systematically perturbed metabolic network. *Science*, 292, pp. 929-934.

Jeffery, C.J. (2003). Moonlighting proteins: old proteins learning new tricks, *Trends in Genetics*, 19, pp. 415-417.

Jeong, H., Tombor, B., Albert, R., Oltvai, Z. N. and Barabasi, A. L. (2000). The large scale organization of metabolic networks, *Nature*, 407, pp. 651-654.

Jones, G. E. (1978). L-asparagine auxotrophs of *Saccharomyces cerevisiae*: genetic and phenotypic characterization, *J. Bacteriol.*, 134, pp. 200-207.

Juhnke, H., Krems, B., Kotter, P. and Entian, K.-D. (1996). Mutants that show increased sensitivity to hydrogen peroxide reveal an important role for the pentose phosphate pathway in protection of yeast against oxidative stress, *Mol. Gen. Genet.*, 252, pp. 456-464.

Kacser, H. and Burns, J. A. (1973). The control of flux, *Symp. Soc. Exp. Biol.*, 27, pp. 65-104.

Kastanos, E. K., Woldman, Y. Y. and Appling, D. R. (1997). Role of mitochondrial and cytoplasmic serine hydroxymethyltransferase isozymes in *de novo* purine synthesis in *Saccharomyces cerevisiae, Biochemistry*, 36, pp. 14956-14964.

Klamt, S. and Stelling, J.(2003). Stoichiometric analysis of metabolic networks, *Tutorial at the 4th International Conference on Systems Biology.*

Krishnan, A., Giuliani, A., Zbilut, J. P. and Tomita, M. (2007). Network scaling invariants help to elucidate basic topological principles of proteins, *Journal of proteome research*, 6, pp. 3924-3934.

Kuepfer, L., Sauer, U. and Blank, L. M. (2005). Metabolic functions of duplicated genes in *Saccharomyces cerevisiae, Genome Res.*, 15, pp. 1421-1430.

Kusano, M., Sakai, Y., Kato, N., Yoshimoto, H., Sone, H. and Tamai, Y. (1998). Hemiacetal dehydrogenation activity of Alcohol dehydrogenases in *Saccharomyces cerevisiae, Biosci. Biotechnol. Biochem.*, 62, pp. 1956-1961.

Lässig, M., Bastolla, U., Manrubia, S. C. and Valleriani, A. (2001). Shape of Ecological Networks, *Phys. Rev. Lett.*, 86, pp. 4418-4421.

Legrain, P., Wojcik, J. and Gauthier, J. M. (2001). Protein-protein interaction maps: a lead towards cellular functions, *Trends Genet.*, 17, pp. 346-52.

Lemke, N., Herédia, F., Barcellos, C. K., dos Reis, A. N. and Mombach, J. C. M. (2004). Essentiality and damage in metabolic network, *Bioinformatics*, 20, pp. 115-119.

Luscombe, N. M., Babu, M. M., Yu, H., Snyder, M., Teichmann, S. A. and Gerstein, M. (2004). Genomic analysis of regulatory network dynamics reveals large topological changes, *Nature*, 431, pp. 308-312.

Ma, H. W. and Zeng, A. P. (2003). Reconstruction of metabolic networks from genome data and analysis of their global structure for various organisms, *Bioinformatics*, 19, pp. 270-277.

Ma, H. W. and Zeng, A. P. (2003). The connectivity structure, giant strong component and centrality of metabolic networks, *Bioinformatics*, 19, pp. 1423-1430.

Ma, H. W., Zhao, X. M., Yuan, Y. J. and Zeng, A. P. (2004) Decomposition of metabolic network into functional modules based on the global connectivity structure of reaction graph. Bioinformatics 20, 1870–1876.

Mangan, S. and Alon, U. (2003). Structure and function of the feed-forward loop network motif. *Proc. Natl. Acad. Sci. USA*, 100, pp. 11980-5.

Masuda, C. A, Xavier, M. A., Mattos, K. A., Galina, A. and Montero-Lomelì, M. (2001). Phosphoglucomutase is an *in vivo* lithium target in yeast, *J. Biol. Chem.*, 276, pp. 37794–37801.

McAlister, L. and Holland, M. J. (1985). Differential expression of the three yeast glyceraldehyde-3-phosphate dehydrogenase genes, *J. Biol. Chem.*, 260, pp. 15019-15027.

McAlister, L. and Holland, M. J. (1985). Isolation and characterization of yeast strains carring mutations in the glyceraldehyde-3-phosphate dehydrogenase gene, *J. Biol. Chem.*, 260, pp. 15013-15018.

McMahon, S. M., Miller, K. H. and Drake, J. (2001) Networking tips fpr social scientists and ecologists, *Science*, 293, pp. 1604-1605.

McNeil, J. B., Bognar, A. L. and Pearlman, R. E. (1996). *In vivo* analysis of folate coenzymes and their compartmentation in *Saccharomyces cerevisiae*, *Genetics*, 142, pp. 371-381.

McNeil, J. B., McIntosh, E. M., Taylor, B. V., Zhang, F.-R., Tangs, S., and Bognar, A. L. (1994). Cloning and molecular characterization of three genes, including two genes encoding serine hydroxymethyltransferases, whose inactivation is required to render yeast auxotrophic for glycine, *J. Biol. Chem.*, 269, pp. 9155-9165.

McNeil, J. B., Zhang, F.-R., Taylor, B. V., Sinclair, D. A., Pearlman, R. E. and Bognar, A. L. (1997). Cloning, and molecular characterization of the GCV1 gene encoding the glycine cleavage T-protein from *Saccharomyces cerevisiae*, *Gene*, 186, pp. 13–20.

Mehler, M. F. and Mattick, J. S. (2007). Noncoding RNAs and RNA editing in brain development, functional diversification, and neurological disease, *Physiol. Rev.*, 87, pp. 799-823.

Monschau, N., Stahmann, K. P., Sahm, H., McNeil, J. B. and Bognar, A. L. (1997). Identification of *Saccharomyces cerevisiae* GLY1 as a threonine aldolase: a key enzyme in glycine biosynthesis, *FEMS Microbiol. Lett.*, 150, pp. 55-60.

Mulquiney, P. J., Bubb, W. A. and Kuchel, P. W. (1999). Model of 2,3-biphosphoglycerate metabolism in the human erythrocyte based on detailed enzyme kinetic equations: in vivo kinetic characterization of 2,3-biphosphoglycerate synthase/phosphatase using ^{13}C and ^{31}P NMR, *Biochem. J.*, 342, pp. 567-80.

Mulquiney, P. J. and Kuchel, P. W. (1999). Model of 2,3-biphosphoglycerate metabolism in the human erythrocyte based on detailed enzyme kinetic equations: equations and parameter refinement, *Biochem. J.*, 342, pp. 581-96.

Mulquiney, P. J. and Kuchel, P. W. (1999). Model of 2,3-biphosphoglycerate metabolism in the human erythrocyte based on detailed enzyme kinetic equations: computer simulation and Metabolic Control Analysis, *Biochem. J.*, 342, pp. 597-604.

Ng, R. K. and Gurdon, J. B. (2008). Epigenetic inheritance of cell differentiation status, *Cell Cycle*, 7, pp. 1173-1177.

Navarro-Avino, J. P., Prasad, R., Miralles, V. J., Benito, R. M. and Serrano, R. (1999). A proposal for nomenclature of aldehyde dehydrogenases in *Saccharomyces cerevisiae* and characterization of the stress-inducible ALD2 and ALD3 genes, *Yeast*, 15, pp. 829–842.

Nielsen, J. (1998). Metabolic engineering: techniques of analysis of targets for genetic manipulations, *Biotechnol. Bioeng.*, 58, pp. 125-132.

Nowak, M. A., Boerlijst, M. C., Cooke, J. and Smith, J. M. (1997). Evolution of genetic redundancy, *Nature*, 388, pp. 167-171.

Opdam, P. (2002). Assessing the conservation potential of habitat networks, In: Gutzwiller K. J. Ed, *Concepts and application of landscape ecology in biological conservation*, Springer Verlag, New York, NY, pp. 381–404.

Overington, J. P., Al-Lazikani, B. and Hopkins, A. L. (2006). How many drug targets are there? *Nat. Rev. Drug. Discov.*, 5, pp. 993-996.

Palumbo, M. C., Colosimo, A., Giuliani, A. and Farina, L. (2005). Functional essentiality from topology features in metabolic networks: a case study in yeast, *FEBS Lett.*, 579, pp. 4642-4646.

Palumbo, M. C., Colosimo, A., Giuliani, A. and Farina, L. (2007). Essentiality is an emergent property of metabolic network wiring, *FEBS Lett.*, 581, pp. 2485–2489.

Papin, J. A., Price, N. D., Wiback, S. J., Fell, D. A. and Palsson, B. O. (2003). Metabolic pathways in the post-genome era, *TRENDS Biochem. Sci.*, 28, pp. 250-258.

Papp, B., Pal, C. and Hurst, L. D. (2004). Metabolic network analysis of the causes and evolution of enzyme dispensability in yeast, *Nature*, 429, pp. 661-664.

Price, N. D., Reed, J. L., Papin, J. A., Wiback, S. J. and Palsson, B. O. (2003). Networkbased analysis of metabolic regulation in the human red blood cell, *J. theor. Biol.*, 225, pp. 185-194.

Rao, F. and Caflisch, A. (2004). The protein folding network, *J. Mol. Biol.*, 342, pp. 299-306.

Ravid, T. and Hochstrasser, M. (2008). Diversity of degradation signals in the ubiquitin-proteasome system, *Nature Rev. Mol. Cell. Biol.*, 9, pp. 679-690.

Rébora, K., Laloo, B. and Daignan-Fornier, B. (2005). Revisiting purine-histidine cross-pathway regulation in *Saccharomyces cerevisiae*: a central role for a small molecule, *Genetics*, 170, pp. 61–70.

Rinaldi, S. (1998). Laura, Petrarch. An Intriguing Case of Cyclical Love Dynamics, *SIAM J. Appld. Math.*, 58, pp. 1205-1221.

Sauro, H. M. and Kholodenko, B.N. (2004). Quantitative analysis of signaling networks, *Prog. Biophys. Mol. Biol.*, 86, pp. 5-43.

Scala, A., Nunes Amaral, L. A. and Barthélémy, M. (2001). Small-world networks and the conformation space of a short lattice polymer chain, *Europhy.s Lett.*, 55, pp. 594-600.

Schaaff-Gerstenschlager, I., Mannhaupt, G., Vetter, I., Zimmermann, F. K. and Feldmann, H. (1993). TKL2, a second transketolase gene of *Saccharomyces cerevisiae*. Cloning, sequence and deletion analysis of the gene, *Eur. J. Biochem.*, 217, pp. 487-492. [published erratum appears in *Eur. J. Biochem.* (1994) Feb 1; 219 (3), 1087]

Schaaff-Gerstenschlager, I. and Zimmermann, F. K. (1993). Pentose-phosphate pathway in *Saccharomyces cerevisiae*: analysis of deletion mutants for transketolase, transaldolase, and glucose 6-phosphate dehydrogenase, *Curr. Genet.*, 24, pp. 373-376.

Schuster, S., Dandekar, T. and Fell, D. A. (1999). Detection of elementary flux modes in biochemical networks: a promising tool for pathway analysis and metabolic engineering, *TIBTECH*, 17, pp. 53-60.

Schuster, S., Dandekar, T. and Fell, D. A. (1999). Detection of elementary flux modes in biochemical networks: a promising tool for pathway analysis and metabolic engineering, *TIBTECH*, 17, pp. 53-60.

Schuster, S., Fell, D. A. and Dandekar, T. (2000). A general definition of metabolic pathways useful for systematic organization and analysis of complex metabolic networks, *Nat. Biotechnol.*, 18, pp. 326-332.

Segrè, D., DeLuna, A., Church, G. M. and Kishony, R. (2005). Modular epistasis in yeast metabolism, *Nat. Genet.*, 37, pp. 77-83.

Segrè, D., Vitkup, D. and Church, G. .M. (2002). Analysis of optimality in natural and perturbed metabolic networks, *Proc. Nat. Acad. Sci. USA*, 99, pp. 15112-15117.

Shen-Orr, S. S., Milo, R., Mangan, S. and Alon, U. (2002). Network motifs in the transcriptional regulation network of Escherichia coli, *Nat. Genet*, 31, pp 64-68.

Shlomi, T., Berkman, O. and Ruppin, E. (2005). Regulatory on/off minimization of metabolic flux changes after genetic perturbations, *Proc. Nat. Acad. Sci. USA*, 102, pp. 7695-7700.

Shlomi, T., Cabili, M. N., Herrgard, M. J., Palsson, B. O. and Ruppin, E. (2008).Network-based predctino of human tissue-specific metabolism, *Nat. Biotech.*, 26, pp. 1003-1010.

Shmulevich, I., and Zhang, W. (2002). Binary analysis and optimization-based normalization of gene expression data, *Bionformatics*, 18, pp. 555-565.

Shmulevich, I., Dougherty, E. R. and Zhang, W. (2002). Gene perturbation and intervention in probabilistic boolean networks, *Bionformatics*, 18, pp. 1319-1331.

Shmulevich, I., Dougherty, E. R., Seungchan, K. and Zhang, W. (2002). Probabilistic boolean networks: a rule-based uncertainty model for gene regulatory networks, *Bionformatics*, 18, pp. 261-274.

Smolen, P., Baxter, D. A. and Byrne, J. (2000). Modeling transcriptional control in gene networks-Methods, recent results, and future directions, *Bull. Math. Biol.*, 62, pp. 247-292.

Somogyi, R. and Sniegoski, C. A. (1996). Modelling the complexity of of genetic networks: Understanding multigenic and pleiotropic regulation, *Complexity*, 1, pp. 45-63.

Sousa, S., McLaughlin, M. M., Pereira, S. A., VanHorn, S., Knowlton, R., Brown, J. R., Nicholas, R. O. and Livi, G. P. (2002). The ARO4 gene of *Candida albicans* encodes a tyrosine-sensitive DAHP synthase: evolution, functional conservation and phenotype of Aro3p-, Aro4p-deficient mutants, *Microbiology*, 148, pp. 1291–1303.

Stelling, J., Klamt, S., Bettenbrock, K., Schuster and S., Gilles, E. D. (2002). Metabolic network structure determines key aspects of functionality and regulation, *Nature*, 420, pp. 190-193.

Stephanopoulos, G. (1999). Metabolic fluxes and metabolic engineering, *Metab. Eng.*, 1, pp. 1-11.

Tibbetts, A S. and Appling, D.R. (2000). Characterization of Two 5-Aminoimidazole-4-carboxamide Ribonucleotide Transformylase/Inosine Monophosphate Cyclohydrolase Isozymes from *Saccharomyces cerevisiae*, *J. Biol. Chem.*, 275, pp. 20920-20927.

Varma, A., Palsson, B. O. (1994). Metabolic Flux balancing – Basic Concepts, Scientific and practical Use. *Bio/Technol*, 12, pp. 994-8.

Verboom, J., Foppen, R., Chardon, P., Opdam, P. and Luttikhuizen, P. (2001) Introducing the key patch approach for habitat networks with persistent populations: an example for marshland birds, *Biol.Conserv.*, 100, pp. 89-101.

Von Dassow, G., Meir, E., Munro, E. M. and Odell, G. M. (2000). The segment polarity network is a robust developmental module, *Nature*, 406, pp. 131-2.

Vos, C. C., Verboom, J., Opdam, P. F. M. and Ter Braak, C. J. F. (2001). Toward ecologically scaled landscape indices, *The Am. Naturalist*, 183, pp. 24-41.

Wagner, A. (2000). Robustness against mutations in genetic networks of yeast, *Nat. Genet.*, 24, pp. 355-361.Watts D.J., and Strogatz, S.H. (1998) Collective dynamics of 'small-world' networks, *Nature,* 393, pp. 440-42.

Whitty, A. (2008). Cooperativity and biological complexity, *Nature Chem. Biol.*, 4, pp. 435-439.

Wiener, N. (1948). *Cybernetics or control and communication in the animal and the machine*, Wiley, New York, NY.

Wilkins, A. S. (2007). For the biotechnology industry the penny drops (at last): genes are not autonomous agents but function within networks! *BioEssays*, 29, pp. 1179-1181.

Wolkenhauer, O. (2002). Mathematical modelling in the post-genome era: understanding genome expression and regulation – a system theoretic approach, *BioSystems,* 65, pp. 1-18.

Zadeh, L. A. and Desoer, C. A. (1963). *Linear System Theory-The State Space Approach*, McGraw-Hill Book Co., New York, NY.

Zhu, D., Qin, Z. S. (2005). Structural comparison of metabolic networks in selected single cell organisms, *BMC Bioinformatics*, 6, 8.

Zotenko E., Mestre J., O'Leary D.P., Przytycka T.M. (2008) Why Do Hubs in the Yeast Protein Interaction Network Tend To Be Essential: Reexamining the Connection between the Network Topology and Essentiality. *PLoS Computational Biology* 4(8): e1000140 doi:10.1371/journal.pcbi.1000140.

PART 2
Brain Networks

Chapter 9

The Human Brain Network

Olaf Sporns

Department of Psychological and Brain Sciences, Indiana University, USA

This article summarizes recent empirical findings and modeling approaches aimed at unraveling the structure of the human brain network. The physical structure of the human brain is still only partially mapped and how structural brain connectivity gives rise to functional brain dynamics is only incompletely understood. Initial mapping studies indicate that the human brain forms a small-world architecture that is structurally organized into modules interlinked by hub regions. This connectivity structure shapes endogenous dynamics as well as responses of the brain to external perturbations. Thus, the topology of the human brain network may be an important ingredient in brain function. We will briefly discuss how complex brain networks may shape human cognition in the healthy and the diseased brain.

9.1. Introduction

Modern neuroscience allows the simultaneous recording and analysis of large numbers of neurons, and even of neural activity across the entire human brain – yet our understanding of how the brain functions as an integrated system organized into a complex biological network is still in its infancy. Nevertheless, the material presented in this chapter argues that complex network approaches (Strogatz, 2001; Amaral and Ottino, 2004; Boccaletti *et al.*, 2006) may provide a fundamental basis for our understanding of brain function (see also Sporns *et al.*, 2004; Sporns and Tononi, 2007). In recent years network science approaches have offered significant new insights into how the structure of the brain shapes its dynamics and how the elements of a neural network can make different contributions to brain function on the basis of how they are interconnected. The dichotomy between brain structure and dynamics presents unique challenges to network approaches. A comprehensive analysis of brain networks needs to address the central issue of how functional brain networks (represented as dynamic linkages between parts of the brain that are transient in nature) can emerge from the physical substrate of structural brain connectivity (the "wires" of the brain which are thought to be relatively time-invariant).

As is the case in other kinds of biological networks, the individual network nodes of the brain are structurally integrated within the overall system, i.e. they maintain specific patterns of connections with other nodes. It is well documented that different nodes in the brain, be they single neurons or brain regions, have different functionalities, expressed in different response preferences or activity patterns in different cognitive tasks. A key idea is that the functionality of an individual neural node is at least in part determined by the pattern of its interconnections with other nodes in the network (Passingham *et al.*, 2002). Thus, the physical interconnections of nodes place important constraints on which dynamic couplings or functional interactions are possible or likely to occur. A corollary of this idea is that nodes with similar connection patterns would tend to share some common functionality, while nodes with dissimilar connections patterns would appear functionally less related and likely be members of different segregated communities. Another basic observation relates to networks as integrated systems. The structured interactions of their constituent nodes result in large-scale patterns of neural activity that can evolve autonomously through time or flexibly respond to exogenous perturbations. Such global states naturally integrate and supervene over the individual functional contributions of network nodes, and yet are shaped by them. Thus, a network perspective naturally reconciles views of local and global brain function.

Before embarking on a brief survey of our current knowledge of the human brain network, we need to introduce a basic distinction between structural, functional and effective brain connectivity (Fig. 9.1; see also a discussion of this distinction in Horwitz, 2003). *Structural connectivity* describes a physical network of connections, which may correspond to fiber pathways or individual synapses. *Functional connectivity* may be defined as the pattern of deviations from statistical independence between distributed and often spatially remote neuronal units. *Effective connectivity* describes a network of the causal effects of one neural system over another. All three modes of brain connectivity can be described as graphs and analyzed using network analysis tools. The distinction itself stems from a long tradition in systems neuroscience and functional neuroimaging to measure patterns of dynamic interactions among recording sites or brain voxels and to observe changes in these interactions in the course of experimental perturbations. From such patterns of observed brain dynamics one may then construct networks that capture either statistical dependencies (functional connectivity) or directed influences (effective connectivity). It turns

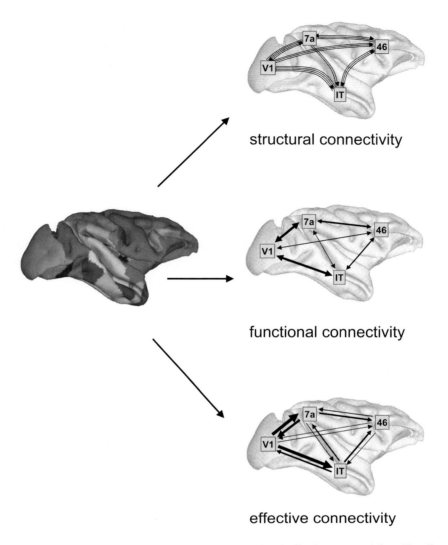

structural connectivity

functional connectivity

effective connectivity

Fig. 9.1. A schematic illustration of structural, functional and effective connectivity. The diagram on the left shows a rendering of the macaque cortical surface partitioned into 47 cortical regions displayed on a gray scale. The column of plots at the right shows schematic diagrams of structural connectivity (fiber pathways linking brain regions), functional connectivity (statistical relations among brain regions) and effective connectivity (directed influences between brain regions). The boxes mark the positions of four specialized cortical regions: V1 = primary visual cortex, 7a = visual cortex in the parietal lobe, IT = visual cortex in the inferior temporal lobe, and area 46 = an area of prefrontal cortex. Note that structural and effective connectivity form directed networks, while functional connectivity is represented as an undirected graph.

out that in most cases these functional or effective patterns are not identical to the patterns of the underlying structural connectivity from which they emerge. This leads to the challenging theoretical issue of how structural, functional and effective connectivity patterns are interrelated. Patterns of structural connectivity are likely to play an important role in shaping emergent patterns of functional and effective interactions between brain regions. Increasingly, these emergent patterns of functional and effective connectivity are thought to be closely associated with human cognition, and their disruption appears to be associated with some forms of mental dysfunction and disease.

The following sections provide an overview of the structure of the human brain network as revealed by network science. The brevity of this chapter and the rapidity with which the field currently advances allow us only to provide a snapshot of some of the current theoretical and computational issues and themes. We will start by considering what we know about the global structural (anatomical) organization of the human cerebral cortex.

9.2.   Structural Connectivity of the Human Cerebral Cortex

The anatomical structure of the brain, in particular the human cerebral cortex, has been a major focus of research since the beginnings of modern neuroscience. The human brain network can be mapped and characterized at multiple spatial scales, ranging from interconnections linking whole brain regions to patterns of connections within a given brain region, for example those between cell populations or even individual cortical neurons (Swanson, 2003). The complete set of structural connections of a brain has been called the "connectome" (Sporns *et al.*, 2005; Hagmann, 2005) which could be represented as a set of nodes and their interconnections, or mathematically in the form of an adjacency or connection matrix. Since there are at least three distinct organizational levels in the human brain (individual neurons at a microscopic level, cell populations at a mesoscopic levels, and brain regions at a macroscopic level) the connectome may also be defined at these three levels of spatial resolution. For species with larger brains, notably humans, a definition of the connectome at the microscopic level currently presents formidable technological challenges. At cellular resolution the connectome would comprise a map of roughly 10^{11} nodes linked by 10^{15} connections, a dataset many orders of magnitude larger than the map of the human genome. While recent advances in cell labeling and optical microscopy have opened up new avenues for tracing individual neurons and connections in three dimensional blocks of neural tissue (Conchello and Lichtman, 2005; Livet *et al.*, 2007), the creation of a map of the entire human brain at cellular resolution still faces seemingly insurmountable methodological obstacles.

While a microscopic map of the human brain would provide connectivity information at an unprecedented level of detail, much can be learned about the global organization of the human brain from applying methods that deliver mesoscopic or macroscopic datasets. Currently among the most promising approaches are those that attempt to construct maps of the entire cerebral cortex that detail the interconnections of segregated cortical regions. Several such maps have been created in the past few years. Increasingly, investigators rely on noninvasive imaging techniques such as diffusion tensor or diffusion spectrum imaging (DTI and DSI, respectively) to obtain *in vivo* structural connectivity. Diffusion imaging records signals from the white matter of the brain and allows the computational reconstruction and estimation of fiber trajectories across the brain (Conturo *et al.*, 1999; LeBihan, 2003).

Outside of diffusion imaging, researchers have utilized observed cross-correlations in cortical thickness or volume across individuals to infer structural connectivity patterns. Although the precise mechanism for this effect is still unknown, previous research indicates that such gray-matter thickness correlations can serve as reliable indicators for the presence or absence of structural connections. He *et al.* (2007) generated a structural connection matrix on the basis of such measurements obtained from 124 individual brain data sets. An analysis of the resulting connection matrices using graph theory methods demonstrated small-world connectivity and the existence of local communities of areas forming structural modules. A more recent study of the connection matrices obtained from cortical thickness correlations has focused in more detail on the modularity structure of cortex (Chen *et al.*, 2008). The study generated modules of brain regions from connectivity data and showed a high degree of overlap between such modules and previously described functional brain systems, e.g. those composed of primarily visual, sensorimotor and auditory/language-related areas. While these studies have provided some of the very first data sets mapping the human cortex in its entirety, it must be noted that the methodological approach of inferring cortico-cortical connections on the basis of cortical thickness or volume correlations has several shortcomings. By its very nature, the method provides relatively indirect information about cortical connection patterns. Furthermore it requires the aggregation of data from large numbers of individuals to derive a single connection data set across a subject group – connection patterns or individual brains cannot be accessed.

Other investigators have attempted to build whole-brain connection matrices from diffusion imaging data. Iturria-Medina *et al.* (2007; 2008) have constructed connectome data sets using diffusion tensor imaging followed by the derivation of average connection probabilities between 70-90 cortical and basal brain gray matter areas. The obtained brain networks were then subjected to graph analysis. All networks were found to have robust small-world attributes and "broad-scale" degree distributions. An analysis of betweenness centrality in these networks demonstrated high centrality for the precuneus, the insula, the superior parietal and the superior frontal cortex. Motif analysis (Milo *et al.*, 2002) revealed that significantly increased motifs were similar to those identified in connection matrices derived by anatomical tract tracing in cat and macaque cortex (Sporns and Kötter, 2004). A particularly innovative aspect of these studies is that all network analyses were carried out while preserving the weighted connectivity structure of the brain network. A separate study using DTI mapped a network of anatomical connections between 78 cortical regions (Gong *et al.*, 2008). This study also identified several hub regions in the human brain, including the precuneus and the superior frontal gyrus.

These diffusion imaging studies were carried out at relatively low spatial resolution (<100 brain nodes) at using DTI which is known to have difficulty identifying connections in parts of the white matter where fiber bundles intersect. Diffusion spectrum imaging can overcome the latter limitation (Wedeen *et al.*, 2005) and was applied in a series of studies by Hagmann *et al.* (2007, 2008). The first study using DSI (Hagmann *et al.*, 2007) constructed a connection matrix from fiber densities measured between homogeneously distributed and equal-sized "regions of interest" (ROIs) numbering between 500 and 4000. A quantitative analysis of connection matrices obtained for approximately 1000 ROIs and approximately 50,000 fiber pathways from two subjects demonstrated an exponential (one-scale) degree distribution as well as robust small-world attributes for the network.

A subsequent study (Hagmann *et al.*, 2008) allowed a detailed analysis of the network structure of human cortical connectivity (Fig. 9.2), using a broad array of network analysis methods including core decomposition, modularity analysis, hub classification and centrality. The study presented evidence for the existence of a structural core of highly and mutually interconnected brain regions, located primarily in posterior medial and parietal cortex (Hagmann *et al.*, 2008). The core is comprised of portions of the posterior cingulate cortex, the precuneus, the cuneus, the paracentral lobule, the isthmus of the cingulate, the banks of the superior temporal sulcus, and the inferior and superior parietal cortex, all located in both cerebral hemispheres. In addition, the analysis detected several additional modules of brain regions in temporal and frontal cortex that maintain

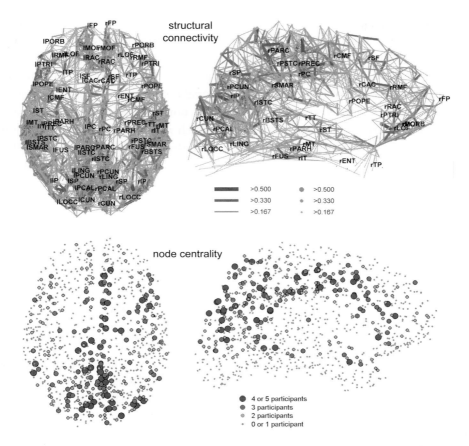

Fig. 9.2. Structural connectivity and node centrality of the human brain network (data replotted from Hagmann *et al.*, 2008). Plots at the top show the backbone of fiber pathways spanning 998 cortical nodes – a dorsal view is shown at the left and a lateral view of the right hemisphere is shown on the right. Note the high proportion of short edges and the high degree of symmetry between left and right hemispheres. Plots at the bottom show nodes that exhibit high betweenness centrality across all five participants imaged in the study of Hagmann *et al.*, 2008. Highly central nodes are found along the cortical midline as well as in selected prefrontal and temporal brain regions. For more information please consult the original publication.

long-range interconnections with regions located in the structural core. Interestingly, core regions are strongly and interhemispherically coupled which suggests that the structure operates as a single functional system.

The studies outlined in this section represent only a first step towards creating comprehensive structural connection data sets for the human brain. More refined datasets and analyses will become available in the near future as brain mapping, imaging and network methodologies become more sophisticated. Future analyses will likely include more comprehensive coverage of subcortical regions and pathways and may aim at achieving higher spatial resolution to capture smaller fiber bundles and anatomical subdivisions. Another advance would be a definition of ROIs (network nodes) according to functional criteria, for example based on the detection of boundaries in functional connectivity patterns. Another particularly promising avenue would be the combination of structural and functional imaging in the same participants, which may provide insight into how structural connectivity of the human brain shapes dynamic brain networks that underlie cognitive function. These latter approaches are discussed in more detail in the next section.

9.3. Dynamic Brain Networks

Structural brain connectivity provides a scaffold for the ongoing dynamics of the brain that unfolds within it. In the adult brain, structural connections are likely to be relatively stable across time (at least over periods of hours and days), while even the most causal analysis of brain dynamics reveals that functional couplings across the brain are highly variable, on a time scale of tens to hundreds of milliseconds. It is these rapid fluctuations of functional connectivity (and by extension, effective connectivity) that are associated with changes in perceptual or cognitive state. Clearly, the static pattern of structural connectivity cannot fully explain these rapid fluctuations, although it may more reliably shape functional connectivity over longer time scales, for example when the brain is cognitively "at rest". The relationship between structural and functional (effective) connectivity represents a major theoretical challenge in cognitive neuroscience.

Network approaches begin to shed light on this relationship. One of the useful attributes of network analysis tools is their applicability to both structural and functional/effective connectivity data sets. For example, the centrality of a vertex can be measured on the basis of its structural connection pattern, as well as on the basis of the pattern of functional or effective connections it maintains. This universality of network analysis approaches invites comparisons between structural and functional data sets.

A rapidly increasing number of studies use network approaches in the analysis of functional or effective connectivity patterns represented as graphs (e.g. Dodel *et al.*, 2002; Salvador *et al.*, 2005a). There are numerous applications of connectivity analyses to EEG, MEG and fMRI data sets. In most of these approaches patterns of cross-correlation or coherence are represented as undirected graphs with edges that represent the existence and, in some cases, the strength of the statistical relationship between the linked vertices. Studies of patterns of functional connectivity (based on coherence or correlation) among cortical regions have demonstrated that functional brain networks, like their structural counterparts (see above), exhibit small-world attributes (Stam, 2004; Salvador *et al.*, 2005b; Achard *et al.*, 2006). A detailed analysis of resting state MEG data across multiple temporal scales has demonstrated the existence of robust small-world architectural features across multiple time scales, resembling a temporally self-similar or fractal mode of organization (Bassett *et al.*, 2006).

As was mentioned briefly above, functional patterns may reflect, at least in part, the underlying structural organization of anatomical connections. However, there are currently only very few empirical comparisons of large-scale structural and functional networks in individual human brains. Such studies are made difficult because charting structural and functional connections in the human brain requires the use of two very different imaging approaches that need to be brought in register. However, such studies would be tremendously useful to determine if nodes in structural and functional neuronal connectivity matrices maintain similar patterns of connectivity and exhibit similar network properties such as clustering or path lengths. Combined structural and functional imaging followed by network analysis in the brains of five participants was carried out by Hagmann *et al.* (2008). The study demonstrated that structural connection patterns estimated from diffusion imaging and functional connection patterns derived from resting state functional MRI are significantly correlated across the entire cortex (Fig. 9.3). The analysis indicated that the presence of a structural connection is highly and quantitatively predictive of a functional connection. However, strong functional connections may also be found where diffusion imaging does not detect a structural connection. This indicates that inference of structural connections from observed functional couplings, for example by simple thresholding of functional connection matrices, may result in an overestimation of structural connections. While the study of Hagmann *et al.* (2008) presents evidence that structural connections can predict functional connections estimated from minute-long BOLD time series in the resting state, it is highly likely that the strength of this correspondence will decrease for functional connections that are estimated over shorter time periods, or across varying conditions of cognitive load. So-called resting state networks have been studied in a variety of contexts but they only represent long-time averages of dynamic couplings in the brain – the underlying short-time dynamics are likely to form a large repertoire of variable patterns across multiple time scales.

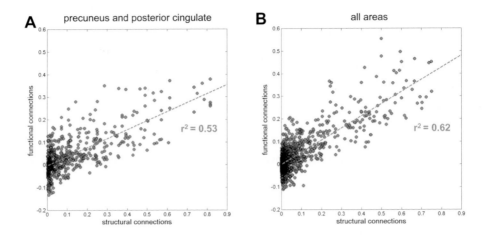

Fig. 9.3. Relationship of structural connectivity and functional connectivity in the human brain (data replotted from Hagmann *et al.*, 2008). Scatter plots show strengths of structural connections versus strengths of functional connections within a lower-resolution network of 66 anatomical subregions. Structural connections were obtained from diffusion imaging, while functional connections were obtained from functional MRI during the resting state. For specific regions (precuneus and posterior cingulated, left panel) as well as across the entire brain (right panel) structural connections are highly predictive of functional connections. For details, please see Hagmann *et al.*, 2008.

Computational models can provide some insights into how structural connections may shape statistical dependencies between parts of the brain at multiple time scales. Because computational models allow the introduction of systematic variations in connectivity, these approaches allow the systematic study of how neural dynamics is shaped by the structure of connection patterns linking individual elements,. Detailed computer simulations of cortical networks have been carried out with heterogeneous (Jirsa and Kelso, 2000; Jirsa 2004; Assisi *et al.*, 2005) and spatially patterned (Sporns, 2004) connection topologies. These studies demonstrated that different connection topologies can generate different modes of neuronal dynamics. For example, as discussed above, connectivity patterns containing locally clustered connections with a small admixture of long-range connections were shown to exhibit robust small-world attributes (Sporns and Zwi, 2004; Sporns, 2004; Kaiser and Hilgetag, 2006), while conserving wiring length. These connectivity patterns also gave rise to functional connectivity of high complexity (Tononi *et al.*, 1994; Sporns *et al.*, 2000) characterized by the coexistence of spatially segregation (regional specialization) and global integration (system-wide interactions that unify processing across segregated clusters). These computational studies suggest the hypothesis that only specific classes of connectivity patterns (structurally similar to cortical networks) simultaneously support short wiring, small-world attributes, clustered architectures (all structural features), and high complexity (combining functional segregation and integration).

The discovery of small-world connectivity patterns in functional connectivity (e.g. Stam, 2004; Bassett *et al.*, 2006) and the demonstration of small-world attributes in structural connection patterns (e.g. He *et al.* 2007; Hagmann *et al.*, 2007, 2008) raises the question of how closely functional connections map onto structural connections. The state- and task-dependence of functional connectivity suggests that a one-to-one mapping of structural to functional connections does not exist. However, it is likely that at least some structural characteristics of individual nodes are reflected in their functional interactions – for example, structural hub regions should maintain larger numbers of functional relations. To examine the relation of structural and functional connectivity directly, Honey *et al.* (2007) implemented and analyzed a large-model of cortical connectivity derived from anatomical tract tracing studies of the macaque monkey visual and somatomotor systems. The neuronal model was a neural mass model, adapted from a classical conductance-based model of neuronal dynamics (Morris and Lecar, 1981) for local population activity. Units of the model describe local populations of densely interconnected inhibitory and excitatory neurons whose behaviours are determined by voltage and ligand-gated membrane channels. Activity in the system arose purely from nonlinear instabilities, generating spontaneous oscillations whose spatiotemporal patterns are shaped through re-entrant excitatory-excitatory internode coupling, provided by the macaque neocortical connectivity matrix. Functional and effective connectivity in the modeled neuronal activity was examined using a range of methods across multiple time scales. Patterns of directed interactions were derived using the information-theoretic measure of transfer entropy (Schreiber, 2000), phase synchrony was measured using the phase locking value (Lachaux *et al.*, 1999) and simulated fMRI responses were derived using a nonlinear Balloon-Windkessel haemodynamic model (Buxton *et al.*, 1998). An analysis of structural and functional connectivity in the model revealed that over long time scales (several minutes of data), functional connectivity patterns closely resembled structural connectivity patterns, and that structural hubs (i.e. nodes with high centrality) also became functional hubs (Fig. 9.4). Closer examination of modeled time series data showed that the model's fast neuronal dynamics exhibited intermittent synchronization and desynchronization at a time scale of hundreds of milliseconds enabling the system to continually explore a repertoire

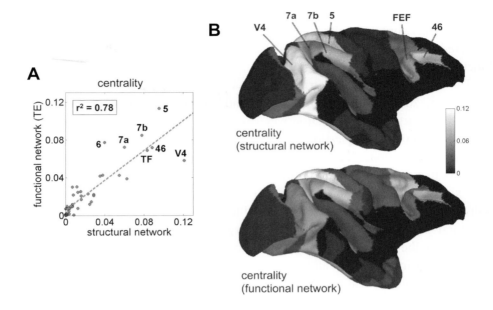

Fig. 9.4. Network measures of structural and functional networks are closely correlated (data replotted from Honey *et al.* 2007). These data are derived from a simulation study of the macaque visual and somatomotor cortex. Structural connections were obtained from a neuroanatomical database and functional connections were derived from long-time samples of nonlinear simulations of endogenous neural activity within this graph. The centrality of nodes in the structural and functional network is highly correlated (for a scatter plot, see A; for a distribution of centrality on the macaque cortical surface, see B).

of functional microstates. Slow variations in the statistics of synchronous coupling gave rise to changes in the strengths of directed interactions between regions at a time scale of seconds (Fig. 9.5). The model suggested that spontaneous cortical dynamics exhibit ongoing changes in the pattern and strengths of functional coupling, and that these coupling events are in turn related to the amplitude of the simulated fMRI response. Thus, functional connectivity patterns exhibit rich and complex spatiotemporal structure at multiple scales. Importantly, these rich dynamics can result even while structural couplings remain constant across time.

Other large-scale models that are capable of relating structural and functional connectivity have been implemented. For example, models of the synchronization dynamics within the cat cortex have revealed significant overlap in clustered synchronization behavior and structurally defined clusters of brain

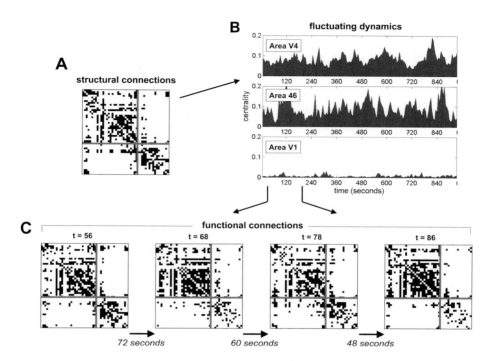

Fig. 9.5. Structural connectivity gives rise to fluctuating neural dynamics that results in time-dependent patterns of functional connectivity (data replotted from Honey *et al.* 2007). The structural network of fiber pathways between 47 macaque cortical regions is shown in panel A – these connections do not change pattern or strength over time. However, samples of functional dynamics (displayed as overlapping windows of 30 seconds of neural activity, panel B) reveal significant fluctuations in centrality for given nodes. For example, nodes with, on average, high centrality (see Fig. 9.4) exhibit significant fluctuations in centrality over time. As a consequence, the precise pattern of functional connectivity will exhibit time-dependent changes on a time scale of seconds to minutes (panel C).

regions (Zemanova *et al.*, 2006), as well as a hierarchical modular organization of functional connectivity that matched that found in structural connectivity (Zhou *et al.*, 2006). An ambitious large-scale model of spiking neural dynamics structurally based on DTI-derived whole-head cortical connectivity was introduced by Izhikevich and Edelman (2008) and showed complex patterns of spike dynamics in spontaneous neural activity. More detailed models of human brain dynamics at multiple time scales are likely to emerge in the near future, incorporating network connectivity data ranging from cells to systems (e.g. Markram, 2006).

9.4. Network Topology and Network Disease

A number of recent studies of natural and technological networks have included analyses that investigated the robustness or vulnerability of such network to disruptions of their connection patterns. The scale-free network of the World Wide Web has been shown to be robust with respect to random deletion of nodes, but vulnerable to targeted attack on heavily connected hubs (Albert *et al.*, 2000; Doyle *et al.*, 2005). Such targeted attacks on hubs often result in disintegration or disconnection of the overall network. The fact that the robustness of social and technological networks is linked to features of network topology raises the interesting question if similar relations may be found for large-scale brain networks.

In the brain, the mapping of functional deficits to underlying structural perturbations is experimentally challenging, but essential for a more complete understanding of brain damage and recovery. Structural perturbations may include deletions of nodes or edges, or the rewiring of edges to new positions in the network. It is currently unknown which structural measures best capture the potential effects of node or edge lesions. A candidate measure of edge vulnerability (Kaiser and Hilgetag, 2004) helps to identify edges whose loss most affects global structural measures. It turns out that such edges often correspond to "bridges", i.e. edges linking segregated clusters of brain regions. More recently, Kaiser *et al.* (2007) studied the effects of removing single areas on the structural integrity of large-scale anatomical networks of the cat and macaque cortex. One main finding indicates that lesions of "bottlenecks", i.e. hubs that link areas across segregated clusters, produce greater effects on structural integrity than lesions of non-hub regions. The clustered architecture of the cortex and the existence of a small number of connector hubs thus may result in greater overall robustness of the network against randomly introduced lesions.

The dynamic consequences of lesioning large-scale brain networks have been investigated in the context of lesion effects on the statistical structure of spontaneous resting state activity (Honey and Sporns, 2008). To understand the effects of a cortical lesion it is necessary to consider not only the loss of local neural function, but also the lesion-induced changes in the larger network of endogenous oscillatory interactions in the brain. To assess these nonlocal effects, a computational model of the macaque visual and somatomotor cortex was designed. The model demonstrated that lesions of single brain regions had effects that extended beyond the immediate neighbors of the lesion site, and the amplitude and dispersal of nonlocal effects are critically influenced by the way areas of the network are organized into clusters. In the model, lesions of hubs that interlink multiple clusters consistently have the most widespread effects, while lesions effects of more local hubs or non-hub regions are spatially more restricted.

Lesions are only one way in which brain networks may become damaged. Another way involves disruptions of connection patterns that occur in the course of a disease process or of degeneration. In the near future, we may expect to see systematic comparisons between connection patterns obtained from clinical populations to those of normal subjects. Diffusion imaging approaches may prove especially useful. When appropriately applied in a clinical context (Johansen-Berg and Behrens, 2006), these approaches can provide important connectional information *in vivo* that may reveal novel associations between network disturbances and brain pathologies. To date, numerous studies of both structural and functional brain connectivity suggest that changes in brain connectivity patterns are associated with mental disorders and brain trauma. Functional imaging has revealed differences in the organization of functional connectivity between normal subjects and patients with Alzheimer's disease (Stam *et al.*, 2007) and schizophrenia (Bluhm *et al.*, 2007; Garrity *et al.*, 2007). In the case of schizophrenia, there is some evidence that the disorder is associated with disorganized cortical white matter (Shergill *et al.*, 2007), supporting the hypothesis that schizophrenia involves cortical disconnection (Friston and Frith, 1995). Future mapping studies of structural connections across the entire cerebral cortex will allow a more detailed characterization of pathological changes affecting brain networks, providing mechanistic links between "network disease" and cognitive or behavioral disturbances. Recovery from a disease state, or from traumatic brain injury, may then be guided by a better understanding of how structural changes affect functional interactions among brain regions.

9.5. Conclusion

Neuroscience stands to gain significant new insights from the systematic application of network approaches across different time and spatial scales. As this brief overview has illustrated, there is rapid progress in mapping structural and functional connections across the brain and emerging new insights into how structural couplings in the brain shape its ongoing dynamics.

An increased understanding and more refined mapping of brain connectivity raises the interesting prospect of large-scale realistic computational models of the human brain. If such models could be designed at a level of scale commensurate with that used in functional neuroimaging they would form important predictive and explanatory tools for cognitive neuroscience. As recent attempts to create such models demonstrate (e.g. Honey *et al.*, 2007; Izhikevich and Edelman, 2008), they appear well within reach of current computational and modeling techniques. A complete map of the brain's structural connections (the brain's connectome) could serve as the coupling matrix that informs a detailed "forward model" of the human brain. Such a model would allow predictions about endogenous brain dynamics, about the responsiveness of the brain system to various exogenous stimuli, and about its changed dynamics in conditions of damage or disease.

Acknowledgement

The author was supported by a grant from the JS McDonnell Foundation.

References

Achard, S., Salvador, R., Whitcher, B., Suckling, J., Bullmore, E. (2006) A resilient, low-frequency, small-world human brain functional network with highly connected association cortical hubs. J. Neurosci. 26, 63-72.

Albert, R., Jeong, H., Barabási, A.-L. (2000) Error and attack tolerance of complex networks. Nature 406, 378-382.

Amaral LAN, Ottino, J.M. (2004). Complex networks. Augmenting the framework for the study of complex systems. Eur. Phys. J. B, 38:147–162.

Assisi, C.G., Jirsa, V.K., Kelso, J.A.S. (2005) Synchrony and clustering in heterogeneous networks with global coupling and parameter dispersion. Phys. Rev. Lett. 94, 018106.

Bassett, D.S., Meyer-Lindenberg, A., Achard, S., Duke, T., Bullmore, E. (2006) Adaptive reconfiguration of fractal small-world human brain functional networks. Proc. Natl. Acad. Sci. USA 103: 19518-19523.

Bluhm, R.L., Miller, J., Lanius, R.A., Osuch, E.A., Boksman, K. (2007) Spontaneous low-frequency fluctuations in the BOLD signal in schizophrenic patients: Anomalies in the default network. Schizophrenia Bull. 33, 1004-1012.

Boccaletti, S., Latora, V., Moreno, Y., Chavez, M., Hwang, D.-U. (2006) Complex networks: Structure and dynamics. Physics Reports 424, 175-308.

Buxton, R.B., Wong, E.C., Frank, L.R. (1998) Dynamics of blood flow and oxygenation changes during brain activation: The balloon model. *Magn Reson Med* 39, 855–864.

Chen, Z.J., He, Y., Rosa-Neto, P., Germann, J., Evans, A.C. (2008) Revealing modular architecture of human brain structural networks by using cortical thickness from MRI. Cerebral Cortex, doi:10.1093/cercor/bhn003

Conchello, J.-A., Lichtman, J.W. (2005) Optical sectioning microscopy. Nature Methods 2, 920-931.

Conturo, T.E., Lori, N.F., Cull, T.S., Akbudak, E., Snyder, A.Z., Shimony, J.S., McKinstry, R.C., Burton, H., Raichle, M.E. (1999) Tracking neuronal fiber pathways in the living human brain. Proc. Natl. Acad. Sci. USA 96, 10422-10427.

Dodel, S., Herrmann, J.M., Geisel, T. (2002) Functional connectivity by cross-correlation clustering. Neurocomp 44, 1065-1070.

Doyle, J.C., Alderson, D.L., Li, L., Low, S., Roughan, M., Shalmov, S., Tanaka, R., Willinger, W. (2005) The "robust yet fragile" nature of the internet. Proc. Natl. Acad. Sci. USA 102, 14497-14502.

Friston, K.J., Frith, C.D. (1995) Schizophrenia: A disconnection syndrome? Clin. Neurosci. 3, 89-97.

Garrity, A.G., Pearlson, G.D., McKiernan, K., Lloyd, D., Kiehl, K.A., Calhoun, V.D. (2007) Aberrant "default mode" functional connectivity in schizophrenia. Am J Psychiatry 164, 450-457.

Gong, G., He, Y., Concha, L., Lebel C., Gross, D.W., Evans, A.C., Beaulieu, C. (2008) Mapping anatomical connectivity patterns of human cerebral cortex using in vivo diffusion tensor imaging tractography. Cerebral Cortex doi:10.1093/cercor/bhn102

Hagmann P. (2005) From diffusion MRI to brain connectomics [PhD Thesis]. Lausanne: Ecole Polytechnique Fédérale de Lausanne (EPFL). 127 p.

Hagmann P., Kurant M., Gigandet X., Thiran P., Wedeen V.J., et al. (2007) Mapping human whole-brain structural networks with diffusion MRI. PLoS ONE 2, e597.

Hagmann, P, Cammoun, L, Gigandet, X, Meuli, R, Honey, C.J., Wedeen, V.J., Sporns, O. (2008) Mapping the structural core of human cerebral cortex. PLoS Biology 6, e159.

He, Y, Chen, Z.J., Evans, A.C. (2007) Small-world anatomical networks in the human brain revealed by cortical thickness from MRI. Cerebr. Cortex 17, 2407-2419.

Honey, C.J., Kotter, R., Breakspear, M., Sporns, O. (2007) Network structure of cerebral cortex shapes functional connectivity on multiple time scales. Proc. Natl. Acad. Sci. USA 104: 10240-10245.

Honey, C.J., Sporns, O. (2008) Dynamical consequences of lesions in cortical networks. Human Brain Mapping 29, 802-809.

Horwitz, B. (2003) The elusive concept of brain connectivity. Neuroimage 19, 466-470.

Iturria-Medina, Y., Canales-Rodriguez, E.J., Melie-Garcia, L., Valdes-Hernandez, P.A., Martinez-Montes, E., et al. (2007) Characterizing brain anatomical connections using diffusion weighted MRI and graph theory. Neuroimage 36, 645-660.

Iturria-Medina, Y., Sotero, R.C., Canales-Rodriguez, E.J., Aleman-Gomez, Y., Melie-Garcia, L. (2008) Studying the human brain anatomical network via diffusion-weighted MRI and graph theory. NeuroImage 40, 1064-1076.

Izhikevich, E.M., Edelman, G.M. (2008) Large-scale model of mammalian thalamocortical systems. PNAS 105, 3593-3598.

Jirsa, V.K., Kelso, J.A.S. (2000) Spatiotemporal pattern formation in continuous systems with heterogeneous connection topologies, Phys. Rev. E 62, 6, 8462-8465.

Jirsa, V.K. (2004) Connectivity and dynamics of neural information processing. Neuroinformatics 2, 183-204.

Johansen-Berg, H., Behrens, T.E.J. (2006) Just pretty pictures? What diffusion tractography can add in clinical neuroscience. Current Opinion in Neurobiology 19, 379-385.

Kaiser, M., Hilgetag, C.C. (2004) Edge vulnerability in neural and metabolic networks. Biol. Cybern. 90, 311-317

Kaiser, M., Hilgetag, C.C. (2006) Nonoptimal component placement, but short processing paths, due to long-distance projections in neural systems. PLoS Comput. Biol. 2, e95.

Kaiser, M., Robert, M., Andras, P., Young, M.P. (2007) Simulation of robustness against lesions of cortical networks. Eur. J. Neurosci. 25: 3185-3192.

Lachaux, J.-P., Rodriguez, E., Martinerie, J., Varela, F.J. (1999) Measuring phase-synchrony in brain signals. *Hum. Brain Mapping* 8, 194-208.

LeBihan, D. (2003) Looking into the functional architecture of the brain with diffusion MRI. *Nature Rev Neurosci* 4, 469-480.

Livet, J., Weissman, T.A., Kang, H., Draft, R.W., Lu, J., Bennis, R.A., Sanes, J.R., Lichtman, J.W. (2007) Transgenic strategies for combinatorial expression of fluorescent proteins in the nervous system. Nature 450, 56-62.

Markram, H. (2006) The Blue Brain project. *Nature Rev. Neurosci.* 7, 153-160.

Milo, R., Shen-Orr, S., Itzkovitz, S., Kashtan, N., Chklovskii, D., Alon, U. (2002) Network motifs: simple building blocks of complex networks. Science 298, 824-827

Morris, C., Lecar, H. (1981) Voltage-oscillations in the barnacle giant muscle fiber. *Biophys. J.* 35, 193-213.

Passingham, R.E., Stephan, K.E., Kötter, R.(2002) The anatomical basis of functional localization in the cortex. Nature Rev Neurosci 3, 606-616.

Salvador, R., Suckling, J., Schwarzbauer, C., Bullmore, E. (2005a) Undirected graphs of frequency-dependent functional connectivity in whole brain networks. Phil. Trans. R. Soc. B 360, 937-946.

Salvador, R., Suckling, J., Coleman, M., Pickard, J.D., Menon, D.K., Bullmore, E.T. (2005b) Neurophysiological architecture of functional magnetic resonance images of human brain. Cereb Cortex 15, 1332-1342.

Schreiber, T. (2000) Measuring information transfer. Phys. Rev. Lett. 85, 461-464.

Shergill S.S., Kanaan R.A., Chitnis X.A., O'Daly O., Jones D.K., *et al.* (2007) A diffusion tensor imaging study of fasciculi in schizophrenia. Am J Psychiatry 164: 467-473.

Sporns, O., Tononi, G., Edelman, G.M. (2000a) Theoretical neuroanatomy: Relating anatomical and functional connectivity in graphs and cortical connection matrices. Cereb. Cortex 10, 127-141.

Sporns, O., Zwi, J. (2004) The small world of the cerebral cortex. Neuroinformatics 2, 145-162.

Sporns, O., Kötter, R.(2004) Motifs in brain networks. PLoS Biology 2, 1910-1918.

Sporns, O., Chialvo, D., Kaiser, M., Hilgetag, C.C. (2004) Organization, development and function of complex brain networks. Trends Cogn Sci 8, 418-425.

Sporns, O. (2004) Complex neural dynamics. In: Coordination Dynamics: Issues and Trends, Jirsa, V.K. and Kelso, J.A.S., (eds.), pp. 197-215, Springer-Verlag, Berlin.

Sporns, O., Tononi, G., Kötter, R. (2005) The human connectome: A structural description of the human brain. PLoS Computational Biology 1, e42.

Sporns, O., Tononi, G. (2007) Structural determinants of functional brain dynamics. In: Handbook of Brain Connectivity, Jirsa, V.K., McIntosh A.R .(eds), pp. 117-147, Springer, Berlin.

Stam, C.J. (2004) Functional connectivity patterns of human magnetoencephalographic recordings: A 'small-world' network? Neurosci. Lett. 355, 25-28.

Stam. C.J., Reijneveld, J.C. (2007) Graph theoretical analysis of complex networks in the brain. Nonlinear Biomed. Physics 1, 3.

Strogatz, S.H. (2001) Exploring complex networks. Nature 410, 268-277.

Swanson, L.W. (2003) Brain Architecture. Oxford University Press, Oxford.

Tononi, G., Sporns, O, Edelman, G.M. (1994) A measure for brain complexity: relating functional segregation and integration in the nervous system. Proc. Natl. Acad. Sci. USA 91, 5033-5037.

Wedeen, V.J., Hagmann P., Tseng, W.Y., Reese, T.G., Weisskoff, R.M. (2005) Mapping complex tissue architecture with diffusion spectrum magnetic resonance imaging. Magn. Reson. Med. 54, 1377-1386.

Zemanova, L, Zhou, C, Kurths, J. (2006) Structural and functional clusters of complex brain networks. Physica D 224, 202-212.

Zhou, C., Zemanova, L., Zamora, G., Hilgetag, C.C., Kurths, J (2006) Hierarchical organization unveiled by functional connectivity in complex brain networks. Physical Rev. Lett. 97, 238103.

Chapter 10

Brain Network Analysis from High-Resolution EEG Signals

Fabrizio De Vico Fallani[1,2] and Fabio Babiloni[1,3]

[1]*IRCCS "Fondazione Santa Lucia", Rome, Italy*
[2]*Research Centre for Models and Information Analysis in Biomedical Systems, University "Sapienza", Rome, Italy*
[3]*Department of Human Physiology and Pharmacology, University "Sapienza", Rome, Italy*

Over the last decade, there has been a growing interest in the detection of the functional connectivity in the brain from different neuroelectromagnetic and hemodynamic signals recorded by several neuro-imaging devices such as the functional Magnetic Resonance Imaging (fMRI) scanner, electroencephalography (EEG) and magnetoencephalography (MEG) apparatus. Many methods have been proposed and discussed in the literature with the aim of estimating the functional relationships among different cerebral structures. However, the necessity of an objective comprehension of the network composed by the functional links of different brain regions is assuming an essential role in the Neuroscience. Consequently, there is a wide interest in the development and validation of mathematical tools that are appropriate to spot significant features that could describe concisely the structure of the estimated cerebral networks. The extraction of salient characteristics from brain connectivity patterns is an open challenging topic, since often the estimated cerebral networks have a relative large size and complex structure. Recently, it was realized that the functional connectivity networks estimated from actual brain-imaging technologies (MEG, fMRI and EEG) can be analyzed by means of the graph theory. Since a graph is a mathematical representation of a network, which is essentially reduced to nodes and connections between them, the use of a theoretical graph approach seems relevant and useful as firstly demonstrated on a set of anatomical brain networks. In those studies, the authors have employed two characteristic measures, the *average shortest path L* and the *clustering index C*, to extract respectively the global and local properties of the network structure. They have found that anatomical brain networks exhibit many local connections (i.e. a high C) and few random long distance connections (i.e. a low L). These values identify a particular model that interpolate between a regular lattice and a random structure. Such a model has been designated as "small-world" network in analogy with the concept of the small-world phenomenon observed more than 30 years ago in social systems. In a similar way, many types of functional brain networks have been analyzed according to this mathematical approach. In particular, several studies based on different imaging techniques (fMRI, MEG and EEG) have found that the estimated functional networks showed small-world characteristics. In the functional brain connectivity context, these properties have been demonstrated to reflect an optimal architecture for the information processing and propagation among the involved cerebral structures. However, the performance of cognitive and motor tasks as well as the presence of neural diseases has been demonstrated to affect such a small-world topology, as revealed by the significant changes of L and C. Moreover, some functional brain networks have been mostly found to be very unlike the random graphs in their *degree-distribution*, which gives information about the allocation of the functional links within the connectivity pattern. It was demonstrated that the *degree distributions* of these networks follow a power-law trend. For this reason those networks are called "scale-free". They still exhibit the small-world phenomenon but tend to contain few nodes that act as highly connected "hubs". Scale-free networks are known to show resistance to failure, facility of synchronization and fast signal processing. Hence, it would be important to see whether the scaling properties of the functional brain networks are altered under various pathologies or experimental tasks. The present Chapter proposes a theoretical graph approach in order to evaluate the functional connectivity patterns obtained from high-resolution EEG signals. In this way, the "Brain Network Analysis" (in analogy with the Social Network Analysis that has emerged as a key technique in modern sociology) represents an effective methodology improving the comprehension of the complex interactions in the brain.

10.1. Cortical Activity Estimation

High-resolution EEG technology has been developed to enhance the poor spatial information of the EEG activity on the scalp and it gives a measure of the electrical activity on the cortical surface. Principally, this technique involves the use of a larger number of scalp electrodes (64-256). In addition, high-resolution EEG uses realistic MRI-constructed subject head models and spatial de-convolution estimations which are commonly computed by solving a linear inverse problem based on boundary-element mathematics. In the present study, the cortical activity was estimated from EEG recordings by using a realistic head model, whose cortical surface consisted of about 5000 triangles disposed uniformly.

Each triangle represents the electrical dipole of a particular neuronal population and the estimation of its current density was computed by solving the linear inverse problem according to techniques described in previous works. In this way, the electrical activity in different Regions Of Interest (ROIs) can be obtained by averaging the current density of the various dipoles within the considered cortical area.

10.1.1. *Head models and regions of interest*

In order to estimate cortical activity from conventional EEG scalp recordings, realistic head models reconstructed from T1-weighted MRIs are employed. Scalp, skull and dura mater compartments are segmented from MRIs and tessellated with about 5000 triangles. Then, the cortical regions of interest (ROIs) are drawn by a neuroradiologist on the computer-based cortical reconstruction of the individual head model by following a Brodmann's mapping criterion.

10.1.2. *Estimation of cortical source current density*

The solution of the following linear system:

$$Ax = b + n \tag{10.1}$$

provides an estimation of the dipole source configuration x which generates the measured EEG potential distribution b. The system includes also the measurement noise n, assumed to be normally distributed. A is the lead field matrix, where each *j-th* column describes the potential distribution generated on the scalp electrodes by the *j-th* unitary dipole. The current density solution vector ξ of Eq. (10.1) was obtained as:

$$\xi = \arg \min_x \left(\| \mathbf{A}x - b \|_{\mathbf{M}}^2 + \lambda^2 \| x \|_{\mathbf{N}}^2 \right) \tag{10.2}$$

where M, N are matrices associated to the metrics of data and source space, respectively; λ is a regularization parameter; $\| \dots \|_M$ represent the M-norm of the data space b and $\| \dots \|_N$ the N-norm of the solutions space x. The formula (10.2) represents a minimization problem also known as *linear inverse* problem.

As a metric of the data space the identity matrix is generally employed. However, the metric in the source space can be opportunely modified when hemodynamic information is available from recorded fMRI data. This aspect can notably improve the localization of the source activity. An estimate of the signed magnitude of the dipolar moment for each one of the 5000 cortical dipoles was then obtained for each time point. The instantaneous average of all the dipoles' magnitude within a particular ROI was used to estimate the average cortical activity in that ROI during the whole time interval of the experimental task.

Figure 10.1 illustrates the effect of the linear inverse problem's solution. From a scalp potential distribution one can estimate accurately the original cortical potential.

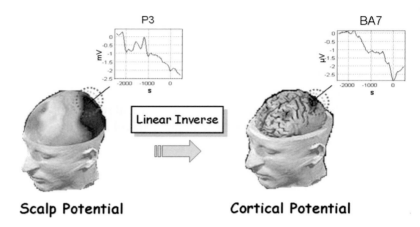

Fig. 10.1. Electrical activity estimation in the Brodmann area 7 from the scalp measurement in the parietal sensor P3.

10.2. Functional Connectivity Estimation

Among the linear and nonlinear methods used to estimate functional brain connectivity, frequency-based methods are particularly attractive for the analysis of EEG or MEG data, since the activity of neural populations is often best expressed in this domain. Many EEG and/or MEG frequency-based methods that have been proposed in recent years for assessment of the directional influence of one signal on another are based mainly on the Granger theory of causality. Granger theory mathematically defines what a "causal" relation between two signals is. According to this theory, an observed time series x(n) is said to cause another series y(n) if the knowledge of x(n)'s past significantly improves prediction of y(n); this relation between time series is not necessarily reciprocal, i.e., x(n) may cause y(n) without y(n) causing x(n). This lack of reciprocity allows the evaluation of the direction of information flow between structures. Kaminski and Blinowska proposed a multivariate spectral measure, called the Directed Transfer Function (DTF), which can be used to determine the directional influences between any given pair of channels in a multivariate dataset. DTF is an estimator that simultaneously characterizes the direction and spectral properties of the interaction between brain signals and requires only one multivariate autoregressive (MVAR) model to be estimated simultaneously from all the time series. The advantages of MVAR modeling of multichannel EEG signals in order to compute efficient connectivity estimates have recently been stressed. Kus *et al.* demonstrated the superiority of MVAR multichannel modeling with respect to the pair-wise autoregressive approach. Another popular estimator, the Partial Directed Coherence (PDC), based on MVAR coefficients transformed into the frequency domain was recently proposed, as a factorization of the Partial Coherence. The PDC is of particular interest because of its ability to distinguish direct and indirect causality flows in the estimated connectivity pattern. If another "true" flow exists from region *x2* to region *x3*, the PDC estimator does not add an "erroneous" causality flow between the signal recorded from region *x1* to region *x3*. This property is particularly interesting in its application to brain signals, where the interpretation of a direct connection between two cortical regions is straightforward.

10.2.1. *MultiVariate AutoRegressive models*

The approach based on multivariate autoregressive models (MVAR) can simultaneously model a whole set of signals. Let X be a set of estimated cortical time series:

$$X = [x_1(t), x_2(t),..., x_N(t)] \tag{10.3}$$

where t refers to time and N is the number of cortical areas considered. Given an MVAR process which is an adequate description of the data set X:

$$\sum_{k=0}^{p} \Lambda(k)X(t-k) = E(t) \tag{10.4}$$

where $X(t)$ is the data vector in time; $E(t) = [e_1(t),...,e_N]$ is a vector of multivariate zero-mean uncorrelated white noise processes; $\Lambda(1), \Lambda(2),...,\Lambda(p)$ are the $N \times N$ matrices of model coefficients $(\Lambda(0) = I)$; and p is the model order. The p order is chosen by means of the Akaike Information Criteria (AIC) for MVAR processes. In order to investigate the spectral properties of the examined process, the Eq. (10.4) is transformed into the frequency domain:

$$\Lambda(f)X(f) = E(f) \tag{10.5}$$

where:

$$\Lambda(f) = \sum_{k=0}^{p} \Lambda(k)e^{-j2\pi f \Delta t k} \tag{10.6}$$

and Δt is the temporal interval between two samples. Eq. (10.5) can then be rewritten as:

$$X(f) = \Lambda^{-1}(f)E(f) = H(f)E(f) \tag{10.7}$$

$H(f)$ is the transfer matrix of the system, whose element H_{ij} represents the connection between the j-th input and the i-th output of the system.

10.2.1.1. *Directed transfer function*

The Directed Transfer Function, representing the causal influence of the cortical waveform estimated in the j-th ROI on that estimated in the i-th ROI is defined in terms of elements of the transfer matrix H, is:

$$\theta_{ij}^2(f) = \left| H_{ij}(f) \right|^2 \tag{10.8}$$

In order to compare the results obtained for cortical waveforms with different power spectra, normalization can be performed by dividing each estimated DTF by the squared sums of all elements of the relevant row, thus obtaining the so-called normalized DTF:

$$\gamma_{ij}^2(f) = \frac{\left| H_{ij}(f) \right|^2}{\sum_{m=1}^{N} \left| H_{im}(f) \right|^2} \tag{10.9}$$

$\gamma^2_{ij}(f)$ expresses the ratio of influence of the cortical waveform estimated in the j-th ROI on the cortical waveform estimated in the i-th ROI, with respect to the influence of all the estimated cortical waveforms. Normalized DTF values are in the interval [0 1], and the normalization condition:

$$\sum_{n=1}^{N} \gamma_{in}^2(f) = 1 \tag{10.10}$$

is applied.

10.2.1.2. *Partial directed coherence*

In order to distinguish between direct and cascade flows, another estimator describing the direct causal relations between signals, the Partial Directed Coherence (PDC), was proposed in 2001. Like DTF, it is defined in terms of MVAR coefficients transformed to the frequency domain. The definition of Partial Directed Coherence (PDC) is:

$$\pi_{ij}(f) = \frac{\Lambda_{ij}(f)}{\sqrt{\displaystyle\sum_{k=1}^{N}\Lambda_{ki}(f)\Lambda_{kj}^{*}(f)}} \qquad (10.11)$$

The PDC from j to i, $\pi_{ij}(f)$, describes the directional flow of information from the activity in the ROI $x_j(t)$ to the activity in $x_i(t)$, whereupon common effects produced by other ROIs $x_k(t)$ on the latter are subtracted leaving only a description that is exclusive from $x_j(t)$ to $x_i(t)$. PDC values are in the interval [0 1] and the normalization condition:

$$\sum_{n=1}^{N}\left|\pi_{ni}(f)\right|^{2} = 1 \qquad (10.12)$$

is verified. According to this condition, $\pi_{ij}(f)$ represents the fraction of the time evolution of ROI j directed to ROI i, as compared to all of j's interactions with other ROIs. Figure 10.2 shows a schematic representation of the functional connectivity estimation from a set of high-resolution EEG signals to the cortical network.

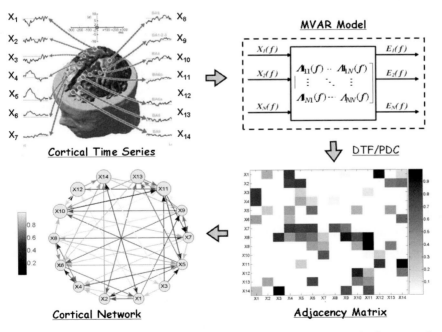

Fig. 10.2.   From a set of cortical time series the MVAR method estimates in the frequency domain a functional connectivity pattern that can be modeled by means of a graph.

10.2.2. *Adaptive MVAR models*

Among the multivariate methods, the Directed Transfer Function (DTF) and the Partial Directed Coherence are estimators characterizing, at the same time, direction and spectral properties of the interaction between brain signals, and require only one MVAR model to be estimated from all the time series. However, the classical estimation of these methods requires the stationarity of the signals; moreover, with the estimation of a unique MVAR model on an entire time interval, transient pathways of information transfer remains hidden. This limitation could bias the physiologic interpretation of the results obtained with the connectivity technique employed. To overcome this limitation, different algorithms for the estimation of MVAR with time dependent coefficients were recently developed. Ding *et al.*, used a short-time windows technique, which requires the stationarity of the signal within short-time windows.

Hesse *et al.*, proposed an application to MVAR estimation of the extension of the recursive least squares (RLS) algorithm with a forgetting factor. This estimation procedure allows for the simultaneous fit of one mean MVAR model to a set of single trials, each one representing a measurement of the same task. In contrast to short-window techniques, the multi-trial RLS algorithm does not require the stationarity of the signals, and involves the information of the actual past of the signal, whose influence decreases exponentially with the time distance to the actual samples. The advantages of this estimation technique are an effective computation algorithm and a high adaptation capability. It was demonstrated in that the adaptation capability of the estimation (measured by its adaptation speed and variance) does not depend on the model dimension.

10.3. Graph Theory

A graph is an abstract representation of a network. It consists of a set of vertices (or nodes) and a set of edges (or connections) indicating the presence of some of interaction between the vertices. The adjacency matrix A contains the information about the connectivity structure of the graph. When a weighted and directed edge exists from the node i to j, the corresponding entry of the adjacency matrix is $A_{ij} \neq 0$; otherwise $A_{ij} = 0$.

10.3.1. *Network density*

The simplest attribute for a graph is its density k, defined as the actual number of connections within the model divided by its maximal capacity; density ranges from 0 to 1, the sparser is a graph, the lower is its value. When dealing with weighted networks, a useful generalization of this quantity is represented by the weighted-density k_w, which evaluates the intensities of the links composing the network. The mathematical formulation of the network density is given by the following:

$$k_w(A) = \sum_{i \neq j \in V} w_{ij} \qquad (10.13)$$

Where A is the adjacency matrix and w_{ij} is the weight of the respective arc from the point j to the point i. $V=1...N$ is the set of nodes within the graph.

10.3.2. *Node strength*

In the same way, the simplest attribute of a node is its connectivity degree, which is the total number of connections with other vertices. In a weighted graph, the natural generalization of the degree of a node i is the node strength or node weight or weighted-degree. This quantity has to be split into in-strength s_{in} and out-strength s_{out}, when directed relationships are being considered. The strength index integrates the information of the links' number (degrees) with the connections' weight, thus representing the total amount of outgoing intensity from a node or incident intensity into it. The formulation of the in-strength index s_{in} can be introduced as follows:

$$s_{in}(i) = \sum_{j \in V} w_{ij} \qquad (10.14)$$

It represents the whole functional flow incoming to the vertex i. V is the set of the available nodes and w_{ij} is the weight of the particular arc from the point j to the point i. In a similar way, for the out-strength:

$$s_{out}(i) = \sum_{j \in V} w_{ji} \qquad (10.15)$$

It represents the whole functional flow outgoing from the vertex i.

10.3.3. *Strength distributions*

For a weighted graph, the arithmetical average of all the nodes' strengths $<s>$ only gives little information about the distributions of the links intensity within the system. Hence, it is useful to introduce $R(s)$ as the fraction of vertices in the graph that have strength equal to s. In the same way, $R(s)$ is the probability that a vertex chosen uniformly at random has weight $= s$. A plot of $R(s)$ for any network can be constructed by making a histogram of the vertices' strength. This histogram represents the strength distribution of the graph and allows a better understanding of the strength allocation in the system. In particular, when dealing with directed graphs, the strength distribution has to be split in order to consider in a separated way the contribution of the incoming and outgoing flows.

10.3.4. *Link Reciprocity*

In a directed network, the analysis of *link reciprocity* reflects the tendency of vertex pairs to form mutual connections between each other [44]. Here we computed the correlation coefficient index ρ proposed by Garlaschelli and Loffredo, which measures whether double links (with opposite directions) occur between vertex pairs more or less often than expected by chance. The correlation coefficient can be written as follows:

$$\rho(A) = \frac{r(A) - k_w(A)}{1 - k_w(A)} \qquad (10.16)$$

In this formula, r is the ratio between the number of links pointing in both directions and the total number of links, while k_w is the connection density that equals the average probability of finding a reciprocal link between two connected vertices in a random network. As a measure of reciprocity, ρ is an absolute quantity that directly allows one to distinguish between reciprocal ($\rho > 0$) and anti-reciprocal ($\rho < 0$) networks, with mutual links occurring more and less often than random, respectively. The neutral or areciprocal case corresponds to $\rho = 0$. Note that if all links occur in reciprocal pairs one has $\rho = 1$, as expected.

10.3.5. *Motifs*

By motif it is usually meant a small connected graph of M vertices and a set of edges forming a subgraph of a larger network with $N > M$ nodes. For each N, there are a limited number of distinct motifs. For $N = 3$, 4, and 5, the corresponding numbers of directed motifs is 13, 199, and 9364. In this work, we focus on directed motifs with $N = 3$. The 13 different 3-node directed motifs are shown in Fig. 10.3. Counting how many times a motif appears in a given network yields a frequency spectrum that contains important information on the network basic building blocks. Eventually, one can looks at those motifs within the considered network that occur at a frequency significantly higher than in random graphs.

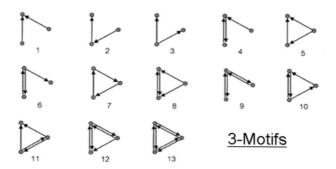

3-Motifs

Fig. 10.3. The 13 possible schemes of connectivity that can be achieved in a graph of 3 nodes.

10.3.6. *Network structure*

Two measures are frequently used to characterize the local and global structure of unweighted graphs: the average shortest path L and the clustering index C. The former measures the efficiency of the passage of information among the nodes, the latter indicates the tendency of the network to form highly connected clusters of vertices. Recently, a more general setup has been examined in order to investigate weighted networks. In particular, Latora and Marchiori considered weighted networks and defined the efficiency coefficient e of the path between two vertices as the inverse of the shortest distance between the vertices (note that in weighted graphs the shortest path is not necessarily the path with the smallest number of edges). In the case where a path does not exist, the distance is infinite and $e = 0$. The average of all the pair-wise efficiencies e_{ij} is the global-efficiency E_g of the graph. Thus, global-efficiency can be defined as:

$$E_g(A) = \frac{1}{N(N-1)} \sum_{i \neq j \in V} \frac{1}{d_{i,j}} \qquad (10.17)$$

where N is the number of vertices composing the graph. Since the efficiency e also applies to disconnected graphs, the local properties of the graph can be characterized by evaluating for every vertex i the efficiency coefficients of A_i, which is the sub-graph composed by the neighbors of the node i. The local-efficiency E_l is the average of all the sub-graphs global-efficiencies:

$$E_l(A) = \frac{1}{N} \sum_{i \in V} E_{glob}(A_i) \qquad (10.18)$$

Since the node i does not belong to the sub-graph Ai, this measure reveals the level of fault-tolerance of the system, showing how the communication is efficient between the first neighbors of i when i is removed. Global- (E_g) and local-efficiency (E_l) were demonstrated to reflect the same properties of the inverse of the average shortest path $1/L$ and the clustering index C. In addition, this new definition is attractive since it takes into account the full information contained in the weighted links of the graph and provides an elegant solution to handle disconnected vertices.

10.4. Application to Real Data

In the following, two applications are presented in order to study the significant features of the functional connectivity networks estimated with the use of advanced high-resolution EEG methodologies.

10.4.1. *Cortical network structure in tetraplegics*

The first study aims at analyzing the structure of cortical connectivity during the attempt to move a paralyzed limb by a group of spinal cord injured patients. Five healthy (CTRL) subjects and five spinal cord injured (SCI) patients participated to the present study. In particular, spinal cord injuries were of traumatic etiology and located at the cervical level (C6 in three cases, C5 and C7 in two cases, respectively); patients had not suffered for a head or brain lesion associated with the trauma leading to the injury. The informed consent statement was signed by each patient after the explanation of the study, which was approved by the local institutional ethics committee. For the EEG data acquisition, subjects were comfortably seated on a reclining chair, in an electrically shielded and dimly lit room. They were asked to perform a brisk protrusion of their lips while they were performing (healthy subjects) or attempting (SCI patients) a right foot movement. By means of the lips protrusion, the SCI patients provided an evident trigger in correspondence of their attempt to move. For each subject, the cortical activity was estimated according to the high-resolution EEG technique (see paragraph X.1). By using the passage through the Tailairach coordinates system, twelve Regions Of Interest (ROIs) were then obtained by segmentation of the Brodmann areas (B.A.) on the accurate cortical model utilized for each subject. Bilateral ROIs considered in this analysis are the primary motor areas for foot (MIF) and lip movement (MIL), the proper supplementary motor area (SMAp), the standard pre-motor area (BA6), the cingulated motor area (CMA) and the associative area (BA7).

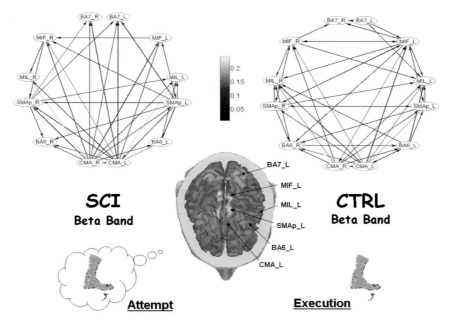

Fig. 10.4. (Up) Average cortical networks in the Beta band for the SCI group and CTRL group. (Centre) Location of the ROIs on the realistic cortex model of a representative subject. (Bottom) The SCI group attempted the foot movement, while the CTRL group executed it.

In order to study the preparation to an intended foot movement, a time segment of 1.5 seconds before the lips pursing was analyzed. The lips movement was detected by means of an EMG. The task was repeated every 6-7 seconds, in a self-paced manner, and the 100 single trials recorded will be used for the estimate of functional connectivity by means of the Directed Transfer Function (DTF, see paragraph X.2) in four frequency bands (Theta 4-7 Hz, Alpha 8-12 Hz, Beta 13-29 Hz, 30-40 Hz). Only the connections that were statistically significant (at $p < 0.001$) after a contrast with a surrogate distribution of one thousand DTF values among the same ROIs were considered for the network to be analyzed with graph theory's tools. Figure 10.4 shows the average cortical network estimated in the Beta frequency band for the SCI group and for the CTRL group, during the motor attempt/execution of the task. The twelve ROIs (the nodes of the cortical network) are indicated on the cortex of one representative subject.

The upper panels of Fig. 10.5 show the average in- and out-degree in the SCI population a) and in the CTRL group b) for the significant Beta band. Direct comparisons of the data show that in the SCI patients the number of links outgoing from both the SMAp areas Left and Right is largely higher than the CTRL subjects. This result puts in evidence the important role of the supplementary motor areas (SMAp Left and Right) that increase their outgoing functional flows to support the diminished activity of their primary motor areas (MIF Left and Right) during the preparation of this motor act.

The panels at the bottom of Fig. 10.5 show the average profiles of the degree distributions for SCI and CTRL group, in the Beta frequency band. An interesting result is that in-degree and out-degree distributions show different trends within each group.

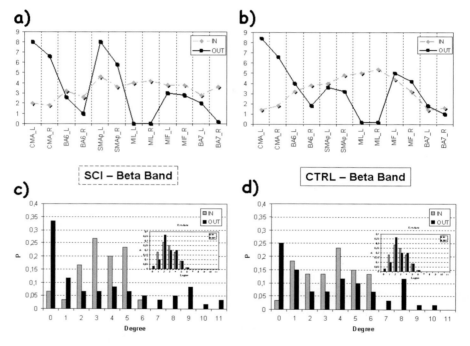

Fig. 10.5. (a) Average in- and out-degrees for the SCI group in the Beta frequency band. (b) Average in- and out-degrees for the CTRL group in the Beta frequency band. (c) Average in- and out-degree distributions for the SCI group in the Beta frequency band. (d) Average in- and out-degrees distributions for the CTRL group in the Beta frequency band.

Right-skew tails of out-degree distributions indicates the presence of few nodes with a very high level of outgoing connections, while for the in-degree distributions there are no ROIs in the network with more than six incoming connections. The inset in each figure illustrates the typical Gaussian profile of the degree-distributions in random graphs, which appears to be different from

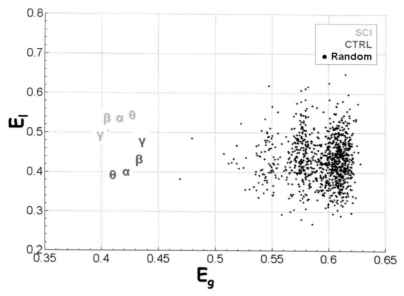

Fig. 10.6. Scatter plot of global- and local-efficiency for SCI networks, CTRL networks and random networks. The Greek symbol codes the average value in a particular frequency band. Black dots identify the values from a distribution of 1000 random graphs.

the estimated cortical networks. Figure 10.6 shows the contrast between the values of global and local efficiency obtained in the two studied populations with those obtained in a set of one thousand random graphs, having the same number of nodes and arcs.

Analysis of variance (ANOVA p = 0.05) was used in order to find significant differences between the indices of efficiency indexes computed in the two groups (SCI, CTRL) for all the frequency bands (Theta, Alpha, Beta and Gamma). ANOVA performed on the global-efficiency E_g variable showed no significant differences for the main factors GROUP and BAND. Instead, the ANOVA performed on the E_l variable revealed a strong influence of the between factor GROUP (F = 32.67, p = 0.00045); while the BAND factor and the interaction between GROUP X BAND were found not significant (F = 0.21 and F = 0.91 respectively, p values equal to 0.891 and 0.457). Post-hoc tests revealed a significant difference between the two examined experimental groups (SCI, CTRL) in Theta, Alpha and Beta band (p = 0.006, 0.01, 0.03 respectively). It can be observed (Fig. 10.6) that the average values of the local efficiency in the SCI subjects are significantly higher than those obtained in the CTRL group, for these three frequency bands. The higher value of local efficiency E_l implies that the network tends to form clusters of ROIs which hold an efficient communication.

These efficient clusters, noticed in the SCI group, could represent a compensative mechanism as a consequence of the partial alteration in the primary motor areas (MIF) due to the effects of the spinal cord injury. Moreover, the estimated cortical networks are not structured like random networks. The statistical contrasts performed by separate Z-tests (Bonferroni corrected for multiple comparisons, p = 0.05) were summarized in the Table 10.1.

By inspecting the data presented in both Table 10.1 and Fig. 10.6, it is clear that in general the cortical networks exhibited ordered and regular properties. In particular, the global efficiency is significantly lower than the random mean value, while the local efficiency of the SCI group is significantly higher than random graphs in each band, meaning fault tolerance is privileged with respect to global communication.

Table 10.1. Z-scores of E_g and E_l from the contrasts with 1000 random graphs.

Z Values	SCI-Theta	SCI-Alpha	SCI-Beta	SCI-Gamma	Healthy-Theta	Healthy-Alpha	Healthy-Beta	Healthy-Gamma
E_g	−237.45	−250.13	−262.88	−267.07	−249.81	−238.21	−225.95	−223.4
E_l	57.714	53.314	57.025	38.936	−15.99	−11.051	7.163	21.674

10.4.2. *Time-varying cortical network during foot movement*

The second study intends to evaluate the dynamics of the cerebral networks during the preparation and the execution of the foot movement in healthy subjects. Five voluntary subjects participated to the study (age, 26-32 years; five males). For the EEG data acquisitions, the participants were comfortably seated on a reclining chair in an electrically shielded and dimly lit room. They were asked to perform a dorsal flexion of their right foot, whose preference was previously attested by simple questionnaires (Chapman 1987). The movement task was repeated every 8 seconds, in a self-paced manner and 200 single trials were recorded by using 200 Hz of sampling frequency. Cortical activity was estimated through high-resolution EEG techniques (see paragraph X.1). The ROIs considered for the left (_L) and right (_R) hemisphere are the primary motor areas of the foot (MF_L and MF_R), the proper supplementary motor areas (SM_L and SM_R) and the cingulate motor areas (CM_L and CM_R). The bilateral Brodmann areas 6 (6_L and 6_R), 7 (7_L and 7_R), 8 (8_L and 8_R), 9 (9_L and 9_R) and 40 (40_L and 40_R) were also considered. In order to inspect the brain dynamics during the preparation and the execution of the studied movement, a time segment of 2 seconds was analyzed, after having

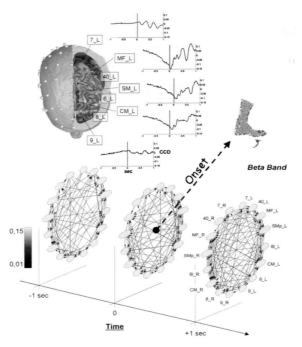

Fig. 10.7. (Up) Realistic head model for a representative subject and cortical activity for the ROIs in the left hemisphere. (Bottom) Three-dimensional representation of the estimated time-varying network in the Beta band for the same subject.

centered it on the onset detected by a tibial EMG. The use of the time-varying Partial Directed Coherence (PDC, see paragraph X.2) to the cortical waveforms obtained from EEG signals returned a cortical network for each selected time sample and frequency band. In order to consider only those task-related connections, a filtering procedure based on statistical validation was adopted. In each trial, a rest period of 2 seconds preceding the movement was selected as an element of contrast (from -4 to -2 s before the onset, i.e. the moment in which the movement occurs). Figure 10.7 illustrates the locations of the regions of interest (ROIs) on the left hemisphere of the cortex model together with their estimated temporal activity. At the bottom, the time-varying cortical network in the Beta frequency band is shown for a representative subject. In particular, three instants are highlighted; one second before the onset, the onset itself and one second after the onset.

Figure 10.8a) shows the in-strength values for the average network during three moments of interest that presented significant differences from random networks. Among all the cortical regions, the supplementary motor areas of both

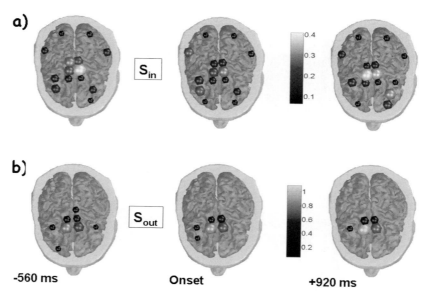

Fig. 10.8. Average in- and out-strength in the Beta band during three significant moments. The little spheres are located in correspondence of each ROI. The size and color of each sphere encodes the degree value.

hemispheres (SM_L and SM_R) show the highest values of in-strength index. In the time points that precedes the onset movement (-560 ms) also the right and left primary motor areas of the foot (MF_L and MF_R) present a considerable number of incoming functional links. Figure 10.8b) shows the average values of out-strength obtained during the three time points of interest. In this particular case, it is evident that the large part of the cortical areas does not produce outgoing edges, while the bilateral cingulate motor region (CM_L and CM_R) presents very high out-strength values. All the indexes calculated on the cortical networks were standardized by considering their Z-score with respect to the distribution obtained from 50 random graphs.

Figure 10.9 shows the average Z values in the analyzed population for the time-varying in- strength R_{in} and out-strength R_{out} distributions in the representative Beta frequency band. An interesting result is that in-strength (R_{in}) and out-strength (R_{out}) distributions show different characteristics. The high Z-scores in correspondence with the high values of S_{out} (i.e. the "right tail" of the distribution) suggest the presence of few ROIs with a very high level of outgoing flows, which makes them act as cortical "hubs". In particular, the intensity of their outgoing links seems to increase as time elapses from the movement

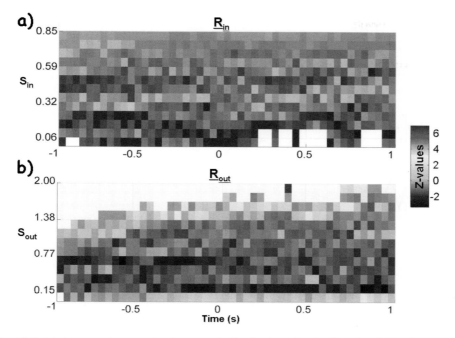

Fig. 10.9. (a) Average time-varying in-strength distributions fro the Beta band. The latency from the movement onset is shown on the x-axes; the in-strength (S_{in}) values on the y-axes. The colour encodes the group-averaged intensity of the R_{in} Z-score. (b) Average time-varying out-strength distributions fro the Beta band. Same conventions as above.

preparation to the movement execution, as revealed by the respective shift of the significant Z values towards high levels of out-strength.

The level of organization in the time-varying cortical networks during the foot movement was analyzed by computing the efficiency indexes E_g and E_l. The E_g and E_l indexes estimated in every subject from the respective cortical networks were contrasted with the ones obtained from the respective random structures. Figure 10.10 shows the average Z-scores of the time-varying E_g (solid line) and E_l - (dotted line) of the connectivity patterns in the Beta frequency band. In particular, one second before the onset (from about -1 to -0.5 s), the cortical networks mostly show low values of E_g and E_l, reflecting a weak pattern of communication characterized by long average distances and few clustering connections between the ROIs. Throughout the period closer to the execution of the movement (from about -0.5 s to the onset), both the global and local properties increase and in correspondence with it, we observe high values of E_g and E_l.

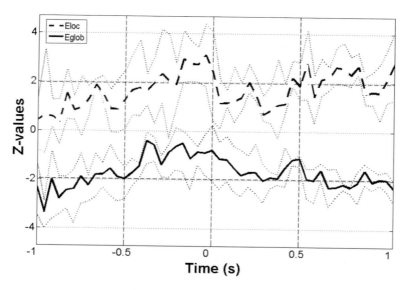

Fig. 10.10. Average time-varying efficiency indexes. The lighter lines around the mean value indicate the time courses of the 25th and 75th percentile. The latency from the movement onset is shown on the x-axes.

Consequently the structure of the cortical networks tends to maximize the interplay between the global integration and its local interactions. This particular structure represents one of the best way in which the cortical areas communicate, since the relevant network presents simultaneously short links between each pair of ROIs and highly connected clusters (i.e. small-world architecture). After the onset (from the onset to +0.5 s), the estimated cortical networks show a typical random organization of the functional links, with a high E_g and a low E_l, reflecting the dense presence of wide-scope interactions among the ROIs, but a low tendency of the same cortical regions to form functional clusters. In the last period of the movement execution (from about +0.5 to +1 s) the estimated cortical networks mainly show high E_l values and low E_g values. The resulting structure is known to reflect the properties of regular and ordered graphs in which the local property of clustering is privileged with respect to the overall communication. Figure 10.11(a) shows the average time-varying course of the weighted-density k_w in the Beta band during the analyzed period of interest.

In particular, the average intensity of the network links during the preparation (from -0.5 s to the onset) is relatively low if compared with its maximum value reached in the following movement execution. In correspondence with this period the network structure presents the most efficient pattern of communication, as revealed by the estimated small-world characteristic.

Therefore, it is interesting to note that the optimal organization of the functional links among the cortical areas during the preparation of the foot movement is not correlated to the need of a high level of overall connectivity.

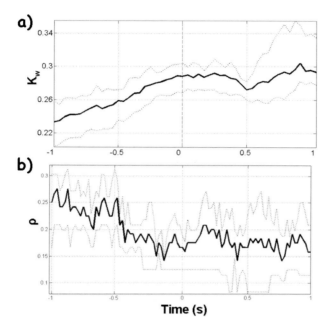

Fig. 10.11. (a) Average time-varying "weighted-density" in the Beta band. (b) Average time-varying "reciprocity" during the period of interest in the Beta band. On y-axes the correlation coefficient ρ while time in seconds on x-axes.

The analysis of the average time-varying reciprocity index revealed an interesting behavior during the preparation (from about -1 to 0 s) of the movement in the Beta frequency band. In such a period, the functional network moved from a reciprocal (r = 0.1) to an anti-reciprocal (r = -0.1) state. This aspect emphasizes the role of the early preparation in which a high level of mutual exchange of information is required to speed up the cortical process in expectation of the execution. Moreover, by tracking the evolving involvement of each single reciprocal connection (see Fig. 10.12(a)) it is possible to observe their "persistence" during the entire period of interest. In particular, the persistent bilateral links between the cingulate motor areas and the supplementary motor areas (they correspond to the rows 58 and 69) in the Beta band reveals a novel aspect of such a connection that anyway was expected in a self-paced modality of movement generation, as in our experimental condition.

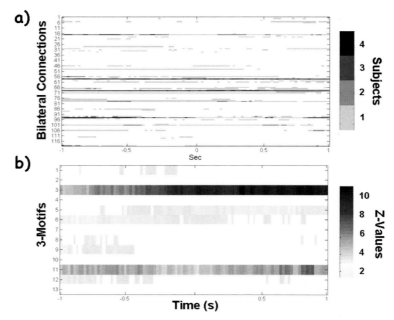

Fig. 10.12. (a) Time-varying persistence of the bilateral connections in the cortical network. On y-axes all the 120 possible reciprocal connections while time in seconds on x-axes. The colour of the line corresponding to a particular link codes the number of subjects that actually hold such a connection. (b) Average time-varying 3-motif spectra. On y-axes all the 13 possible directed 3-motifs are listed while time in seconds is displayed on x-axes.

In Figure 10.12(b), we compared the 3-motif properties of real brain networks with random networks and we identified some motif classes that occurred more frequently during particular stages of the movement. Of particular interest is the involvement of the feed-forward-loop motif (the fifth in the Figure 10.4) that tends to significantly ($p < 0.01$) increase during the proper movement execution (from about 0 to +1 s). This type of building block is known to play an important functional role in information processing. In fact, one possible function of this circuit is to activate output only if the input signal is persistent and to allow a rapid deactivation when the input goes off. In the cortical context, a possible interpretation of such a motif would make a particular ROI act as a "switch" for the communication between the others two ROIs composing the triad. Another interesting aspect was revealed by the significant ($p \ll 0.01$) "persistence" of the single-input motif (the third in the Figure 10.4) that represented the highest recurrent pattern of interconnections during the entire evolution of the foot movement. The main function of this motif is known to involve the "activation" of several parallel pathways by a single activator. Thus, since the single-input only differs from the feed-forward-loop motif for the functional link between the two areas activated, we can claim the privileged scheme of communication within the functional networks estimated consists in a parallel activation from a particular ROI of two other distinct areas, whose communication seems to increase significantly only during the proper movement execution.

10.5. Conclusions

One of the interesting characteristics of the brain networks presented in this Chapter is that such networks have no precise anatomical support, i.e. there is no particular cerebral structure that implements the network itself. Thus, those brain networks represent functional networks, which could change in topology and properties according to the specific subject's behavior. Another attractive characteristic is that these functional networks are estimated from high-resolution EEG signals. This allows the representation of the graph nodes as particular regions of interest on the cortex. This approach gives to the researcher a "window" to access the brain functions in a different perspective than the usual techniques encountered in the neuroscience literature.

In fact, the development of brain imaging devices (such as the functional Magnetic Resonance Imaging (fMRI), but also the high-resolution EEG technology) often give to the scientist a series of colored hot-spots in the brain that sub-serve the functions performed by the subject during a particular task. Actually, if we look at thousands of fMRI studies a possible impression is that a specific cortical area gets "activated" during the performance of whatever cognitive or motor operation. In this scenario of modern "color phrenology", the study of functional cortical connectivity suggests an image of the brain as a system of objects that rapidly changes the way in which they are interconnected, according to the complexity and to the dynamic of the task proposed to the subject. It is opinion of the Authors that the perspective offered by the use of graph theory to the functional cortical connectivity networks estimated from high-resolution EEG recordings could be a promising way to approach the brain functioning from a modern point of view.

References

Achard S., Salvador R., Whitcher B., Suckling J. and Bullmore, Ed. A Resilient, Low-Frequency, Small-World Human Brain Functional Network with Highly Connected Association Cortical Hubs. The Journal of Neuroscience, 26(1):63–72, 2006.

Akaike H. (1974) A new look at statistical model identification. IEEE Trans Automat Control AC-19:716-723.

Astolfi L., Cincotti F., Mattia D., De Vico Fallani F., Tocci A., Colosimo A., Salinari S., Marciani M.G., Hesse W., Witte H., Ursino M., Zavaglia M., Babiloni F.Tracking the time-varying cortical connectivity patterns by adaptivemultivariate estimators. IEEE Trans Biomed Eng. 55(3):902-13, 2008.

Astolfi L., Cincotti F., Mattia D., Marciani M.G., Baccalà L., De Vico Fallani F., Salinari S., Ursino M., Zavaglia M., Ding L., Edgar J.C., Miller G.A., He B. and Babiloni F. A comparison of different cortical connectivity estimators for high resolution EEG recordings. Human Brain Mapping; 28(2):143-57, 2006.

Babiloni F., Babiloni C., Locche L., Cincotti F., Rossini P.M., Carducci F. High resolution EEG: source estimates of Laplacian-transformed somatosensory-evoked potentials using a realistic subject head model constructed from magnetic resonance images. Med. Biol. Eng. Comput., 38:512-9, 2000.

Babiloni F., Cincotti F., Babiloni C., Carducci F., Basilisco A., Rossini P.M., Mattia D., Astolfi L., Ding L., Ni Y., Cheng K., Christine K., Sweeney J., He B. Estimation of the cortical functional connectivity with the multimodal integration of high resolution EEG and fMRI data by Directed Transfer Function. Neuroimage, 24(1):118-3, 2005.

Baccalà L.A., Sameshima K.., Partial Directed Coherence: a new concept in neural structure determination. *Bio.l Cybern.*, 84: 463-474, 2001.

Barabasi A.L., Albert R. Emergence of scaling in random networks. Science, 286: 509-512, 1999.

Bassett D.S., Meyer-Linderberg A., Achard S., Duke Th., Bullmore E. Adaptive reconfiguration of fractal small-world human brain functional networks. PNAS, 103:19518-19523, 2006.

Bartolomei F., Bosma I., Klein M., Baayen J.C., Reijneveld J.C., Postma T.J., Heimans J.J., van Dijk B.W., de Munck J.C., de Jongh A., Cover K.S., Stam C.J. Disturbed functional connectivity in brain tumour patients: evaluation by graph analysis of synchronization matrices. Clin Neurophysiol; 117:2039-2049, 2006.

Boccaletti S., Latora V., Moreno Y., Chavez M., Hwang D.U. Complex networks: structure and dynamics. Physics Reports, 424:175-308, 2006.

Chávez M., Martinerie J., Le Van Quyen M. Statistical assessment of nonlinear causality: application to epileptic EEG signals. J Neurosci Methods, 124(2):113-28.

De Vico Fallani F., Astolfi L., Cincotti F., Mattia D., Marciani M.G., Salinari S., Kurths J., Gao S., Cichocki A., Colosimo A., Babiloni F. Cortical functional connectivity networks in normal and spinal cord injured patients: Evaluation by graph analysis. Hum Brain Mapp, 28:1334-6, 2007.

De Vico Fallani F., Astolfi L., Cincotti F., Mattia D., Marciani M.G., Tocci A., Salinari S., Witte H., Hesse W., Gao S., Colosimo A., Babiloni F. Cortical network dynamics during foot movements. Neuroinformatics. Spring, 6(1):23-34. 2008.

Eguiluz V.M., Chialvo D.R., Cecchi G.A., Baliki M., Apkarian A.V. Scale-free brain functional networks, Phys. Rev. Lett. 94:018102, 2005.

Garlaschelli D. and Loffredo M.I. Patterns of Link Reciprocity in Directed Networks Phys. Rev. Lett. 93, 268701, 2004.

Hesse W., Möller E., Arnold M., Schack B. The use of time-variant EEG Granger causality for inspecting directed interdependencies of neural assemblies. Journal of Neuroscience Methods 124: 27-44, 2003

Hilgetag C.C., Burns G.A.P.C., O'Neill M.A., Scannell J.W., Young M.P. Anatomical connectivity defines the organization of clusters of cortical areas in the macaque monkey and the cat. Philos. Trans. R. Soc. Lond. B Biol. Sci., 355:91-110, 2000.

Kaminski M., Blinowska K. A new method of the description of the information flow in the brain structures. *Biol. Cybern.* 1991, 65: 203-210.

Kaminski M., Ding M., Truccolo W.A., Bressler S. Evaluating causal relations in neural systems: Granger causality, directed transfer function and statistical assessment of significance. *Biol. Cybern.* 2001, 5, 145-157.

Kus R., Kaminski M., Blinowska K.J. Determination of EEG activity propagation: pair-wise versus multichannel estimate. IEEE Trans. Biomed. Eng. Sep;51(9):1501-10, 2004.

Lago-Fernandez L.F., Huerta R., Corbacho F., Siguenza J.A. Fast response and temporal coherent oscillations in small-world networks, Phys. Rev. Lett.; 84: 2758–61, 2000.

Latora V and Marchiori M. Efficient behaviour of small-world networks. Phys. Rev. Lett. 87:198701, 2001

Latora V. and Marchiori M. Economic small-world behaviour in weighted networks. Eur. Phys. JB 2003; 32:249-263.

Le J. and Gevins A. A method to reduce blur distortion from EEG's using a realistic head model. IEEE Trans Biomed Eng., 40:517-28, 1993.

Micheloyannis S., Pachou E., Stam C.J., Vourkas M., Erimaki S., Tsirka V. Using graph theoretical analysis of multi channel EEG to evaluate the neural efficiency hypothesis. Neuroscience Letters, 402:273-277, 2006.

Milgram, S. *The Small World Problem, Psychology Today* 60-67, 1967.

Milo R., Shen-Orr S., Itzkovitz S., Kashtan N., Chklovskii D. and Alon U. Network motifs: simple building blocks of complex networks Science, 298 824-7, 2002.

Newman M.E.J. The structure and function of complex networks. SIAM Review; 45:167-256, 2003.

Nunez P.L. Neocortical dynamics and human EEG rhythms. New York: Oxford University Press, 708 p., 1995.

Pfurtsheller G., Lopes da Silva F.H. Event-related EEG/EMG synchronizations and desynchronization: basic principles. Clin Neurophysiol. 110:1842–1857, 1999

Salvador R., Suckling J., Coleman M.R., Pickard J.D., Menon D., Bullmore E. Neurophysiological Architecture of Functional Magnetic Resonance Images of Human Brain. Cereb Cortex; 15(9):1332-42, 2005.

Shen-Orr S., Milo R., Mangan S. and Alon U. Network motifs in the transcriptional regulation network of Escherichia coli Nature Genetics, 31 64-8, 2002.

Sporns O., Honey C.J., Kötter R. Identification and classification of hubs in brain networks. PLoS ONE.;2(10):e1049, 2007.

Sporns O., Honey C.J. Small worlds inside big brains. Proc. Natl. Acad. Sci. USA. 19, 103(51):19219-20. 2006.

Stam C.J., Jones B.F., Manshanden I., van Cappellen van Walsum A.M., Montez T., Verbunt J.P., de Munck J.C., van Dijk B.W., Berendse H.W., Scheltens P. Magnetoencephalographic evaluation of resting-state functional connectivity in Alzheimer's disease. Neuroimage; 32:1335-44, 2006.

Stam CJ, Jones BF, Nolte G, Breakspear M, Scheltens Ph. Small-world networks and functional connectivity in Alzheimer's disease. Cereb Cortex; 17:92-99, 2007.

Strogatz SH. Exploring complex networks. Nature; 410:268-76, 2001.

Tononi G., Sporns O., Edelman G.M. A measure for brain complexity: relating functional segregation and integration in the nervous system. Proc. Natl. Acad. Sci. USA, 91:5033-7, 1994.

Watts D.J. and Strogatz S.H. Collective dynamics of 'small-world' networks. Nature, 393:440-2, 1998.

Chapter 11

An Optimization Approach to the Structure of the Neuronal Layout of *C. elegans*

Alex Arenas, Alberto Fernández and Sergio Gómez

Departament d'Enginyeria Informàtica i Matemàtiques,
Universitat Rovira i Virgili, E-43007 Tarragona, Spain

11.1. Introduction

Between 1899 and 1904, in the masterpiece *Textura del sistema nervioso del hombre y de los vertebrados* (published in separated folded sheets during these years), S. Ramón y Cajal established the linchpin of modern neuroscience [1]. Among its capital contributions, he stated the *law of maximum economy in space, time and inter-connective matter*, that explicitly hypothesizes about an optimization of the structure and function of the nervous systems during evolution, reflected in an economical principle for informational driving processes in neuronal circuitries. This fascinating elucidation of the complex structure of nervous systems, has been however very difficult to quantify with real data. The topological mapping of each one of the neurons of a vertebrate's brain is still out of nowadays technical possibilities. However, it exists an invertebrate organism for which the complete neuronal layout is known, the nematode *C. elegans*, syee Fig. 11.1. The current computational capabilities and the disposal of such a connectivity data set allow us to explore the conjecture of S. Ramón y Cajal about the "wiring economy principle".

In this chapter we will review a recent optimization approach to the wiring connectivity in *C. elegans*, discussing the possible outcomes of the optimization process, its dependence on the optimization parameters, and its validation with the actual neuronal layout data. We will follow the main procedure described in the work by Chen *et al.* [2–4]. The results show that the current approach to optimization of neuronal layouts is still not conclusive, and then the "wiring economy principle" remains unproved.

11.2. The Dataset

The nematode *Caenorhabditis Elegans* has become in biology the experimental organism par excellence to understand the mechanisms underlying a whole animal's

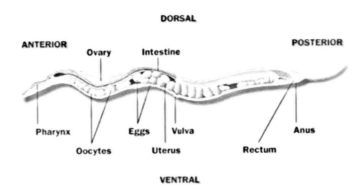

Fig. 11.1. Sketch of *C. elegans* anatomy.

behavior, at the molecular and cellular levels [5, 6]. It has been extensively studied to understand particular biological phenomena, with the expectation that discoveries made in this organism will provide insight into the workings of other organisms. It is one of the usually called model organisms. In particular, model organisms are widely used to explore potential causes and treatments for human disease when human experimentation would be unfeasible or unethical. This approximation is supported by the fact of common descent of all living organisms, and the conservation of metabolic and developmental pathways and genetic material over the course of evolution. In fact, this goal was a primary motivation behind the development of *C. elegans* as an experimental organism 40 years ago. Yet it has proven surprisingly difficult to obtain a mechanistic understanding of how the *C. elegans* nervous system generates behavior, despite the existence of a "wiring diagram" that contains a degree of information about neural connectivity unparalleled in any organism. Studying model organisms can be informative, but generalizations should be carefully considered.

The structural anatomy of *C. elegans* is basically that of a cylinder around 1 millimeter in length and 0.1 millimeter in diameter, see Fig. 11.1. In the following, we will use the common hypothesis of study of a one dimensional entity. We are interested in its neuronal system, in particular in the position along the body of the different neurons and its interconnections. The current work uses the public data found in [7]. The construction of this data set started with the work by Albertson *et al.*, and White *et al.* [8, 9], and has been contributed by many authors since then, in the multimedia project *Wormatlas* [7]. The particular wiring diagram we use was revised and completed by Chen *et al.* [3] using other valuable sources [11, 12]. The wiring information we have used is structured in four parts: connectivity data between neurons, neuron description, neuron connections to sensory organs and body muscles, and neuronal lineage. The architecture of the nervous system of *C. elegans* shows a bilaterally symmetric body plan. With a few exceptions,

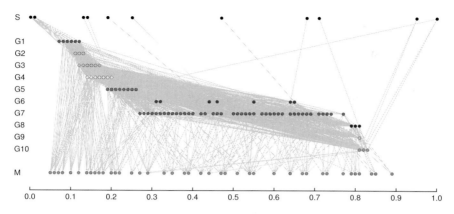

Fig. 11.2.   *C. elegans* layout of sensorial (S), motor (M) and non-pharyngeal neuronal cells, and its connectivity. Neurons in the same ganglia are given the same vertical offset for clarity: G1) anterior ganglion, G2) dorsal ganglion, G3) lateral ganglion, G4) ventral ganglion, G5) retrovesicular ganglion, G6) posterolateral ganglion, G7) ventral cord neuron group, G8) pre-anal ganglion, G9) dorsorectal ganglion, G10) lumbar ganglion. The bottom ruler shows the longitudinal assigned coordinates, with values 0.0 and 1.0 for the head and tail of the worm respectively.

neurons in *C. elegans* have a simple uni- or bipolar morphology that is typical for invertebrates. Synapses between neurons are usually formed *en passant* and each cell has multiple presynaptic regions dispersed along the length of the axon.

The neuronal network connectivity of the *C. elegans* can be represented as a weighted adjacency matrix of 279 nonpharyngeal neurons, out of a total of 302 neurons (pharyngeal neurons are not considered in this work because they are not reported in the above mentioned database). The abstraction at this point consists in to assume that the nervous system of the *C. elegans* can be modeled as a network, where nodes represent the center of the cell bodies, and the links represent synapses, see Fig. 11.2. The order and nomenclature of the neurons in the matrix follows that of [12], for a detailed biological record of the dataset see [7]. The position of neurons has been defined in the data set as follows: i) neuron location is considered at the center of the cell body projected onto the anterior–posterior axis of the worm, ii) a neuron is assumed to make a single connection to a given sensory organ, iii) the position of each muscle is defined as the midpoint between anterior and posterior extremities of the sarcomere region, and iv) there is a lack of data specifying the location of individual synapses in the worm.

11.3. Optimization Model Problem

Assuming that the neuronal wiring of *C. elegans* has been optimized by natural evolution, it seems plausible to formulate an optimization model problem whose results will reproduce a neuronal layout in agreement with the real data, and support then the hypothesis of the wiring economy principle. The model must be robust

to the variation of the parameters, and statistically significant. Here we follow
the general formulation presented in [2, 3]. The optimization problem, is stated
as a cost function that must be minimized. The cost is separated in two different
contributions: a connection cost between neurons, and a connection cost between
neurons and sensorial organs or muscles. This difference arises from the different
connectivity patterns and also because its mathematical convenience, note that the
position of sensors and muscles is taken as input data, constraining then the optimization problem. We will work in the scope of the dedicated-wire model (following
the terminology in [3]) which simply means that every synapse has its own wire.
At difference, the shared-wire model introduced also in [3] considers a neuron as a
wire (segment) with multiple synapses; we do not include its analysis here because
it does not provide betters results.

Mathematically, the general formulation of wiring cost is:

$$C^{\text{tot}} = C^{\text{int}} + C^{\text{ext}} \tag{11.1}$$

where

$$C^{\text{int}} = \frac{1}{2\alpha_A} \sum_i \sum_j A_{ij} (x_i - x_j)^2 \,, \tag{11.2}$$

$$C^{\text{ext}} = \frac{1}{\alpha_S} \sum_i \sum_k S_{ik} (x_i - s_k)^2 + \frac{1}{\alpha_M} \sum_i \sum_r M_{ir} (x_i - m_r)^2 \tag{11.3}$$

and

- $\mathbf{A} = (A_{ij})$ neuron-neuron connectivity matrix ($N_A \times N_A$)
- $\mathbf{S} = (S_{ik})$ neuron-sensor connectivity matrix ($N_A \times N_S$)
- $\mathbf{M} = (M_{ir})$ neuron-motor connectivity matrix ($N_A \times N_M$)
- $\mathbf{x} = (x_1, \ldots, x_{N_A})^T$ neurons positions vector
- $\mathbf{s} = (s_1, \ldots, s_{N_S})^T$ sensors positions vector
- $\mathbf{m} = (m_1, \ldots, m_{N_M})^T$ motors positions vector
- α_A, α_S, and α_M are parameters to be determined by normalization constraints

Everything will be treated as input data except the position of the neurons $\mathbf{x}$.
This quadratic expression of the cost is just a convention, other exponents have been
considered in [3] with no improvement on the optimization results. Moreover, it is
mathematically convenient because in this form the system is analytically solvable.
The optimization of the total cost (11.1) is obtained by imposing

$$\frac{\partial C}{\partial x_a} = 0\,, \quad a = 1, \ldots, N_A\,. \tag{11.4}$$

They yield a system of linear equations whose solution, in matrix form, reads

$$\mathbf{x} = \mathbf{Q}^{-1} \left(\frac{1}{\alpha_S} \mathbf{S}\mathbf{s} + \frac{1}{\alpha_M} \mathbf{M}\mathbf{m} \right) \tag{11.5}$$

where

$$Q_{ij} = \left(\frac{1}{\alpha_A} \sum_{\ell} \tilde{A}_{i\ell} + \frac{1}{\alpha_S} \sum_{k} S_{ik} + \frac{1}{\alpha_M} \sum_{r} M_{ir} \right) \delta_{ij} - \frac{1}{\alpha_A} \tilde{A}_{ij} \qquad (11.6)$$

being δ_{ij} the elements of the identity matrix, and

$$\tilde{\mathbf{A}} = \frac{1}{2}(\mathbf{A} + \mathbf{A}^T). \qquad (11.7)$$

Note that the symmetrization of matrix $\mathbf{A}$ is not an initial constraint but a consequence of the optimization process.

It is convenient to define also the following constants for the determination of the α parameters:

$$\tau_A = \frac{1}{2N_A} \sum_{i} \sum_{j} A_{ij}, \qquad (11.8)$$

$$\tau_S = \frac{1}{2N_A} \sum_{i} \sum_{k} S_{ik}, \qquad (11.9)$$

$$\tau_M = \frac{1}{2N_A} \sum_{i} \sum_{r} M_{ir}. \qquad (11.10)$$

The actual neuronal layout of *C. elegans* is represented in Fig. 11.2. In the plot we have separated neurons at different areas and also differentiated them from sensor neurons and muscles. The first interesting observation is that the connectivity is clearly biased towards the head of the animal, where more sensorial connections are established. The muscular connectivity is also more dense in the anterior part of the body although not as dense than for the sensors. Realizing that our constraints are the positions of sensor neurons and muscles, we expect a significant scatter of the predictions in the region between the middle and the posterior part of the body. Moreover, the scatter will result in an under-prediction of the position of neurons, the predicted positions will be biased towards the anterior part.

11.4. Results and Discussion

Equation (11.5) gives the position of the neurons in the abstracted optimization model. To measure the success of this method, the mean absolute difference between the actual ($\mathbf{x}$) and predicted ($\mathbf{x}'$) neuron positions is used:

$$E = \frac{1}{N_A} \sum_{i} |x_i - x_i'|. \qquad (11.11)$$

We have prepared several experiments in the scope of the current optimization problem, to check the reliability of the current approach to get support on the hypothesis of wiring optimization in the neuronal layout of *C. elegans*. Here we expose the set up and results of each experiment. Finally a summary is presented in Table 11.1.

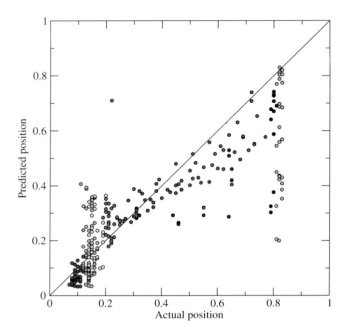

Fig. 11.3.   Experiment 1. Predicted versus real neuronal layout of *C. elegans*, setup detailed in text. Colors correspond to the different ganglia shown in Fig. 11.2.

11.4.1. *Experiment 1*

First, we reproduce the results presented in [3] for the dedicated-wire model. In this case, the matrices **A**, **S** and **M** correspond to the weighted connectivity matrices of the real data. The parameters α_A, and α_M are set to normalize the neuron-neuron, and neuron-muscle interaction by the average number of synapses per neurite, 29.44 (or equivalently 58.88 synapses per neuron divided by two neurites per neuron). In the referenced work α_S is set to 1, with no apparent reason, so we did it. The results are depicted in Fig. 11.3, different colors corresponds to the different ganglia shown in Fig. 11.2. The error E in this approximation is 9.69%.[a] The authors contrasted this result with the positioning of neurons uniformly at random along the body of the worm, which raises an $E \approx 34\%$, moreover they computed that the probability of obtaining by chance the results of the optimization is of order 10^{-68}. They also compare the optimization results against the null hypothesis that more related neurons are positioned closer to each other. To this end they use instead of **A** the "relatedness" matrix, which is a matrix connecting neurons by number of jumps in the lineage tree, substituting non existent connections in this

---

[a]In the original paper the authors find an error of 9.71%, we attribute this difference to the use of different data, since we are using the last update of the data in [7].

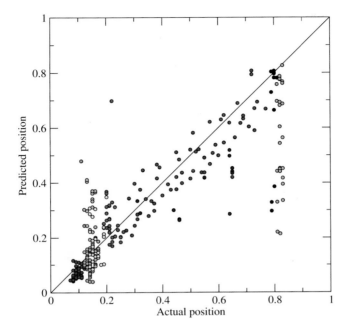

Fig. 11.4.   Experiment 2. Predicted versus real neuronal layout of *C. elegans*, setup detailed in text. Colors correspond to the different ganglia shown in Fig. 11.2.

matrix by a uniform repulsive force.[b] We will reproduce this null hypothesis in Experiment 6. The authors find an error in this case of $E = 26.1\%$ far bigger than the optimization result, supporting then the optimization process as a meaningful description of the relationship between neuronal arrangement and connectivity in the worm.

11.4.2. *Experiment 2*

We noticed that the results presented in [3] can be quantitatively improved by a separate normalization of the neuron-neuron interaction, and the neuron-muscle interaction. Given that the adjacency matrix $\mathbf{A}$, and $\mathbf{M}$ are distinguishable, it makes sense to compute the normalization $\alpha_A = \tau_A$, and $\alpha_M = \tau_M$. In doing this slight change in the parameters, the error reduces to $E = 8.75\%$ which is noticeable in this scenario (nearly a 10% relative improvement with respect to the result in Experiment 1). However, qualitatively the results do not seem to change, see Fig. 11.4.

---

[b]This last prescription seems to us unnecessary because the relatedness matrix even with the zero values for non existent connections is non singular.

11.4.3. *Experiment 3*

If the modification in the normalization is a key factor in the optimization process, one expects this to hold also for the normalization of the neuron-sensor connections. Mathematically, there is no handicap for this consideration because it simply implies $\alpha_S = \tau_S$, and it provides consistency with the prescription above. We used this normalization, plus the one in Experiment 2, and obtained the unsatisfactory result of an increasing in the error to $E = 10.15\%$, see Fig. 11.5. This surprising effect raises doubts on the approach, because this sensibility to the normalization parameters was unexpected.

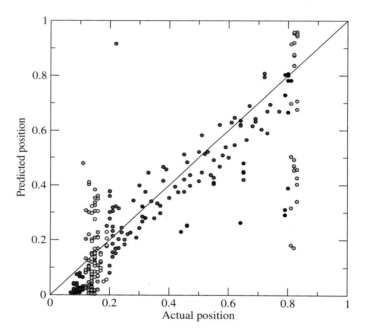

Fig. 11.5. Experiment 3. Predicted versus real neuronal layout of *C. elegans*, setup detailed in text. Colors correspond to the different ganglia shown in Fig. 11.2.

11.4.4. *Experiment 4*

After the last observation, we decided to change the input data considering that the neuron-neuron connectivity matrix is an unweighted matrix. This seems a more realistic approach, since multiple synapses between neurons are not built upon multiple wires but only two neurites with multiple *en passant* synapses. We filtered **A** assigning 1 if there is a connection between the neurons, disregarding its number, and 0 otherwise. The matrices **M** and **S** were remained unchanged. Using the normalization exposed in Experiment 2, $\alpha_A = \tau_A$, and $\alpha_M = \tau_M$, the error now is reduced to $E = 8.33\%$, see Fig. 11.6. Now the results are more striking because they point out not only that the normalization factor notably affects the results, but also that the consideration of the weights (number of synapses) between neurons makes no real difference in the outcome of the optimization process.

11.4.5. *Experiment 5*

Equivalently to Experiment 3, we introduce the normalization of the neuron-sensor connections $\alpha_S = \tau_S$ in the set up exposed in Experiment 4, i.e. unweighted connectivity matrix. Again the results of the optimization process are worst, with an error $E = 9.62\%$, see Fig. 11.7. Definitely the inclusion of the normalization of the neuron-sensor connections provide worst results in terms of global error. The main reason is that the normalization of neuron-sensor connections enhances even more the connectivity towards the anterior part of the animal. However, we think that its elimination is not mathematically consistent, and then it must be preserved in the same way we preserve the normalization of the other type of connections.

11.4.6. *Experiment 6*

At the light of the previous results, we think that the contribution of the actual neuronal connectivity in the global error of the optimization process is very low. Following the idea in [3], we used the relatedness of neurons, which is a measure of distance in the lineage tree (not linearly correlated with the connectivity), instead of the adjacency matrix. The authors in [3] did something similar and provided an error significantly larger than the error considering the actual connectivity. We simply substituted the adjacency matrix by the relatedness matrix, and normalized according to the previous Eq. (11.8). The error in this case is $E = 11.76\%$, slightly larger than the error in Experiment 1 but absolutely comparable, moreover the scatter of the neuron positions is qualitatively equivalent to that in Experiments 1 to 5, see Fig. 11.8.

11.4.7. *Experiment 7*

Finally, we want to show the results of a really out-minded experiment where the true adjacency matrix is replaced by an all-to-all connectivity between neurons. The result obtained in this particular case, raises an error $E = 11.60\%$ which is comparable to the results so far obtained using the true connectivity matrix, see Fig. 11.9. This experiment raises serious doubts about the validity of the optimization procedure proposed so far, as a way to test the wiring economy principle in neural networks.

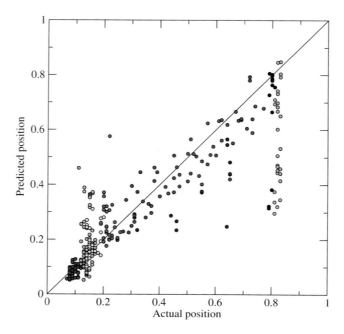

Fig. 11.6. Experiment 4. Predicted versus real neuronal layout of *C. elegans*, setup detailed in text. Colors correspond to the different ganglia shown in Fig. 11.2.

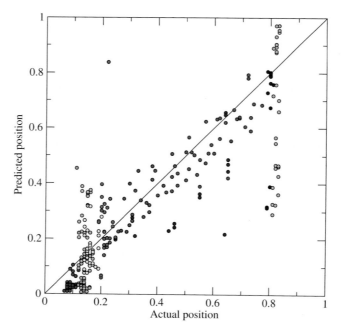

Fig. 11.7. Experiment 5. Predicted versus real neuronal layout of *C. elegans*, setup detailed in text. Colors correspond to the different ganglia shown in Fig. 11.2.

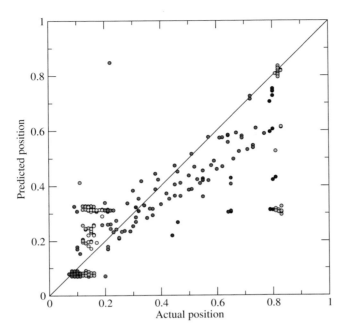

Fig. 11.8.   Experiment 6. Predicted versus real neuronal layout of *C. elegans*, setup detailed in text. Colors correspond to the different ganglia shown in Fig. 11.2.

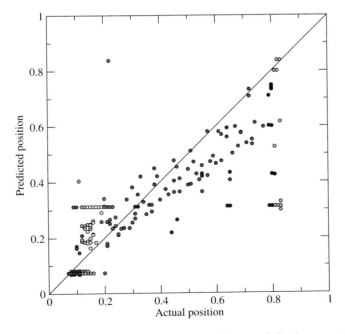

Fig. 11.9.   Experiment 7. Predicted versus real neuronal layout of *C. elegans*, setup detailed in text. Colors correspond to the different ganglia shown in Fig. 11.2.

Table 11.1.   Summary of the different experiments and results.

Exp.	Parameters	E
1	$A_{ij} >= 0$ (see Ref. 3) $\alpha_A = \alpha_M = \tau_A + \tau_M = 29.44$ $\alpha_S = 1$	9.69%
2	$A_{ij} >= 0$ $\alpha_A = \tau_A = 26.10$ $\alpha_M = \tau_M = 3.34$ $\alpha_S = 1$	8.75%
3	$A_{ij} >= 0$ $\alpha_A = \tau_A = 26.10$ $\alpha_M = \tau_M = 3.34$ $\alpha_S = \tau_S = 0.15$	10.15%
4	$A_{ij} \in \{0,1\}$ $\alpha_A = \tau_A = 8.20$ $\alpha_M = \tau_M = 3.34$ $\alpha_S = 1$	8.33%
5	$A_{ij} \in \{0,1\}$ $\alpha_A = \tau_A = 8.20$ $\alpha_M = \tau_M = 3.34$ $\alpha_S = \tau_S = 0.15$	9.62%
6	A_{ij} replaced by relatedness R_{ij} (lineage distance) $\alpha_A = \tau_A = 2255.97$ $\alpha_M = \tau_M = 3.34$ $\alpha_S = \tau_S = 0.15$	11.76%
7	$A_{ij} = 1$, $\forall i, j$ $\alpha_A = \tau_A = 139.60$ $\alpha_M = \tau_M = 3.34$ $\alpha_S = \tau_S = 0.15$	11.60%

11.4.8. *Summary*

The summary of the experiments results are presented in Table 11.1. These show that the outcome of the optimization problems are basically driven by the information we provide *a priori*, i.e. the position of the sensor and muscle neurons, which determine the main anchorage of neurons' connections, with very low impact of the neuron-neuron connectivity in the process. We also present a persuasive and forceful comparison of the resulting layouts of the experiments at Fig. 11.10. Note that the kernel of the prediction success is in the ventral cord neuron group (G7) in the central area of the worm, which is not surprising given the biological nature of these neurons, see [10]. The ventral cord is spatially coherent; neurons, particularly interneurons running in the cord, maintain their positions relative to their nearest neighbors in spite of local distortions produced by intrusions of cell bodies,

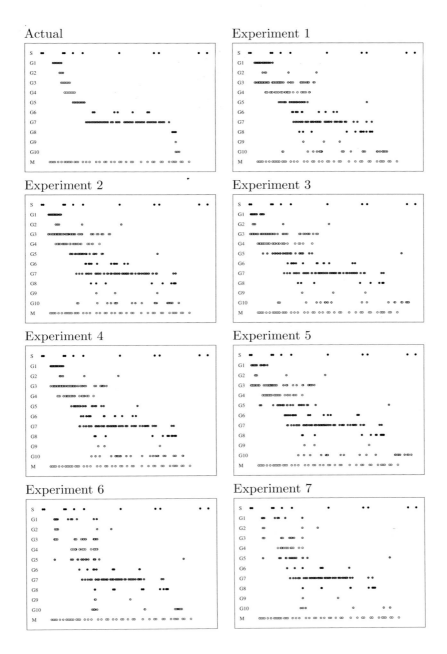

Fig. 11.10.   Comparison of the *C. elegans* real neuronal layout and predictions of the experiments described in the text.

which means that any optimization procedure where the motor neurons are fixed will provide, given the motor-neuron connectivity, a good balance in the positioning of neurons of G7. This is pretty uninformative about the actual position of the rest of neurons but eventually gives a bound for the error that dominates the out-coming results. In our opinion, these findings arise important concerns about the method to support the wiring economy principle in neuronal networks, and encourage the scientific community to find new approaches to reveal the mechanisms governing the biological topology of neuronal networks.

References

[1] Ramón y Cajal, S. (1899). Textura del Sistema Nervioso del Hombre y de los Vertebrados. Translation: Texture of the Nervous System of Man and the Vertebrates. Vol. 1. 1899, New-York: Springer (1999) 631.

[2] Chklovskii, D. B. (2004). Exact Solution for the Optimal Neuronal Layout Problem, *Neural Computation* **16**(10), pp. 2067–2078.

[3] Chen, B. L., Hall, D. H. and Chklovskii, D. B. (2006). Wiring optimization can relate neuronal structure and function, *Proc. Natl. Acad. Sci. USA* **103**, pp. 4723–4728.

[4] Chen, B. L. and Chklovskii, D. B. (2006). Placement and routing optimization in the brain, *Proc. Int. Symposium on Physical Design (ISPD'06)* (San José, California, USA), pp. 136–141.

[5] Wood, W.B., ed. (1998). The nematode Caenorhabditis elegans, Cold Spring Harbor Laboratory: Cold Spring Harbor, New York. 25.

[6] Schafer, W.R. (2005). Deciphering the Neural and Molecular Mechanisms of C. elegans Behavior, *Current Biology* **15**, pp. R723–R729.

[7] http://www.wormatlas.org/handbook/nshandbook.htm/nswiring.htm

[8] Albertson, D.G. and Thomson, J.N. (1976). The Pharynx of Caenorhabditis elegans, *Phil. Trans. R. Soc. London, series B* **275**, pp. 299–325.

[9] White, J.G., Southgate, E., Thompson, J.N., Brenner, S. (1986). The structure of the nervous system of the nematode caenorhabditis elegans, *Phil. Trans. Royal Soc. London. Series B* **314**, pp. 1–340.

[10] White, J.G., Southgate, E., Thompson, J.N., Brenner, S. (1976). The structure of the ventral nerve cord of the nematode caenorhabditis elegans, *Phil. Trans. Royal Soc. London. Series B* **275**, pp. 327–348.

[11] Durbin, R.M. (1987). Studies on the Development and Organisation of the Nervous System of Caenorhabditis elegans. PhD Thesis, University of Cambridge.

[12] Achacoso, T.B., Yamamoto, W.S. (1992). AY's Neuroanatomy of C. elegans for Computation. CRC Press, Boca Raton.

Chapter 12

Cultured Neuronal Networks Express Complex Patterns of Activity and Morphological Memory

Nadav Raichman[1], Liel Rubinsky[2], Mark Shein[1,3], Itay Baruchi[1], Vladislav Volman[4,5] and Eshel Ben-Jacob[1,4]

[1]*School of Physics and Astronomy, Raymond & Beverly Sackler Faculty of Exact Sciences, Tel-Aviv University, Tel-Aviv 69978, Israel*
[2]*Department of Anesthesia and Perioperative Medicine University of California at San Francisco, San Francisco, CA 94143-0542, USA*
[3]*Department of Physical Electronics, School of Electrical Engineering Tel Aviv University, Tel Aviv 69978, Israel*
[4]*Center for Theoretical and Biological Physics, University of California San Diego,La Jolla, California 92093*
[5]*Computational Neurobiology Laboratory, The Salk Institute for Biological Studies, La Jolla, California*

The mammalian brain is perhaps the most complex biological network in nature. It is an intricate system composed from billions of interconnected neuron and glia cells, each with sophisticated intracellular mechanisms and rich intercellular communication dynamics (Levitan, 1991; Nicholls *et al.*, 2001; Kandel *et al.*, 1995). Understanding the mechanisms underlying neuronal network activity is essential for deeper understanding of higher brain functions such as memory and learning (Yuste *et al.*, 2005).

12.1. Cultured Neuronal Networks

Cultured neuronal networks cultivated on micro-electrode arrays (MEAs) offer a well controlled setup for this purpose (Maeda *et al.*, 1995; Marom and Shahaf 2002; Segev *et al.* 2002; Raichman *et al.* 2006; Wagenaar *et al.* 2006; Rubinsky *et al.* 2007; Shein *et al.*, in press). These cultures are widely used as a tool for non-invasive investigation of cellular and network neuronal systems, as these simplified "brain" models show a rich neuronal dynamics of bioelectrical activity and cell communication. The cultures are simple to handle and to design in various formats, and by using a MEA setup, simultaneous parallel recordings of many individual neurons, both under normal conditions and undergoing chemical and electrical manipulations, are enabled.

In the studies presented here, cortical cells that are extracted from rat embryos are dissociated and uniformly spread over a planar glass substrate combined with a grid of 60 recording sites (electrodes). The cells are maintained for days and weeks in a life-supporting environment with proper biochemical conditions, ensuring functionality over long periods of time. During the time of incubation the individual cells grow extensions (neurite growth) and establish connectivity (synapse formation) between cells. After 10-20 days in vitro (DIV) the cell assembly evolves into a mature planar neuronal network with functional synaptic connections and a unique pattern of bioelectrical activity.

12.2. Recording the Network Activity

Using a set of amplifiers and appropriate software enables the recording and saving of the electrical signals generated by the cells. Single action potentials (spikes), which are the basic form of neuronal signal transmission, are characterized by higher amplitudes in respect to noise and are detected using a predefined threshold. The relatively large size of electrodes (30 µm in diameter, as compared to the 10 µm diameter of neuronal cell body) allows the recording of several cells from the same electrode. Therefore, further software tools are applied in order to classify the recorded spikes from into clusters of similar waveforms, leading to the association of each detected spike with a unique neuron (a process known as spike-sorting, see Refs. Hulata *et al.*, 2000, 2002 for details). The resulting data is comprised of the relative timing of action-potential occurrences (spike time-series) of all single neurons. With a common system of 60 electrodes per plate, it is possible to reliably monitor the simultaneous activity of several tens of cells.

In Fig. 12.1 we show a microscope image of a network grown over a micro-electrode array. The two-dimensional network lies over a glass plate placed in the bottom of a well filled with a life-supporting liquid medium (Fig. 12.1A). Figure 12.1B shows the entire recording area, covering 60 electrodes (distance between neighboring electrodes is 500 µm). The area in view, of around 30 mm^2, covers a small section of the entire network, which is 2 cm in diameter and includes 2 million cells. Figure 12.1C is a closer view showing the network densely surrounding a single electrode.

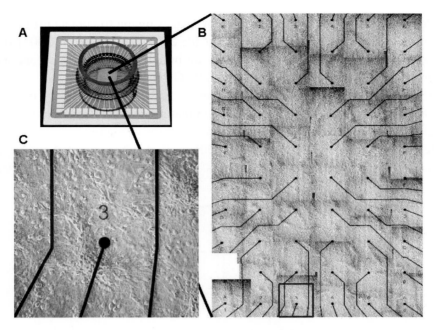

Fig. 12.1. Neuronal network on an electrode array. (A) A micro-electrode array on which a cultured network is cultivated. The network is kept in a liquid medium inside the central well, where there are 60 electrodes uncovered on the upper surface. Conductive leads connect the electrodes to external pads on which amplifiers transfer the signals into a computer. Plate size is 4 cm in width and the well diameter is 2 cm. (B) A photo montage of the central part of a cultured network cultivated on a micro-electrode array with a rectangular 6 × 10 design. Distance between electrodes is 500 μm. (C) Zoom-in on one electrode (marked by a blue rectangle in (B)). Electrode diameter is 30 μm.

12.3. Network Engineering

The laboratory process of culture generation enables great flexibility in the patterning of the network. Several parameters may affect the final outcome of network connectivity and cell organization. For example, the physical location of the cells may be controlled by using advanced lithography methods. The plated cells typically do not attach to the glass, and a "gluing" polymer (poly-D-lysine) is first necessary to be evenly spread over the substrate before the placement of the cells. Segev *et al.* (2002) have applied a lithography technique in which the precise design of the polymer location is determined to the scale of micrometers. This method enables the shaping of the culture into different sizes, such as limited 2D or 1D cultures. In Raichman and Ben-Jacob (2008) cultures were designed to take shape of a connected ring of four sub-networks by placing poly(dimethylsiloxane) (PDMS) strips that act as barriers. Other novel methods enable a finer resolution of cell localization such as by micro-contact printing and lithography (Segev *et al.* 2002; Sorkin *et al.* 2006; Wilson *et al.* 2007) or by synthesizing carbon nanotube islands that are both cell adhesive and conductive so signals can be recorded (Gabay *et al.* 2005; Gabay *et al.* 2007).

Another important factor in determining the network's morphology is the cell density. In Segev *et al.* (2003) it was shown that networks with different number of cell per unit area show a transition from uniform spreading of the cells to a clustered network where neurons join together into dense ensembles. In Fig. 12.2 we show an example of clustered networks, in which grouping of cells are clearly

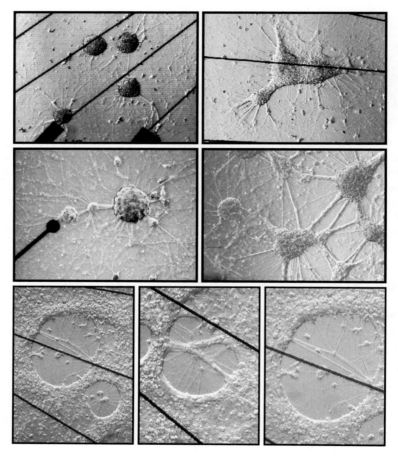

Fig. 12.2. Inverted microscope images of cultures that exhibited dense clusters of cells. The cultures grow clear extensions of interconnecting neurite bundles. In the bottom images are cases of "holes" in the culture. Note that in all of the images the extensions are completely straight, with no curves, due to the tension forces working on both ends of each bundle.

seen. Tension forces hold tightly the neuron extensions between clusters. As is also seen in the bottom of Fig. 12.2, in some cases the tension between cells may lead to "holes" within the network connectivity.

The topology of the neuronal network had been the focus of many studies. It has been mentioned that connectivity between neurons in culture takes the form of a "small-world" network, consisting long-range connections that make short-cuts between distant regions of the culture (Netoff *et al.* 2004). Highly clustered networks, as the ones shown in Fig. 12.2, can take the form of a scale-free network that includes hubs of dense connections. As was previously mentioned, the scale-free parameter that defines small-world connectivity may be strongly related to the density of the plated cells (Segev *et al.* 2003). The main causes for a culture to collapse into a clustered network are the strong tension forces along the axons, which grow as the number of cells increases, compared to the adhesive forces to the surface. However, recent work by Breskin *et al.* (2006), in which electrical stimulation was used in order to engage network bio-electrical activity, had shown, by applying percolation theory, that the scale-free parameter found is closer to a Gaussian distribution of connectivity.

12.4. The Formation of Synchronized Bursting Events

Synchronization is a fundamental phenomenon that is commonly observed in network dynamics and takes a special form in cultured networks (Baruchi *et al.*, in press). In fact, the basic mode of network activity, appearing as early as in day 10 in culture, is composed of synchronized eruptions of neuronal firing. Traditionally, this activity mode is termed as a "synchronized bursting event" (SBE) (Maeda *et al.* 1995), though other titles such as "neuronal avalanches" (Beggs and Plenz 2004), "network spikes" (Eytan and Marom 2006), "network bursts" (van Pelt *et al.* 2004) or "giant depolarizing potentials" (Menendez de la Prida *et al.* 1998) are also used. Figure 12.3 shows network activity recorded over 24 hours. The synchronized basic nature of network firing is clearly visible as simultaneous events that involve the activation of most of the neurons. In general, SBEs are a momentary outbreak of most of the neurons in the network, usually lasting for several hundreds of milliseconds. As is seen in Fig. 12.3, in between each event the network remains relatively silent for periods lasting several seconds.

The occurrence of SBEs in the network activity is the major trademark of cultured networks, and therefore requires a precise definition and a proper measure to characterize it. A common definition for a SBE is a limited time

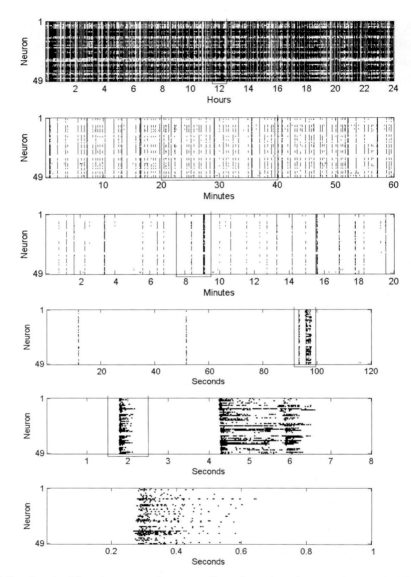

Fig. 12.3. Synchronized bursting events in a recording of a cultured neuronal network. The top image shows the activity of 49 neurons recorded over 24 hours by a micro-electrode array. Each row indicates the spikes (black marks) of each of the neurons. Lower images are zoom-ins into smaller time windows, where each image is an enlargement of a smaller section from its above image, marked by a rectangle. Synchronized bursting events (SBEs) are distinguished as vertical lines of intense activity of most of the recorded neurons. The bottom image is of a 1 second window showing a single SBE.

firing event that includes at least 90% of the recorded neurons (Maeda *et al.* 1995; Marom and Shahaf 2002; Segev *et al.* 2002; Mukai *et al.* 2003). However, to our view, an important aspect of a SBE is not the relative number of activated neurons, i.e. the scale on which the collective firing takes place, but the fact that the SBE represents an initiated *activation* of a sub-network in the culture (assumed by a localized initiation zone) followed by a *propagation* of the activity from the initiating (leading) sub-network towards the rest of the network. These two terms, activation and propagation, lead together to the phenomenon of synchronization of cells when the activity cascades along the network map of connectivity (Raichman and Ben-Jacob 2008). By our definition a SBE is any synchronized event in the network that is a result of an activated sub-network and not a random occurrence of nearly-timed spikes. We also do not limit the size of SBEs to include only a large percentage of the recorded neurons, which enables us to analyze synchronized events of just a few neurons as well.

12.5. The Characterization of the SBEs

The dynamical state of a network at a certain time window can be measured by several key parameters, such as the rate of SBE generation, fraction of active neurons and SBE intensity. In Raichman and Ben-Jacob (2008) we had shown that the intensity of a SBE, either measured by the number of active neurons or the SBE duration, is positively correlated to the time passed from the preceding event. Thus, the intensity of the SBE reflects the network "readiness" to fire after some time integral of silenced activity. Other two important parameters are the activation time τ_{act}, which is the time it takes to trigger the neurons (from the first firing neuron to the initiation of the last one neuron), and the spike-train duration time τ_{st}, which is the average duration of the spike-trains of the firing neurons in the burst. The activation time reflects the time it takes to activate the entire network from end-to-end of the recording area, disregarding the intensity of firing of the individual neurons, and the spike-train duration time reflects the intensity of firing of the neurons, disregarding the temporal order in which neurons were triggered. The exact method of calculating τ_{act} and τ_{st} is described in Raichamn and Ben-Jacob (2008).

Early developing networks show dramatic changes in several key parameters, reflecting their ongoing maturation and increase in connectivity. Developing networks begin to exhibit spiking activity at around 3-5 DIV. Synchronized bursting with an increasing number of neurons shortly follows the activation of the network. In Fig. 12.4 we show the changes in the SBE frequency and intensity in an immature developing culture, as reflected by several key parameters. As is shown in the figure, there is a monotonic, nearly linear, increase in the rate of SBE. As the network matures, it eventually reaches a steady state of a constant average rate. In addition, there is a gradual increase in the intensity of SBEs, as reflected by all three measures of SBE intensity: number of neurons in SBE, duration of activation time τ_{act} and spike-train duration τ_{st}, which increase monotonically with the culture age.

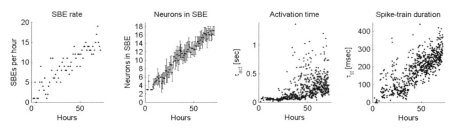

Fig. 12.4. Increase of SBE intensity in an immature developing network. The plots show the SBE firing rate and three measures of SBE intensity for a developing network. This network began to show activity at around 3-4 DIV, and the first synchronized events (within a window of 100 msec) of at least 3 neurons at 5 DIV. The plot on the far left is of the firing rate of SBE following 3 days from the first appearing SBE. The rest of the plots show the measures of SBE intensity for the same time period. The measures are (from left to right): number of neurons in SBE, duration of activation time (τ_{act}) and spike-train duration (τ_{st}).

After few more days the network reaches a mature state, in which the activity shows complex patterns of SBE firing. In Segev *et al.* (2002) it was shown that a mature network, monitored over long periods of time (days and weeks), will show non trivial statistics of SBE rate (measured by inter-burst-interval, *IBI*) and its derivatives. In their article, Segev *et al.* (2002) show that the derivative of the *IBI*s follow a Lévy distribution (Bertoin, 1996) which is characterized by the distribution's tail that decreases on many time scales, and the *IBI*s power spectrum covers a wide range of frequencies. This latter result hints to a complex dynamics of the network and the ability of the culture to serve as a template for long and short term memory. In Wagenaar *et al.* (2006) it was also shown that a network never really reaches a steady state behavior with stationary statistics.

The above experimental observations of scale-free collective activity can be explained within the framework of generative network modeling, as detailed in Volman *et al.* (2004). Bursting events in these model networks emerge as some spikes are spontaneously generated by background noise current, and spread to other neurons through synapses that adjust their strength in an activity-related way. As soon as inhibitory neurons are recruited, the collective activity is terminated. It was shown that the Levy distribution in individual neuronal intraburst activities is related to the nonlinear dynamical properties of neuronal membrane and to the time-dependent coupling between neurons. On the other hand, in order to obtain scale-free statistics of synchronized bursting events timing, the background current has to be rendered with temporal correlation. It was further suggested in Volman *et al.* (2004) that such currents can, at least in part, represent the signaling from the network of glial cells that surrounds the neurons.

12.6. Highly-Active Neurons

In contrast to the ongoing amplification in the activity of individual neurons during network development, no dramatic changes are observed in their synchronization to the network. By synchronization we mean the tendency of a neuron to fire spikes solely during a SBE, where a non-synchronized neuron will fire spikes during the network silence periods (between SBEs) as well (See definition and measure of synchronization in Shein *et al.*, in press). In Fig. 12.5A we show a network activity, where most neurons are synchronized with the network SBEs, while two neurons, marked by a "*" symbol on the left axis, are unsynchronized. In Shein *et al.* (in press) we name these neurons as highly-active (HA) neurons.

Most neurons are synchronized to the network from the onset of their activity. They do not generate spikes endogenously, and their activity is always network-dependent. In this sense, the synchronized firing is an innate property of most neurons in the network. Despite the above, a small subset of HA neurons exhibit an unsynchronized activity pattern between SBEs, in addition to firing during SBEs. In Shein *et al.* (in press) it was shown that these neurons were the first active neurons in the network during development, and that as synchronized activity emerged, they were precursors of SBEs. This is also shown in Fig. 12.5B, where a higher temporal resolution on one SBE shows that the two HA neurons identified in Fig. 12.5A initiate spike-trains before the rest of the neurons. Statistics over 3 different cultures is shown in Fig. 12.5C show consistency. It was suggested that the HA neurons are involved in the regulation of activity in the network (Volman *et al.* 2004; Shein *et al.*, in press).

12.7. Function—Form Relations in Cultured Networks

Closer inspections of the SBEs, such as the one shown in Fig. 12.3, reveal that each network burst could be described as a momentary spatio-temporal image in

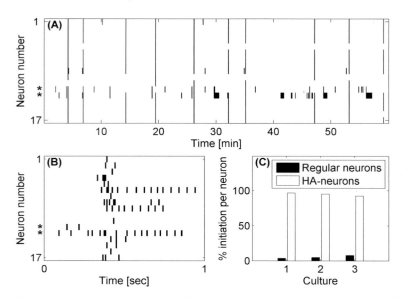

Fig. 12.5. Highly-active neurons in culture. (A) Network activity showing SBEs. Two HA neurons are identified by "*" symbols on the left axis. These neurons fire a substantial percentage of their activity between SBEs. See Ref. Shein *et al.* (2008) for specific details and measures of HA neurons. (B) A closer view of one SBE showing that the same two neurons from (A) are precursors to the rest of the neurons and are the first to initiate their spike-train firing. (C) Statistics over 3 different cultures show that in most cases, the first neurons to fire during a SBE are HA type. Figure taken with permission from Shein *et al.*, in press.

which neurons at different locations fire spike-trains at different delays relative to each other. A single culture will usually show a few different SBEs, each with its own characteristic pattern of neuronal firing time ordering (activity propagation) and neuron temporal correlations. Nevertheless, in many of the cultures it is possible to detect SBEs that share a similar propagation profile between the recording sites, i.e. a repeating motif. For example, Fig. 12.6 shows four different spatio-temporal patterns of SBE propagation in one single culture. This culture was divided into four sub-networks connected in a ring-like shape. The figure clearly shows that the activity in each mode of SBE begins in one region, propagates to the neighboring regions and ends in the region at the opposite side of the initiation cluster. Each image in Fig. 12.6 is composed from several tens of SBEs, showing a clear repetition of the pattern for each mode.

Several works suggested explanations for the burst repetition in propagation profile, such as network morphology of substructures of cell clusters (Gross and Kowalski, 1999), existence of endogenously active neurons (Latham *et al.*,

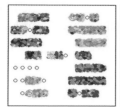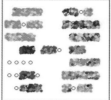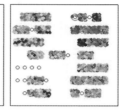

Fig. 12.6. Four motifs of spatio-temporal propagation of activity in a clustered network. The four figures show the spread of activity during SBEs taken from a single culture that had architecture of four interconnected sub-networks. The SBEs were clustered into groups of similar propagation motifs, and for each SBE the neurons were color-coded according to their order of firing (blue – first firing neuron, red – last neuron). The plots show the overlapped neuron order of firing of all SBEs belonging to each motif. Figure taken with permission from Baruchi *et al.*, in press.

2000a; Latham *et al.*, 2000b; Shein *et al.*, in press) or neurons with an enhanced sensitivity (Eytan and Marom, 2006; Shein *et al.*, in press), interplay between inhibition and excitation activity in the network (Streit *et al.*, 2001), or the presence of localized initiation zones characterized by high neuronal density and by recurrent and inhibitory network connections (Feinerman *et al.*, 2007). The fact that cultures spontaneously generate repeating motifs is of significant importance as it proves that neuronal networks in-vitro are capable of maintaining long-term memory. In a recent paper it was shown that new motifs can also be imprinted into the culture by localized chemical stimulation (Baruchi and Ben-Jacob, 2007). The number of motifs and how they correlate among themselves may shed light on the morphology and architecture of the culture. This may become apparent through the relationship between excitatory and inhibitory neurons (Volman *et al.* 2005), the type of connectivity within the culture, or the geometry of the culture sections (Blinder *et al.*, 2005; Raichman *et al.*, 2006).

One modeling approach to study the emergence of different SBEs motifs utilizes the "function follows form" principle, based on the realization that different network architectures might support different types of dynamics. Consequently, it was shown in Volman *et al.* (2005) how synchronized bursting events with different internal spatio-temporal organization arise in a network with non-homogeneous connectivity. A network composed of two homogeneous sub-networks that had a small overlap, supported two types of SBEs, each one corresponding to the activity initiating in one sub-networks and spreading, through the overlap region, to the other one. Moreover, the temporal appearance of different SBEs could be controlled by imposing regulatory constraints on the background noise current. The predictions of the model were later tested in experimental system of two coupled cultured networks (Baruchi *et al.*, in press).

12.8. Analyzing the SBEs Motifs

Identification of clusters of repeating burst motifs can serve as an important tool for investigating the inter-neuronal relations in the neuronal network (Baruchi *et al.*, in press). Successful clustering of burst motifs based on different measures of burst similarity has been previously achieved. For example, in the works by Mukai *et al.* (2003) and Madhavan *et al.* (2005) SBEs were clustered according to the firing intensity of individual neurons, with disregards to the temporal delays between neurons. We have developed a novel measure of SBE similarity that inspects the changes in the temporal delays between neuron activation in the SBE (Raichman and Ben-Jacob, 2008). By applying a standard clustering algorithm, one can determine the number of observed motifs in a culture.

In our method, we reduce the image of the SBE motif to include only the first spike of each neuron, and thus capture its initiation profile. Our assumption to include only the first spike of the neuronal spike-train is in accordance with results showing that spike timing is more accurate in the beginning of the spike-train, both in spontaneous firing and in bursts generated as a response to electric stimuli (Jimbo and Robinson, 2000; Bonifazi *et al.* 2005; Luczak *et al.* 2007).

In order to identify sets of SBE motifs of neuronal activation, we looked at the first spike of each neuron during the activation time τ_{act} for each burst (as previously defined). We represented the activation of each SBE by an activation matrix, A, where $A(i,j)$ is the delay in milliseconds between the first spike of neuron i and the first spike of neuron j in the SBE. Neurons that do not fire in the SBE received a NULL value in the activation matrix. Obviously, the matrix is anti-symmetric: $A(i,j) = -A(j,i)$ and $A(i,i) = 0$.

We then defined the similarity, $S(A_p,A_q)$, between SBEs p and q as:

$$S\left(A_p, A_q\right) = \frac{1}{n(n-1)} \sum_{i \neq j} H\left(th - \left|A_p\left(i,j\right) - A_q\left(i,j\right)\right|\right)$$

where H is the Heaviside step function ($H(x) = 0$ if $x < 0$ and $H(x) = 1$ if $x \geq 0$) and th is a time-threshold parameter. According to this formula, $S(A_p,A_q)$ is the fraction of neuron pairs (i,j) that obey the condition that the accuracy in delays between the two bursts is less than th: $|A_p(i,j) - A_q(i,j)| < th$. The summation is made only on neuron pairs that did not receive NULL values in any of the two SBEs (i.e. both neurons fired a spike in both SBEs). In our analysis we set *th* = 30 msec to reflect the average spike precision in bursts, as was shown in Harris *et al.* (2003) and Bonifazi *et al.* (2005). We also tested other values of *th*, proving the robustness of our classification method to this parameter where the results are shown in Raichman and Ben-Jacob (2008).

The computed similarity matrix contains the complete data on the SBE-to-SBE relations. However, in order to detect and define the number *M* of well-defined sets of SBEs with similar activation pattern, we need to develop a measure that evaluates the goodness of each set of SBE motif. Such a proposed method is described in length in Raichman and Ben-Jacon (2008), introducing a simple procedure to detect centers of SBE clusters with high similarity, by applying a two-stage method that uses a hierarchical clustering algorithm followed by an iterative search for independent cluster centers. By re-ordering the similarity matrix, one can visualize the clustered data, where similar SBEs are grouped into square areas along the matrix diagonal with high similarity values.

12.9 Network Repertoire

In Fig. 12.7 we show a clustering process of one culture. As is shown in the figure, the similarity matrix of the examined culture breaks into five distinctive clusters of SBEs with high similarity of activation profile (Fig. 12.7A and B). The different motif sets are marked on the right of the reordered similarity matrix (Fig. 12.7B). In this example we have chosen a set size of 20 SBEs, from a sample of 300 consecutive SBEs. In Fig. 12.7C plotted are the similarity value between every SBE that belongs to one of the five sets and the mean similarity matrix that represents each of the sets. The values show that inside the clusters the SBEs are highly similar to their representative mean activation matrices, and are not similar to any of the other sets. Figure 12.7D shows the raster of the activation pattern (first spikes only of each neuron) for each of the identified motifs, where each motif has a different propagation profile across the culture. Figure 12.7E plots the appearance of each motif along time. The order of appearance of the different motifs is rather random and almost uniform. Thus, there is no dominant motif that takes over the network in this specific culture.

We have applied our motif detecting method on different cultures with different spatial morphologies (Raichman and Ben-Jacob, 2008). In large uniform cultures, 50% showed to have at least two distinguished motifs. Motifs were also detected in smaller cultures and in clustered ones, where the clustered cultures showed an increased number of motifs, proportional to the number of clusters.

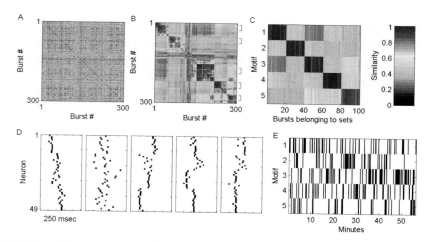

Fig. 12.7. Identifying clusters of SBEs with similar activation patterns. (A) Similarity matrix between 300 consecutive bursts in a culture with an architecture of four interconnected clusters. The resulting matrix shows no special features that depend on the temporal ordering of the SBEs. The color-coding for (A), (B), and (C) is according to the scale on the far-right. (B) The same matrix shown in (A), after re-ordering of the rows and columns according to a dendrogram algorithm. The resulting new matrix reveals a clear division of the data into blocks of square areas along the diagonal containing high similarity values, separated by lower values between blocks. Five independent motifs were identified in the matrix, marked by leaning "U" symbols on the right of the image. (C) Similarity values between 20 bursts of each motif and the mean representation of each motif. The matrix clearly shows the high similarity within each motif, relative to the low values of similarity between motifs. (D) Raster plot of the activation of each mean SBE representing each of the 5 motifs identified in (B). The raster plots show only the mean relative timing of the first spike of each neuron. Neurons are grouped according to their physical location in the 4 clusters of the network architecture. (E) Raster plot of the SBEs according to the time of appearance in the recording, and sorted according to the identified motifs. Each bar marks the timing of each burst (bin=10 min.). Each of the 300 bursts was associated with one of the 5 motifs identified in (B) according to its maximal similarity. According to the image, it appears that the expression of the motifs is in a random, nearly uniform, temporal manner. Figure taken with permission from Raichman and Ben-Jacob, 2008.

12.10. Network under Hypothermia

The fact that cultured networks exhibit repeating motifs may help us in monitoring changes taking place within the network. For example, the number of motifs is related to the large scale structure, and an elimination of a motif might hint at some process of deterioration. Another factor is the activation path of each motif, which may be affected as well by external or internal processes. As an example of using network motifs as a tool for monitoring network reaction to stimuli, we conducted several experiments in which hypothermia was applied on the culture. Hypothermia of the brain is commonly used to reduce metabolism for the protection of the brain immediately after trauma, during surgical procedures and in recovery of patients (Sessler, 2001; Varathan *et al.* 2001; McIntyre *et al.* 2003). We studied the effect of hypothermia on network dynamics, by observing the apparent changes in the network activity during and following application of cold (Rubinsky *et al.* 2007; Rubinsky *et al.*, in press).

In our experiments, we exposed cultured neuronal networks to deep hypothermia by pumping chilled water through channels in the walls of the life-supporting chamber. This effectively reduced the temperature of the culture to 19°C. The cultures were maintained in this hypothermic state for 20 hours. Following 20 hours of deep hypothermic conditions the cooling system was shut down and the temperature was returned to 37°C while continuously recording for additional 5 hours. This was repeated on 3 separate cultures.

The most pronounced and immediate observation was the drastic decrease in the SBE rate (Fig. 12.8, top) seen within a few minutes after application of hypothermia. This low level of activity remained constant throughout the whole duration of hypothermia application. Moreover, as is shown in Rubinsky *et al.* (2007), during hypothermia there was a gradual decrease in burst width and burst intensity.

Immediately following hypothermia the network exhibited a dramatic increase in activity, with a temporary overshoot reaching 3 times the SBE rate compared to the activity level prior to hypothermia. This overshoot lasted for over an hour and was also accompanied by a slight elevation in the averaged inter-neuron correlation, which afterwards gradually returned to initial levels. This overshoot could be explained by an increase in the network excitability, as compensation for the strong inhibition of the network activity during hypothermia. Approximately two hours after termination of hypothermia application the network activity gradually returned to its pre-hypothermia SBE rate and activity level.

Homeostasis of the network activity before and after hypothermia was also seen in the classification of the SBEs into motifs. In this example the measure of similarity between SBEs was based on correlation analysis, as explained in Segev *et al.* (2004) and Rubinsky *et al.* (2007). In the bottom of Fig. 12.8 we see the clustered SBE similarity matrix prior to hypothermic exposure in one of the cultures. The resulting clustered matrices exhibit a clear block organization of SBEs with high similarity within the blocks and low levels in between blocks, where each block represents a separate motif of spatio-temporal pattern of propagation. In the figure we detect three such motifs prior to hypothermia application. These three motifs are quite evident when seen in the plots of the

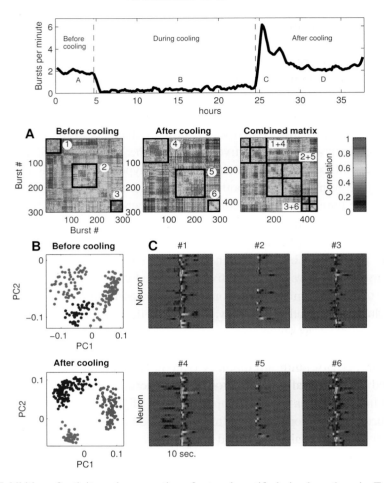

Fig. 12.8. Inhibition of activity and preservation of network motifs during hypothermia. Top: A plot of the burst rate of the network before, during and after hypothermia stimulation. Dashed lines mark the onset (left) and offset (right) of stimuli. Bottom: Classification of SBEs from before and after hypothermia. (A) Inter-SBEs correlation matrices where calculated for two data sets consisting of 300 SBEs taken from before hypothermia and 300 SBEs taken from 5 h after hypothermia. Clustering analysis was then applied over the two correlation matrices. The resulting clustered matrices show a clear separation of the data into three sub-groups that contain high values of correlation. Areas of the matrix in between the three sub-groups have lower values of correlation. The third matrix shows the inter-SBEs correlation values as calculated from the combined data of the six sub-groups identified in the previous matrices. Sub-groups #1 and #4 and sub-groups #2 and #5 share high values of correlation. (B) PCA plots of the three sub-groups identified in each of the two clustered matrices. Distinction between SBEs of different sub-groups is color-coded. After hypothermia the separation between the sub-classes becomes more pronounced. (C) Raster plots of the average neuronal activity in each SBE sub-group shown in (A). Red/blue colors represent high/low spike rate. Figures taken with permission from Rubinsky et al. 2007.

two first principal components (PCA) of the correlation matrix, emphasizing the separation between the different groups. Following hypothermia we again see the same three motifs of SBEs with the same spatio-temporal pattern. Since the spatio-temporal patterns of SBEs are strongly related to the network connectivity, this last result may indicate that the neuronal connections did not change significantly during the 20 hours of hypothermia. We also observed that the separations between the SBE sub-classes in the PCA plots became more pronounced after hypothermia application, with a narrower statistical distribution of each cluster around its mean. Similar findings were observed in the two other cultures.

All the above results clearly demonstrate that the network survived hypothermia with little if any detectable changes to the underlying neural network connectivity, though there was evidence of some cell death. The strong inhibition of our network activity during hypothermia may support and may help explain a recent experiment in which cooling was used to arrest epileptic seizures in both human and animal experiments (Yang *et al.* 2003). This cooling of the brain might have reduced the neuronal activity while leaving the general connectivity intact, as seen in our model.

In addition this model may be of use in the study of the effect of hypothermia on memory. Our findings show that the network maintained its spatio-temporal patterns before and after hypothermia. Each spatio-temporal pattern could be regarded as representing a different active "state" of the network. The maintenance of these patterns suggests the possibility that memory in neuronal networks may be retained even following strong activity inhibition (Rubinsky *et al.*, in press).

12.11. Summary

We have shown here analysis results on the observed activity in homogeneous and structured cultured neuronal networks, with emphasis on the investigation of the characteristics modes of network activity and on the expression of repeating motifs and their relation to the network structure. It is evident that cultured networks exhibit a rich spontaneous dynamical activity. The networks show complex patterns in all aspects: In the developmental process from a collection of isolated cells to a fully connected and functional network; In the spontaneous emergence of bio-electrical activity; In the formation and structure of the network topology; In the long-term characteristics of the network activity under "steady state" conditions; And in the appearance of synchronized bursts.

The SBEs, as a trade mark of cultured networks activity, reflect the diversity and complexity of the neuronal network. In most cases SBEs are a collective phenomenon that involves the participation of almost all of the recorded neurons and it is reasonable to assume that the whole network is activated. The internal and spontaneous initiation of SBEs may be regarded as "natural" stimulation of the network, as opposed to artificial stimulation such as by electrical pulses. When there are multiple motifs of SBEs, such as often observed in large or dissected cultures, each SBE motif may be treated as an outcome of a different natural stimulation.

As was noted, most of the recorded cultures exhibited at least two sets of propagation motifs. Patterned networks with large distances between culture sections distinctively exhibit more burst motifs, proportional to the number of sub-sections (Raichman and Ben-Jacob, 2008; Baruchi et al. In press). One clear result of our analysis is the understanding of the contribution of spatial distance between electrodes and the attribution of SBEs to different motifs, where clustered networks significantly exhibit more motifs.

Previous works have shown that mature cultured networks exhibit a rich repertoire of bursting patterns, where bursts vary largely in duration, intensity and in the number of participating neurons (Kamioka et al. 1996; Mukai et al. 2003; Wagenaar et al. 2006). Large variations in network activity may also be achieved by stimulation of the network (Gross et al. 1997; Jimbo et al. 1998; Jimbo et al. 1999; Bi and Poo, 1999; Eytan et al. 2003; Bonifazi et al. 2005; Chiappalone et al. 2007). This diverse form of activity may indicate that the networks fire bursts repeatedly in different states, which are associated with the different relaxation times and amount of accumulated neurotransmitters between bursts.

We have also applied cold on our cultures in order to investigate methods for neuro-protection. Low temperatures are a common tool used to protect the brain during traumatic injury, patient recovery and surgical procedures. However, it is known that some damage is caused to the cells under these procedures. Our cultured networks offer a well controlled system for testing the effect of cold.

Our experiments, in which cold was applied to cultures for several hours, show major and important results. The experiments clearly demonstrate that the network survived hypothermia with little if any detectable changes to the underlying neural network connectivity, exhibited by the protection of SBE motifs by number and by structure. The same motifs that appeared before the application of cold were also seen after the network stimulation. The strong inhibition of our network activity during hypothermia may support and may help to explain recent experiments in which cooling was used to arrest epileptic seizures in both human and animal experiments (Yang *et al.* 2003). This cooling of the brain might have reduced the neuronal activity while leaving the general connectivity intact, as seen in our model.

All of the above findings describe our cultured neuronal network as a complex dynamical biophysical system that has some forms of intrinsic memory, information coding and self-regulation. By applying different forms of manipulations, such as morphology constrains and thermal stimulations, we discovered repeating activity motifs and observed the network's long term adaptation to a changing environment. These kinds of network characteristics fit well to the notion that the brain is composed of such complex living network systems.

Future experiments on cultured networks may continue to reveal higher levels of network plasticity and communication forms. For example, we already mentioned our recent work (Baruchi and Ben-Jacob, 2007) in which we have shown that new "memories" may be imprinted into a given neuronal network. By carefully placing microscopic droplets of chemicals, new regions in the network could be activated to consistently generate new SBE motifs. An interesting continuation to this research will be to determine the maximal number of SBE motifs that may be imprinted into a given network, and how are they related to the network morphology, topology and excitation state.

The observed SBE motifs in our cultures suggest that a random ensemble of neuron cells will not produce a random "jumble-mumble" of signals, but are rather forced into a small number of repeating patterns or attractors. Though the SBE motifs shown in our cultures are not necessarily associated with real brain "memories", they can still hint that the brain may work in specific highly regulated patterns, in which memories are pulled out and stored in controlled processes.

Acknowledgments

We are thankful to Inna Brains for her devoted technical assistance in extraction of the cells and the networks plating and monitoring. This research has been supported in part by the Tauber Foundation and te Maguy-Glass chair in Physics of Complex Systems at Tel Aviv University. Vladislav Volman acknowledges the support of the U.S. National Science Foundation I2CAM International Materials Institute, grant DMR-0645461.

References

Baruchi, I. and Ben-Jacob, E. (2007). Towards neuro-memory-chip: Imprinting multiple memories in cultured neural networks, *Phys. Rev. E* 75, pp. 050901.

Baruchi I., Volman, V., Shein, M., Raichman, N. and Ben-Jacob, E. (in press). The emergence and properties of mutual synchronization in in-vitro coupled cortical networks. *Eur. J. Neurosci.*

Beggs, J. M. and Plenz, D. (2004). Neuronal avalanches are diverse and precise activity patterns that are stable for many hours in cortical slice cultures, *J. Neurosci.* 24(22), pp. 5216-5229.

Bertoin, J. (1996). Levy processes , Cambridge University Press, Cambridge, UK.

Bi, G. and Poo, M. (1999). Distributed synaptic modification in neural networks induced by patterned stimulation, *Nature* 401, pp. 792-796.

Blinder, P., Baruchi, I., Volman, V., Levine, H., Baranes, D. and Ben Jacob, E. (2005). Functional topology classification of biological computing networks, *Neural Computing* 4, pp. 339-361.

Bonifazi, P., Ruaro, E. and Torre, V. (2005). Statistical properties of information processing in neuronal networks, *Eur. J. Neurosci.* 22, pp. 2953-2964.

Breskin, I., Soriano, J., Moses, E. and Tluski, T. (2006). Percolation in living neural networks, *Phys. Rev. Lett.* 97, pp. 188102.

Chiappalone, M., Vato, A., Berdindini, L., Koudelka-hep, M. and Martinoia, S. (2007). Network dynamics and synchronous activity in cultured cortical neurons, *Int. J. Neur. Sys.* 17(2), pp. 87-103.

Eytan, D., Brenner, N. and Marom, S. (2003). Selective adaptation in networks of cortical neurons, *J. Neurosci.* 23(28), pp. 9349-9356.

Eytan, D. and Marom, S. (2006). Dynamics and effective topology underlying synchronization in networks of cortical neurons, *J. Neurosci.* 26(33), pp. 8465-8476.

Feinerman, O., Segal, M. and Moses, E. (2007). Identification and dynamics of spontaneous burst initiation zones in unidimensional neuronal cultures, *J. Neurophysiol.* 97, pp. 2937-2948.

Gabay, T., Ben-David, M., Kalifa, I., Sorkin, R., Abrams, Z. R., Ben-Jacob, E. and Hanein, Y. (2007). Electro-chemical and biological properties of carbon nanotube based multi-electrode arrays, *Nanotechnology* 18, pp. 035201.

Gabay, T., Jakobs, E., Ben-Jacob, E. and Hanein, Y. (2005). Engineered self-organization of neural networks using carbon nanotube clusters, *Physica A* 350, pp. 611-621.

Gross, G. W., Harsch, A., Rhodes, B. K. and Göpel, W. (1997). Odor, drug and toxin analysis with neuronal networks in vitro: extracellular array recording of network responses, *Biosensors and Bioelectronics* 12(5), pp. 373-393.

Gross, G. W. and Kowalski, J. M. (1999). Origins of activity patterns in self-orgaizing neuronal networks in vitro, *J. Intel. Mat. Sys. Str.* 10, pp. 558-564.

Harris, K. D., Csicsvari, J., Hirase, H., Dragoi, G. and Buzsáki, G. (2003). Organization of cell assemblies in the hippocampus, *Nature* 424, pp. 552-556.

Hulata, E., Segev, R. and Ben-Jacob, E. (2002). A method for spike sorting and detection based on wavelet packets and shannon's mutual information, *J. Neurosci. Meth.* 117, pp. 1-12.

Hulata, E., Segev, R., Shapira, Y., Benveniste, M. and Ben-Jacob, E. (2000). Detection and sorting of neural spikes using wavelet packets, *Phys. Rev. Lett.* 85(21), pp. 4637-4640.

Jimbo, Y. and Robinson, H. (2000). Propagation of spontaneous synchronized activity in cortical slice cultures recorded by planer electrode arrays, *Bioelectrochemistry* 51, pp. 107-115.

Jimbo, Y., Robinson, H. P. C. and Kawana, A. (1998). Strengthening of synchronized activity by tetanic stimulation in cortical cultures: applications of planer electrode arrays, *IEEE T. Bio-Med. Eng.* 45(11), pp. 1297-1304.

Jimbo, Y., Tateno, T. and Robinson, H. P. C. (1999). Simultaneous induction of pathway-specific potentiation and depression in networks of cortical neurons, *Biophys. J.* 76, pp. 670-678.

Kandel, E., Schwartz, J. and Jessel, T. (1995). Essentials of Neural Science and Behavior, Appleton & Lange.

Kamioka, H., Maeda, E., Jimbo, Y., Robinson, H. P. C. and Kawana, A. (1996). Spontaneous periodic synchronized bursting during formation of mature patterns of connections in cortical cultures, *Neurosci. Lett.* 206, pp. 109-112.

Latham, P. E., Richmond, B. J., Nelson, P. G. and Nirenberg, S. (2000). Intrinsic dynamics in neuronal networks. i. theory, *J. Neurophyiol.* 83(2), pp. 808-827.

Latham, P. E., Richmond, B. J., Nirenberg, S. and Nelson, P. G. (2000). Intrinsic dynamics in neuronal networks. ii. Experiment, *J. Neurophysiol.* 83(2), pp. 828-835.

Levitan, I. B. (1991). The Neuron: Cell and Molecular Biology, Oxford University Press, NY.

Luczak, A., Barthó, P., Marguet, S. L., Buzsáki, G. and Harris, K. D. (2007). Sequential structure of neocortical spontaneous activity in vivo, *P. Natl. Acad. Sci. U.S.A.* 104(1), pp. 347-352.

Madhavan, R., Chao, Z. C. and Potter, S. M. (2006). Spontaneous bursts are better indicators of tetanus-induced plasticity than responses to probe stimuli, *Proc. 2ⁿᵈ Int. IEEE EMBS conf. on Neural Eng.* pp. V-VIII.

Maeda, E., Robinson, H. P. C. and Kawana, A. (1995). The mechanisms of generation and propagation of synchronized bursting in developing networks of cortical neurons, *J. Neurosci.* 15(10), pp. 6834-6845.

Marom, S. and Shahaf, G. (2002). Development, learning and memory in large random networks of cortical neurons: lessons beyond anatomy, *Quart. Rev. Biophys.* 35(1), pp. 63-87.

McIntyre, A. L., Fergusson, A. D., Hebert, C. P., Moher, D. and Hutchison, S. J. (2003). Prolonged therapeutic hypothermia after traumatic brain injury in adults: A systematic review, *JAMA* 22(289), pp. 2992-2999.

Menendes de la Prida, L., Bolea, S. and Sanchez-Andres, V. (1998). Origin of the synchronized network activity in the rabbit developing hippocampus, *Eur. J. Neurosci.* 10, pp. 899-906.

Mukai, Y., Shina, T. and Jimbo, Y. (2003). Continuous monitoring of developmental activity changes in cultured cortical networks, *Electr. Eng.* 145(4), pp. 28-37.

Netoff, T. I., Clewley, R., Arno, S., Keck, T. and White, J. A. (2004). Epilepsy in small-world networks, *J. Neurosci.* 24(37), pp. 8075-8083.

Nicholls, J. G., Martin, A. R., Wallace, B. G. and Fuchs, P. A. (2001). From Neuron to Brain, Sinauer Association, Inc., Sunderland, Massachusetts USA.

Raichman, N. and Ben-Jacob, E. (2008). Identifying repeating motifs in the activation of synchronized bursts in cultured neuronal networks, *J. Neurosci. Meth.* 170, pp. 96-110.

Raichman, N., Volman, V. and Ben-Jacob, E. (2006). Collective plasticity and individual stability in cultured neuronal networks, *Neurocomputing* 69, pp. 1150-1154.

Rubinsky, L., Raichman, N., Baruchi, I., Shein, M., Lavee, J., Frenk, H. and Ben-Jacob, E. (2007). Study of hypothermia on cultured neuronal networks using multi-electrode arrays, *J. Neurosci. Meth.* 160, pp. 288-293.

Rubinsky, L., Raichman, N., Lavee, J., Frenk, H. and Ben-Jacob, E. (in press). Spatio-temporal motifs 'remembered' in neuronal networks following profound hypothermia, *Neural Networks*.

Segev, R., Baruchi, I., Hulata, E. and Ben-Jacob, E. (2004). Hidden neuronal correlations in cultured networks, *Phys. Rev. Lett.* 92(11), pp. 118102.

Segev, R., Benveniste, M., Shapira, Y. and Ben-Jacob, E. (2003). Formation of electrically active clusterized neural networks, *Phys. Rev. Lett.* 90, pp. 168101.

Segev, R., Benveniste, M., Hulata, E., Cohen, N., Palevski, A., Kapon, E., Shapira, Y. and Ben-Jacob, E. (2002). Long term behavior of lithographically prepared in vitro neuronal networks, *Phys. Rev. Lett.* 88(11), pp. 118102.

Sessler, I. D. (2001). Complications and treatment of mild hypothermia, *Anesthesiology* 95, pp. 531-543.

Shein, M., Volman, V., Raichman, R., Hanein, Y. and Ben-Jacob, E. (in press). Management of synchronized network activity by highly active neurons, *Phys. Biol.*

Sorkin, R., Gabay, T., Blinder, P., Baranes, D., Ben-Jacob, E. and Hanein, Y. (2006). Compact self-wiring in cultured neural networks, *J. Neur. Eng.* 3, pp. 95-101.

Streit, J., Tscherter, A., Heuschkel, M. O. and Renaud, P. (2001). The generation of rhythmic activity in dissociated cultures of rat spinal cord, *Eur. J. Neurosci.* 14, pp. 191-202.

van Pelt, J., Corner, M. A., Wolters, P. S., Rutten, W. L. C. and Ramakers, G. J. A. (2004). Longterm stability and developmental changes in spontaneous network burst firing patterns in dissociated rat cerebral cortex cell cultures on multielectrode arrays, *Neurosci. Lett.* 361, pp. 86-89.

Varathan, S., Shibuta, S., Shimizu, T. and Mashimo, T. (2001). Neuroprotective effect of hypothermia at defined intraischemic time courses in cortical cultures, *J. Neurosci. Res.* 65, pp. 583-590.

Volman, V., Baruchi, I. and Ben-Jacob, E. (2005). Manifestation of function-follow-form in cultured neuronal networks, *Phys. Biol.* 2, pp. 98-110.

Volman, V., Baruchi, I., Persi, E. and Ben-Jacob, E. (2004). Generative modelling of regulated dynamical behavior in cultured neuronal networks, *Physica A* 335, pp. 249-278.

Wagenaar, D. A., Pine, J. and Potter, S. M. (2006). An extremely rich repertoire of bursting patterns during the development of cortical cultures, *BMC Neurosci.* 7, pp. 11.

Wilson, N. R., Ty, M. T., Ingber, D. E., Sur, M. and Liu, G. (2007). Synaptic reorganization in scaled networks of controlled size, *J. Neurosci.* 27(50), pp. 13581-13589.

Yang, X. F., Chang, J. H. and Rothman, S. M. (2003). Long-lasting anticonvulsant effect of focal cooling on experimental neocortical seizures, *Epilepsia* 44, pp. 1500-1505.

Yuste, R., MacLean, J. N., Smith, J. and Lansner, A. (2005). The cortex as a central pattern generator, *Nat. Rev. Neurosci.* 6, pp. 477-483.

Chapter 13

Synchrony and Precise Timing in Complex Neural Networks

Raoul-Martin Memmesheimer[1,2,3] and Marc Timme[1,2]

[1] *Network Dynamics Group,*
Max Planck Institute for Dynamics and Self-Organization (MPI DS)
Bunsenstr. 10, 37073 Göttingen, Germany
[2] *Bernstein Center for Computational Neuroscience (BCCN) Göttingen*
Bunsenstr. 10, 37073 Göttingen, Germany
[3] *Center for Brain Science, Faculty of Arts and Sciences Harvard University*
Northwest Lab Building, 52 Oxford Street, Cambridge, MA02138 USA

13.1. Precise Timing in the Brain?

Dynamical processes in the human brain underlie all our perceptions, actions and intellectual abilities, such as the formation of memories and the construction of high-level knowledge about the world. It is generally assumed that the large number of neurons that collectively interact in networks – and not the high diversity of the individual neurons – support the various functions of the brain. Understanding these collective processes thus constitutes a fundamental current problem of research, with key consequences also in disciplines outside neuroscience, ranging from psychology to the social sciences and philosophy. Moreover, most theoretical investigations of the collective dynamics of neural networks require new and advanced methods originally used in similar form in theoretical physics, computer science and mathematics. Often these methods are to be newly developed and thus also initiate complementary lines of research in these disciplines.

Neurons are connected to networks via synapses and communicate with each other using short-lasting ($\approx$ 1ms) electrical pulses called action potentials or spikes [79, 126]. These spikes are predominantly sent via chemical synapses to other neurons which process the incoming signals, change their internal state in response, and in turn send spikes at state-dependent times. The chemical processes involved are comparatively slow (several milliseconds) and cause a significant delay in signal transmission [79]. Chemical synapses are also highly adaptive and their strengths can be changed during learning processes.

Several basic functional features of neuronal networks, such as the selectivity of visual cortical neurons to oriented bar stimuli [126], are well characterized by spatially and temporally coarse-scale quantities, e.g. by the number of spikes emitted by a local population of neurons in a larger time interval (their spike rate) [33]. Nevertheless, there is accumulating evidence [6, 44, 45, 55, 121, 130] that the timing of spikes may be highly coordinated between neurons and play a role in neural processing as well. Neurons that spike coincidentally within a few milliseconds, or with a precise time lag between them have been observed in different neuronal systems [6, 44, 67, 87, 109, 119, 121, 130]. Coincident spiking can occur with high statistical significance correlated to internal states of the brain [121, 130] and patterns of spikes may re-occur repeatedly and in a particular order (second order spike patterns) [44, 45]. Patterns of precisely timed spikes and synchronization in the millisecond range are therefore discussed to be essential for information processing in the brain [4–6, 17, 31, 96, 117, 124, 130]. For a number of physiological experiments [6, 67, 87, 130] however, the statistical significance of some of the findings is currently highly debated [14, 99, 109, 119]. It has been argued that in some experiments [6, 87, 130] the significance of the occurrences of spike patterns highly depends on the underlying statistical assumptions about the spike trains [14, 119]. Further, [109] shows that the occurrence of repeated dynamical motifs of the membrane potential (which were assumed to indicate and generate spike patterns) is equally likely in random or randomized sub-threshold dynamics if the randomly generated membrane potential has similar coarse statistical properties (such as the power spectrum) as the actually measured one. It is thus still an open problem whether and how neurons may precisely coordinate their spiking activity across complex networks, and which role the identity of individual neurons and their inter-connectivity actually play.

Below we present two classes of hypotheses that may explain the dynamical origin of patterns of precisely timed spikes and microscopic, inter-neuronal synchronization, i.e. non-random, coincident spiking. One hypothesis states that feed-forward anatomical structures are embedded in cortical circuits and support the propagation of synchronous spiking activity of groups of neurons that constitute the layers of the feed-forward architecture [4, 5, 38, 62]. This kind of dynamics was termed 'synfire chain' activity [4]. As in current physiological experiments only small subsets of neurons are observed, the synfire chain hypothesis permits the occurrence of spiking activity that is synchronized with millisecond precision as well as the persistence of spike patterns over longer time periods. A second, alternative hypothesis states that recurrent networks may collectively organize patterns of precisely timed spikes without the need of specific feed-forward anatomy. We will give more emphasis to this latter hypothesis as it is more recent and its theoretical aspects are only marginally described so far in standard references.

This chapter is organized as follows: In Section 13.2, we briefly present the key ideas underlying synfire chain dynamics and state the main results on this topic. The remainder of this chapter is devoted to recurrent network models. In Section 13.3 we introduce a class of analytically tractable models of spiking neural networks that serves as our guide throughout; we also list related model classes as well as biophysically more detailed models. Section 13.4 gives an overview of basic states and important collective phenomena in recurrent spiking neural networks. In Section 13.5 we present recent approaches to characterize the emergence of patterns of coordinated, precisely timed spikes in neural network models. Finally, in Section 13.6, we conclude and highlight some open questions. To keep this overview concise, we focus on conceptual questions and theoretical challenges throughout, sometimes passing over technical subtleties and smaller (yet not unimportant) problems.

13.2. Feed-Forward Mechanisms: Synfire Chains

The synfire chain hypothesis states that precisely timed spiking in cortical networks is due to the existence of anatomical feed-forward structures that are part of cortical circuits [4, 5]. Thinking abstractly, such a feed-forward structure can be separated into groups of neurons or 'layers', such that neurons in one layer receive many synaptic connections from neurons in the previous layer (Fig. 13.1a). In its simplest setting, the connectivity between layers is uni-directional and global, i.e. each neuron in a layer receives an excitatory synapse from every neuron in the previous layer. In general, this connectivity between layers is diluted and only predominantly excitatory; still, when many spikes are received from a sufficiently synchronized presynaptic group of neurons, the likelihood that a post-synaptic neuron generates a spike is increased.

If now some initial layer of neurons emits spikes synchronously, i.e. with only small inter-neuronal variations (on the order of one millisecond) each of the postsynaptic neurons in the next layer receives an almost synchronous collection of spikes ('volley') after an effective transmission delay [79]. Collectively, this may initiate synchronous spiking activity generated by that next layer. Depending on the temporal spread of the spikes, on the inter-connectivity between layers and on the total number of synchronously firing neurons in a group, this may lead to the persistent (or decaying) propagation of synchronous activity along the chain [38, 62], (Fig. 13.1b). The feed-forward anatomy underlying synfire chains is viewed as an embedded part of a larger recurrent circuit such that each neuron receives in addition many synaptic inputs from outside the chain; as the basic state of the entire circuit is often asynchronous and irregular [33, 138, 140, 162], this additional input is typically regarded as noise that adds to the propagating synchronous activity [4, 5, 38, 53, 61, 62, 83, 101, 159].

Already in 1963 Griffith [56] suggested that densely coupled feed-forward anatomy may have a functional role in the brain. He investigated the capability of what he calls 'transmission lines' to reliably transmit information in a nontrivial way along feed-forward chains of abstract units. This idea was refined by Abeles [4, 5] to account for the appearance of precise spiking sequences of neurons. Diesmann and others showed that both fully connected and randomly diluted

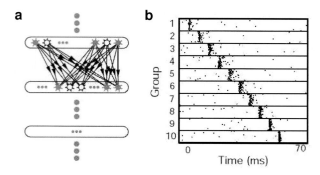

Fig. 13.1. Synfire chain anatomy (a) and dynamics (b). (a) Chain of groups of neurons connected in a feed-forward manner. (b) Activity synchronizes further as it propagates along the chain. (Modified from [38].)

feed-forward structures permit the stable propagation of synchrony [7, 38, 53, 62, 145, 159]. Model studies of synfire chains that are actually embedded in recurrent networks (rather than treating the influence of the embedding network just as additional noise) show that persistent propagation of localized synchronous events within recurrent cortical circuits is not simple to realize [83, 101, 146]. In particular, extensive numerical studies show that although propagation of synchrony can be achieved, often pathological dynamics occur, for instance synchronous activity which spreads and covers the entire network after a short time ('synfire explosion'), or synchronous activity in the embedded chain dies out quickly due to inhibition from the embedding network [12, 101, 146]. It has been shown recently that depending on the features of individual neurons and on the network architecture, stable propagation of synchrony along chains embedded in recurrent networks may also be achieved in a robust way [83]. Another possibility to construct networks with feed-forward pathways is to strengthen those connections that are already present in a recurrent random network [164]. However, strong amplifications of synapses and specific changes in the response properties of neurons along the pathway are required to enable the propagation of synchrony over a few groups. A mechanism that might enable persistent synchronous activity in embedded architectures with moderately strong pathway structure is nonlinear enhancement of synchronous inputs due to dendritic spikes that was recently found in neurophysiological experiments ([9, 46, 47, 116, 122], cf. also [106]).

Successive excitation of neurons in groups with distributed transmission delays can generate spiking activity that is not synchronous, but precisely time-lagged (with the lag defined by the transmission delays), resembling synfire chain dynamics. Works by Izhikevich and coworkers [68, 69] show that such groups of neurons with strong coupling can spontaneously form in a random network due to spike timing dependent plasticity (see, e.g., [33, 79]) and that they generate detectable spike patterns with millisecond precision although embedded in a larger network. Taken together, current theoretical knowledge supports the synfire hypothesis in systems with specific constraints on synaptic dynamics, single neuron features, neural inter-actions, and inter-connectivity between groups.

The current stage of experimental research is inconclusive. Despite some inter-esting studies which might support the synfire chain hypothesis [94, 113, 125, 139], there is no key experiment that directly proves – or disproves – the existence of synfire chain anatomy or dynamics. Such an experiment would require either a large-scale structural investigation of local cortical anatomy, proving or excluding the necessary non-random feed-forward connectivity; or a large-scale dynamical study, recording spikes of a large number of neurons simultaneously and repeatedly under controlled conditions, such as to explicitly show (or exclude) the existence of synchronous activity propagating along fixed paths.

13.3. Recurrent Neural Networks

Alternatively, in recurrent networks without specifically embedded feed-forward structures, mechanisms other than synchronous excitation along feed-forward anatomy might generate spikes that are precisely coordinated in time and among different neurons. For recurrent networks, however, theoretical investigations that take into account individual neurons' spike times and thus go beyond mean-field de-scriptions are often highly non-standard. Thus up to date precise timing of spikes in recurrent networks is far less understood than synfire chain dynamics. Con-ceptual challenges include the nonlinear features of individual neurons and their interactions, the complex recurrent connectivity of the networks, the existence of transmission delays that make the dynamical systems formally infinite-dimensional, and strong heterogeneities that might be present among the neurons and their in-teractions.

To cope with these challenges, many studies have focused on networks of ide-alized model neurons, e.g. of integrate-and-fire type [33, 51, 71, 86, 120]. In the following, we introduce a class of spiking neural network models for which a wide range of dynamical phenomena becomes analytically accessible. We briefly list re-lated model classes and biophysically more detailed models at the end of this section and describe some basic and more involved dynamical states of spiking activity in the subsequent sections.

13.3.1. *An analytically accessible class of models*

Consider a network of $N \in \mathbb{N}$ neurons that interact by sending and receiving spikes (see, e.g., [40, 60, 108, 149]). The state of each neuron j at time s is specified by a single real variable, the membrane potential $V(s)$ that evolves according to

$$\frac{d}{ds}V_j(s) = g(V_j(s)) + \sum_{i=1}^{N}\sum_{m\in\mathbb{Z}}\varepsilon_{ji}K(s-(s_i^m+\tau_{\mathrm{V}})) \qquad (13.1)$$

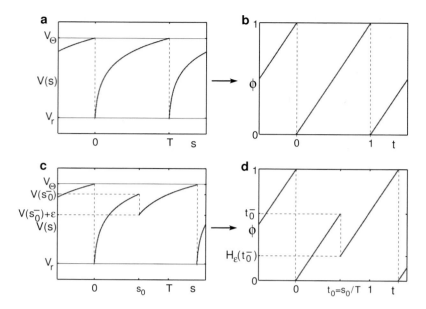

Fig. 13.2.   Dynamics of membrane potentials and phases. Upper panels show free evolution of potential (a) and corresponding phase (b). Lower panels show evolution of potential (c) and phase (d) with an inhibitory spike arriving at time s_0. (Modified from [81].)

where $g(.) > 0$ specifies the local dynamics of neuron j, ε_{ji} denotes the strength of synaptic coupling from neuron i to neuron j, and s_i^m specifies the time neuron i sends its mth spike. When a neuron i reaches a potential threshold $V_i((s_i^m)^-) = V_\Theta$, its potential is reset to $V_i(s_i^m) = V_r$ and it sends a spike which is received by the postsynaptic neurons j after a delay time $\tau_V > 0$. Here, $K(.)$ is a response kernel that determines the post-synaptic current in response to an incoming spike signal. Such a kernel satisfies $\int_{-\infty}^{\infty} K(s)ds = 1$ and $K(s) = 0$ for $s < 0$. Often one considers the limiting case of fast synaptic response, $K(s) = \delta(s)$. For such systems the smooth dynamics of the neurons is interrupted by two kinds of events that occur at discrete times only: sending of spikes (and reset) and receiving of spikes. This results in a hybrid dynamical system [13, 19, 133] with continuous-time dynamics interrupted at discrete times where maps are applied [10, 23].

A universal representation of the network dynamics provides elegant analytical access to state space trajectories. The network of spiking neurons (13.1) with $K(s) = \delta(s)$ is equivalently described by the dynamics of phase variables $\phi_i(t) \leq 1$ with rescaled time variable $t = s/T$: The free solution $\tilde{V}(s)$ of Eq (13.1) in the absence of coupling (all $\varepsilon_{ji} = 0$) through the initial condition $\tilde{V}(0) = V_r$ increases monotonically and is assumed to reach the threshold after a time T such that $\tilde{V}(T^-) = V_\Theta$. This free solution defines a bijective map (Fig. 13.2a,b)

$$U : (\phi_-, 1] \to (V_-, V_\Theta]; \phi \mapsto \quad U(\phi) := \tilde{V}(\phi T), \tag{13.2}$$

between potential and phase representation via a continuously differentiable 'rise function' $U(\phi)$ that is monotonic increasing because $g(.) > 0$ in Eq. (13.1). In Eq. (13.2), V_- and ϕ_- are possible lower bounds of potential and phase; if there are no bounds in potential or phase, $V_- = -\infty$ or $\phi_- = -\infty$, cf. [74, 102, 103]. In the absence of interactions, the phases increase uniformly and obey $d\phi_i/dt = 1$. When ϕ_i reaches its phase threshold $\phi_i(t^-) = 1$ it is reset to $\phi_i(t) := 0$ and a spike is sent such that $t = t_i^m = s_i^m/T$. This spike is now received by the post-synaptic neurons j after a rescaled delay time $\tau = \tau_V/T$, where it causes an instantaneous phase jump (Fig. 13.2c,d) according to

$$\phi_j(t + \tau) = H_{\varepsilon_{ji}}(\phi_j((t + \tau)^-)) \qquad (13.3)$$

mediated by the transfer function

$$H_\varepsilon(\phi) = U^{-1}(U(\phi) + \varepsilon) \qquad (13.4)$$

that is strictly monotonic increasing both as a function of ε and ϕ, because $U' > 0$ [74, 102, 103]. If $U(\phi) + \varepsilon$ may exceed the codomain of $\tilde{V}$ and thus the domain of U^{-1}, this has to be accounted for by case distinctions in the definition of H_ε [40, 102, 103, 151]; for the sake of conciseness, we will not discuss this complication here. If the coupling is excitatory ($\varepsilon > 0$), it is phase-advancing, $H_\varepsilon(\phi) > \phi$, enabling the neuron to emit its next spike earlier than without that coupling; if it is inhibitory ($\varepsilon < 0$), it is phase-retarding, $H_\varepsilon(\phi) < \phi$, such that the neuron will emit its next spike later than without that coupling.

As demonstrated before [40, 74, 103, 104, 108, 150, 151] this phase representation allows for exact numerical integration and provides elegant analytic access to trajectories of the network dynamics, even if the local neuron dynamics is not characterized by a simple, e.g. linear, differential equation [108].

We remark that the above models are current-based, i.e. the coupling strengths ε_{ji} do not explicitly depend on the state of the post-synaptic neuron. Under certain conditions, networks with conductance based synapses (where the coupling strength depends on the state as $\alpha(V_j - V_{\text{rev}})\varepsilon_{ji}$), or others where the interaction is explicitly state-dependent, can also be modeled using a phase description together with a modified rise function [132, 151, 160]. Often models of spiking neural networks are formulated in a generalized way, with various kinds of heterogeneities in the neuron dynamics, in the delay times and with temporally extended interaction kernels $K(.)$ that may depend on the synaptic connection and the state of the post-synaptic neuron at reception time. Moreover, additional noise and driving forces may add to the recurrent network dynamics (13.1). For simplicity of presentation, we will not describe these features here in detail, but only refer to the respective literature where appropriate.

The model class includes, among others, the simple non-leaky integrate-and-fire (IF) model where $g(V) = \text{const.}$ and thus $U(\phi) = (V_\Theta - V_r)\phi + V_r$ [50, 100, 110], standard leaky IF models $(g(V) = I - \gamma V)$ [1, 27, 86, 155], quadratic integrate-

and-fire (or theta-) neurons ($g(V) = V^2 + I$) [39, 43, 71, 114], abstract neural
oscillator models where $U(\phi) = b^{-1}\ln(1 + (\exp(b) - 1)\phi)$ [36, 108], and exponential
IF neurons where $g(V) = I - \gamma V + \exp(\beta V)$ [42, 43, 129]. The model class is limited
by the idealization that the sub-threshold neuron dynamics is well characterized by
a single variable; moreover, whereas an analytic approach is conceptually simple
in the limit of infinitely fast response (where the kernel is a delta-distribution),
it typically becomes restricted for certain post-synaptic response kernels that are
temporally extended.

13.3.2. *Related models*

Besides the standard model class defined via (13.1), several variants are widely used
as well. Often, additional degrees of freedom are introduced. One phenomenological
class of models has a spike-triggered adaptation variable [22, 70, 71, 77]. In dependence of the parameters, neurons of this class show a wide spectrum of qualitative
features observed in biological neurons and at the same time allow fast numerical
simulations of large networks if the individual neurons' features are well understood [70, 71]. It has the disadvantages that its dynamics is analytically accessible
only in rare special cases, and in numerical simulations, though they are much faster
than, say for Hodgkin Huxley neurons (see below), the dynamical parameters are
similarly hard to restrict. The 'spike response model' works in the original potential
representation and includes additional refractoriness or adaptation, modeled as a
threshold dynamics that is not present in (13.1). Recent works [22, 76, 77] suggest
that certain representatives of spike response model neurons with adaptation well
reproduce the response of real neurons to specific random current inputs.

Spiking neural network models with temporally extended interactions often also
characterize the response dynamics by one additional degree of freedom per neuron, e.g. by a second differential equation, which is, however, usually chosen to be
solvable in closed form such that Eq. (13.1) is regained (see, e.g., [3, 154, 168]).
Biophysically more detailed models, such as the Hodgkin-Huxley, Morris-Lecar,
Fitzhugh-Nagumo, or Hindmarsh-Rose models ([64, 65, 111], see [71] for a comprehensive review) require several dynamical variables and many physiological parameters for each neuron. As such they are appropriate for modeling dynamical network
aspects of well-known systems (see, e.g., [35]); at the same time, they typically
preclude analytical arguments and for many systems it is unclear how to suitably
restrict all model parameters or even whether the chosen model is appropriate at
all [115].

Whereas all deterministic models have their variants that include additional
stochastic influences, modeling, e.g., synaptic failure [80] or local noise induced by
ion channels [15], intrinsically stochastic models may sometimes be more appropriate for the description of single neuron or network dynamics [49, 78, 88, 127, 128,
143, 144, 156]. For instance, Levina *et al.* [88, 89] have recently shown that the
dynamics of branching processes under certain conditions well describe the stochas-

tic dynamics of large recurrent networks. Of course in all the above-mentioned frameworks, also additional features may be studied, including synaptic plasticity on short [2, 88, 93, 153] and long time-scales [63, 68, 90, 112], ion channel co-operativity [115], compartmental or spatially extended structure [20, 82, 135] and non-additive features of the interactions [9, 88, 93, 106, 115, 153].

Moreover, various models of abstract rate-coded often binary-state or discrete-time neurons [16, 66, 91, 92, 131, 165] exist that are valuable for studying conceptual problems of computation or information processing in neural systems, but by their very nature generically do not capture the precise timing of spikes. Very recently the link between continuous-time and discrete-time models has been reconsidered with the interesting resulting suggestion [29, 30] that under certain conditions specific discrete time models may actually more appropriately describe the spiking dynamics of recurrent networks.

13.4. Basic Collective States of Recurrent Networks

Here we provide a brief overview on basic collective states of deterministic recurrent networks. This should pave the way to a better understanding of the concepts and the complex spatio-temporal dynamics presented in the next section.

13.4.1. *Quiescence and synchrony*

Obviously, the simplest dynamical state of a spiking neural network is global quiescence, where no neuron is emitting any spike. This is a trivial network state because its dynamics is just the collection of all individual neuron dynamics, even in the presence of driving signals and fluctuations. Still it is sometimes valuable to know under which conditions the quiescent state exists and is stable, for instance if the emergence of a complex, persistent state from the quiescent one (or from an almost quiescent one) is to be understood [57, 59].

In fact, single neurons are intrinsically excitable systems and typically quiescent if not driven by synaptic inputs, external currents, fluctuations or by other means. Therefore, the modeling of single spiking neurons ranges essentially between two extreme limits (see, e.g., [155, 156]). One limit is stochastic: neurons receive independent stochastic sequences of spikes (e.g. Poisson spike trains) and therefore also generate stochastic spiking dynamics theirselves (e.g. [27, 28, 156]). The second limit is deterministic: neurons receive sufficiently strong temporally uniform input currents such that their dynamics becomes tonic periodic spiking. We here focus on deterministic models of spiking neural networks. As we will see below, however, these neurons collectively still often exhibit dynamics that resembles random processes [58, 74, 106, 162, 163].

Arguably the simplest non-trivial and truly collective state is the fully synchronous state, a periodic orbit in which every neuron emits spikes periodically and at the same times as all the other neurons in the network. It is dominant in globally

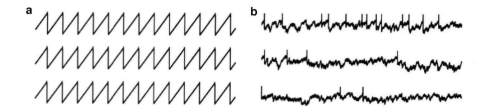

Fig. 13.3. Synchronous (a) and irregular (b) dynamics of three neurons in a sparse random network [149]. Both dynamics may coexist in the same network and external stimulations can induce switching between them, cf. also Fig. 13.4. (Modified from [149].)

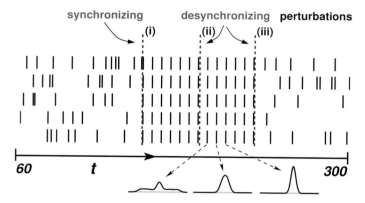

Fig. 13.4. Transitions between irregular and synchronous states due to external input signals. Synchronized excitatory external input (i) causes the dynamics to assume the synchronous state. The synchronous state is stable: A sufficiently small desynchronizing perturbation (ii) does not lead to the irregular state. Only after strong desynchronizing signals (iii), e.g. induced by a large number of random inhibitory and excitatory input spikes, the system switches back to the irregular state. (Modified from [149].)

excitatorily coupled networks of leaky (and non-leaky [136]) IF-like neurons if the interactions exhibit zero delay ([108], see also [52]), but it is unstable in the presence of arbitrarily small delays in globally and more complex connected excitatory networks [40, 41, 148, 149]. If the coupling is inhibitory, however, synchrony might occur and even be predominant [106, 149, 152, 161] in the presence of interaction delays as well, cf. Figs. 13.3a, 13.4. Noise in such systems affects synchrony in a non-trivial way [26].

    In globally coupled or spatially extended homogeneous networks of spiking units also less symmetric solutions exist, including waves [21], periodic localized activity [134] and cluster states [40] exist that are well known from smoothly coupled systems.

13.4.2. *Away from synchrony:*
first hints towards spike patterns in recurrent networks

Systems that exhibit full synchrony, with all units obeying identical dynamics, necessarily are constraint, e.g. they exhibit some invariance. For instance, when considering networks of identical inhibitory neurons, the total input strength to each neuron in a complex network needs to be the same for the fully synchronous state to exist [148, 149]. Such an idealized condition is atypical for biological networks in which the total synaptic strengths may be roughly the same due to homeostasis [32, 157, 158], but at least weak inhomogeneities are prevalent. Naturally, weak inhomogeneities and the presence of other less idealized features will induce states similar to the fully synchronous one [3, 36, 154]. Stronger heterogeneities typically lead to states that are very distinct from synchrony and sometimes completely asynchronous [36].

Already in large networks of all-to-all and homogeneously coupled excitatory neurons with temporally extended synaptic responses, a partially synchronous state exists for a certain range of temporal extent [160]. In this partially synchronous state the total network firing rate oscillates periodically whereas the individual neurons send spikes quasi-periodically. This result by van Vreeswijk provides one possible mechanism for oscillations in neural circuits and at the same time is of interest mathematically as the local quasi-periodic activity adds up to global periodic activity. Brunel *et al.* [24, 25, 48] showed that also sufficiently strong inhibitory interactions can lead to high frequency network oscillations where the individual neurons fire irregularly and with low frequency. This type of dynamics has recently been proposed to underlie high frequency oscillations of Purkinje cells in the cerebellum [34].

Tsodyks, Mitkov and Sompolinsky uncovered a different interesting state similar to synchrony [154]; for globally and excitatorily coupled neurons with temporally extended synaptic responses arbitrarily weak inhomogeneities in the individual neurons' intrinsic time scales may split the neurons into two sub-populations (Fig. 13.5), one sub-population consisting of the slower neurons, that stay identically synchronized forever, and a second consisting of the intrinsically faster spiking neurons, which also have collective frequencies that are different from each other and larger than in the synchronized sub-population. We remark that already in such states, the timing of spikes of the synchronized sub-population is highly precise, despite inhomogeneities in the individual neuron features; moreover, the timing of spikes of neurons in the unlocked sub-population is relatively precise and close to that of the locked sub-population, for repeated, long stretches of time.

Networks with more complex connectivity may exhibit additional collective dynamical features induced by heterogeneities. Recent work [36] has shown that in networks of inhibitorily coupled IF-like neurons with delayed interactions, weak inhomogeneities in the coupling strengths (or equivalently, in other system parameters) induce a state close to full synchrony that exhibits well-defined patterns of

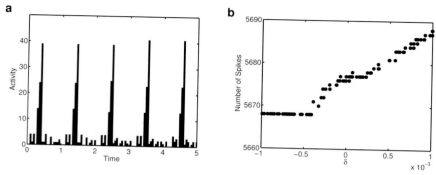

Fig. 13.5.   Network activity and rates of individual neurons in a network of excitatory neurons with weak inhomogeneities as studied in [154]. The neurons split into two populations, neurons that are constantly phase-locked with zero lag and neurons which are not phase-locked. The network activity (a) shows large peaks of well synchronized firing together with some non-synchronous activity. The rate profile (b), displaying the rate of the neurons versus the perturbation of their driving, shows that there is a driving I_c separating phase locked neurons (driving $I < I_c$) and non-phase locked neurons (driving $I > I_c$).

spiking activity coordinated between the neurons. The exact analysis of spike times revealed also the transition point at which inhomogeneities become too strong such that states close to full synchrony (short patterns) cease to exist. Furthermore, the same work considered the occurrence of patterns of precisely timed spikes as an inverse problem (see sec. 13.5.2): For any given, predefined pattern of spikes that spreads over a sufficiently short time interval, Denker *et al.* [36] showed how to find the set of networks that exhibit that pattern as an invariant solution of its collective dynamics. Interestingly, in homogeneous sparsely connected random networks (or in those with sufficiently weak inhomogeneities) a synchronous (or almost synchronous) state may coexist [148, 149] with highly irregular asynchronous states (see Figs. 13.3 and 13.4 and also the next subsection).

Numerical investigations of inhomogeneous networks of inhibitory and excitatory sub-populations with delayed, temporally extended interactions [18] have shown that the two sub-populations may send spikes phase-locked but out-of-phase with each other, with all neurons in the separate sub-population close to synchronous with each other. As a sideline, that work suggests that patterns of locked spikes may occur also in neural circuits with a mixture of excitatory and inhibitory neurons; the mechanism underlying this phenomenon is similar to that described in Ref. [36] for purely inhibitory recurrent interactions.

13.4.3. *Asynchrony: Irregular, chaotic, and balanced activity*

Besides simple synchronous states, asynchronous states provide a second type of basic activity in spiking neural networks. Depending on the features of the network considered, asynchronous states may predominantly emerge (i) as states in which neurons emit spikes individually and periodically and phase-locked to all other neu-

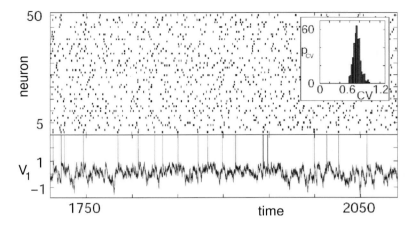

Fig. 13.6. Balanced irregular neural activity. Upper part shows irregular spiking dynamics characteristic for the balanced state. Lower part displays the normalized, highly irregular membrane potential of neuron 1. Inset displays the distribution of the coefficients of variation of the neurons' spike trains. This dynamics is stable against small perturbations, the irregularity does not result from chaos. (Modified from [74].)

rons in the network, see, e.g., [168], (ii) as non-periodic states with periodic oscillatory network rate dynamics (e.g. splay states [24, 25, 160] and cf. section 13.4.2) and (iii) and as aperiodic irregular states with constant network rate and no apparent coordination of spike times [24, 163].

The balanced state constitutes a key example of asynchronous irregular spiking as it is considered as a possible ground state of cortical activity. In such a state, excitatory and inhibitory synaptic input to each neuron balances such that the average membrane potential is sub-threshold and large fluctuations generate spikes at low rate and at seemingly random times [24, 162, 163]. Whereas the original assumption was that chaotic dynamics causes these apparently random spiking sequences, it was recently found ([73, 74], (cf. also [167]) that dynamics with the same irregularity is prevalent also in systems which do not exhibit chaotic, but rather stable microscopic dynamics. This raises the question which dynamical features actually generate asynchronous irregular spiking dynamics characteristic for the balanced state. A further question of current research is how balanced irregular activity may persist after a transient external stimulus has initiated it [57, 58, 84].

In theoretical investigations, the analysis of asynchronous states may also serve as a starting point to reveal mechanisms that underlie more coordinated neural activity by studying bifurcations away from asynchrony, see, e.g., [59].

13.5. Precise Timing in Recurrent Networks

As some results presented in the previous section already suggest, patterns of spikes that are precisely timed and coordinated among neurons may also emerge in the collective dynamics of recurrent neural networks. Compared to patterns generated by feed-forward networks, this possibility so far is much less explored. Below, we present some recent developments where in part a detailed understanding of the collective phenomena is possible.

13.5.1. *Spike patterns as attractors of recurrent networks*

In certain networks that are dominated by inhibitory non-delayed interactions, the collective network dynamics converges to periodic spike patterns [75, 98]. These spike patterns are typically of low period and reached quickly such that they dominate the network dynamics on the relevant time scales. As shown before already for globally coupled networks [40, 41], interaction delays may have a drastic influence on the collective network dynamics. This is even more so if the network topology is complex and local dissipation becomes relevant, compare, e.g., [169] vs. [50, 108]. For instance, in inhibitorily coupled units delays strongly enhance the transient times towards periodic spike patterns [73, 74] such that stable irregular transients dominate the dynamics. Moreover, periodic orbits in these systems typically are long. In another example, very long delays induce switching between sequences [54] that recur several times and afterwards are non-recurrent. The results of Ref. [74] strongly suggest that the dynamics is nevertheless stable. An explanation for the switching phenomenon is detailed in [102].

13.5.2. *Realizing spike patterns in complex networks –*
An inverse problem

Nearly all the above studies considered certain pre-specified networks of spiking neurons and studied what kinds of dynamics they may exhibit. In recent years, an inverse perspective was introduced [36, 97, 98, 103, 104, 123, 147, 166] where now the central question becomes "What kind of networks exhibit a given dynamics?". Such questions have been conceptually addressed two decades ago in abstract networks of non-spiking neurons [37, 66, 91, 92].

Prinz and coworkers [123] presented an extensive numerical analysis of three-neuron circuits and identified broadly distinct networks that exhibit nearly the same spiking dynamics. Makarov *et al.* [97] used stochastic optimization to find networks of given model neurons that most closely match observed spiking data. An analytical deterministic framework of network design was introduced recently [103, 104] to find the set of all possible networks that exhibit a predefined, e.g. periodic, spike pattern (Fig. 13.7). If a network solution exists at all (which it does under mild constraints), there typically is a high-dimensional set of networks that exhibit the same spike pattern as a possible dynamics. This set is parametrized, for instance, by the coupling strengths. The set is restricted by spike timing conditions, equations that impose the constraint that a given spike occurs where predefined, and by silence conditions, inequalities that ensure that a neuron does not emit a spike when none

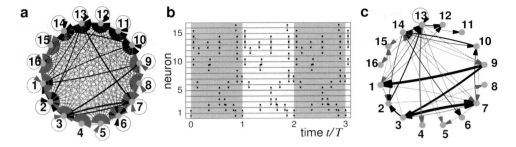

Fig. 13.7. Using the methods derived in [103, 104], networks with very different coupling statistics (a), (c) can be designed to generate the same pattern of spiking activity (b). Network (a) minimizes the L_2-norm ($\sqrt{\sum_{i,j} \varepsilon_{ij}^2}$) of the coupling matrix, network (c) minimizes the L_1-norm ($\sum_{i,j} |\varepsilon_{ij}|$). (Modified from [103].)

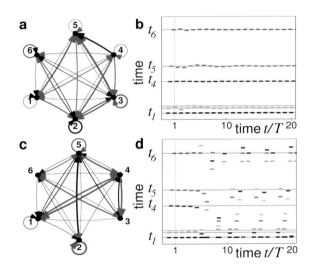

Fig. 13.8. The same periodic pattern of spikes can be realized as stable invariant dynamics (b) in one network (a) and as unstable invariant dynamics (d) in another network (c). (Reproduced from [103].)

is predefined. The equations and inequalities involve all parameters available in the class of model systems considered, including the network topology and the coupling strengths, the delay times and the local individual neuron dynamics [103, 104].

The high dimensionality of a typical solution space has important conceptual consequences. For instance, the same pattern may exist in networks with very distinct topologies and coupling types, cf. Fig. 13.7. Moreover, the same pattern may exist in networks with statistically similar topology but may be stable in one and unstable in another network, cf. Fig. 13.8. Besides numerical analyses, for large classes of networks even general analytical statements on the stability of such spike patterns have been derived [102–105]. The stability properties are of particular interest, e.g. because they determine the computational capabilities of a network (where computation is not only possible with stable states [10, 11, 72, 95]) and because motifs that exhibit patterns which are stable (or unstable) might, if embedded in a larger network, influence the entire network's function in a specific way.

The theory of coupled phase oscillators suggests so-called chimera states as a possible link [118] between a fully synchronous state and asynchronous states. Chimera states were originally found [85] in rings of coupled identical limit-cycle oscillators with translation-invariant, non-local coupling. In chimera states, one sub-population of neighboring oscillators is phase-locked whereas oscillators in a second sub-population are asynchronous and neither locked with the first sub-population nor with each other. Such chimera states are thus similar to the partially synchronized states [154] described above. However, they are significantly different as they occur also in homogeneous, translation invariant systems ([8, 85, 137], Ref. [118] introduces a location-dependent stimulation that breaks this symmetry), whereas the partial locking found in pulse-coupled systems [154] was induced by inhomogeneities [154].

Since in each biological neural system one particular network is selected that generates a desired dynamics (and function) an open question is whether and in which aspects networks may be optimized, for instance structurally. First examples [103, 104] show that even very sparse heterogeneously coupled networks and very dense, homogeneously coupled networks may be capable of generating the same predefined pattern, cf. Fig. 13.7.

13.6. Conclusions and Open Questions

The currently debated question under which conditions and how patterns of precisely timed spikes and microscopic synchrony may emerge in neural circuits is still far from being answered and also their potential functional role is explored further. On the path towards a final conclusion, experimental and theoretical findings need to be jointly evaluated in a critical way in order to generate and confirm (or reject) key hypotheses. We have here presented two major hypotheses for the mechanism underlying the generation of spike patterns; one where feed-forward anatomy is crucial, and one asserting that spike patterns may emerge collectively in recurrent networks. Up to now, both hypotheses have been neither confirmed nor rejected experimentally and on theoretical grounds, both seem possible.

The hypotheses on synfire and recurrent mechanisms mark only the extreme starting points for such investigations, working in the limits of strong feed-forward chains and of no particular coarse structure, respectively. Intermediate possibilities need to be explored as well. For instance, what is the impact of observed non-random topology, such as motifs present in otherwise apparently randomly connected, spatially extended circuits [103, 107, 123, 141, 142]?

On the one hand, one may be tempted to argue that in model studies on the emergence of spike patterns more and more biological details need to be taken into account. For instance, only few works so far take into account additional dynamical features such as synaptic plasticity, synaptic failure and intrinsic noise or dendritic non-linearities (e.g. [68, 106, 112]), among others. On the other hand, model reduction is essential to isolate potential mechanisms that underlie any hypothesis and thus to restrict a hypothesis as strongly as possible to make it experimentally testable. The theoretical analyses of feed-forward chains of spiking neurons already raise many non-trivial problems, and even the idealized recurrent neural network models considered above typically exhibit highly complex dynamics. Idealized models enable us, nevertheless, to understand mechanisms in feed-forward and even in recurrent networks in a systematic way.

And indeed, recent studies of reduced models for instance revealed that non-additive dendritic integration can support the propagation of synchronous spiking activity in recurrent networks even if they are purely randomly connected and do not contain additional feed-forward connectivity [106]. Conceptually, this constitutes another cornerstone for bridging the gap between feed-forward and recurrent perspectives because propagation of synchronous activity similarly underlies the emergence of spike patterns in both hypotheses. At the same time, propagation of synchronous activity in random recurrent networks might be experimentally distinguished from that along feed-forward chains: In networks containing groups of neurons with specific feed-forward connections between them, synchronous activity propagates along paths that are predefined by that anatomy such that propagation takes place with high probability along the same paths if the experiment is repeated with the same inital group of neurons synchronized. In contrast, in random recurrent networks without specific feed-forward structures but with non-additive dendritic features, synchrony propagates along self-organized paths that may vary among repetitive trials of an experiments. This mechanistic difference implies different types of possible patterns of precisely timed spikes and thus a dynamics that offers an experimental distinction between the two hypotheses.

The balanced activity described in section 13.4.3 constitutes another example where simple models of neural circuits substantially helped to understand the biological network dynamics. The balanced state was first investigated for binary-state neuron models [162] and the mechanism that generates its irregular spiking behavior points to a collective network effect that is due to simultaneously strong and only weakly correlated inhibitory and excitatory inputs. Due to this basic network mechanism, such balanced activity robustly occurs across different systems; it is prevalent also for biophysically more detailed models and provides the first consistent explanation of the irregularity observed in biological neural circuits. Perhaps reduced models will similarly help to better understand the conditions under which spike patterns emerge in feed-forward and in recurrent networks.

When physiological experiments in the near future pin down in which systems patterns of precisely timed spikes definitely carry information that is not contained in the spike rate, the question of their functional role and their actual origin becomes more specific – and even more urgent. In particular if linked with further extensive theoretical studies on all levels, on stochastic and deterministic, abstract and biophysically detailed models, on mechanistic single neuron aspects and on network effects, such experimental investigations are likely to give key answers to the question how the brain computes.

Acknowledgments

We are grateful to our colleagues and collaborators, who provided many insights to the theory of spiking neural networks, including in particular Ad Aertsen, Markus Diesmann, Theo Geisel, Sonja Grün, Sven Jahnke, Christoph Kirst and Fred Wolf. We also thank Hinrich Kielblock for support with manuscript preparation. This work has been supported by the Ministry of Education and Research (BMBF) Germany via the Bernstein Center for Computational Neuroscience (BCCN) Göttingen, under Grant No. 01GQ0430, by a grant of the Sloan-Swartz foundation to RMM, and by a grant of the Max Planck Society to MT.

References

[1] Abbott, L. (1999). Lapicque's introduction of the integrate-and-fire model neuron (1907), *Brain Res. Bull.* **50**, pp.303–304.
[2] Abbott, L. and Regehr, W. (2004). Synaptic computation, *Nature* **431**, pp. 796–803.
[3] Abbott, L. and van Vreeswijk, C. (1993). Asynchronous states in networks of pulse-coupled oscillators, *Phys. Rev. E* **48**, pp. 1483–1490.
[4] Abeles, M. (1982). *Local Cortical Circuits: An Electrophysiological Study* (Springer, Berlin).
[5] Abeles, M. (1991). *Corticonics: Neural Circuits of the Cerebral Cortex* (Cambridge University Press, Cambridge).
[6] Abeles, M., Bergman, H., Margalit, F. and Vaadia, E. (1993). Spatiotemporal firing patterns in the frontal cortex of behaving monkeys, *J. Neurophysiol.* **70**, pp. 1629–1638.
[7] Abeles, M., Hayon, G. and Lehmann, D. (2004). Modeling compositionality by dynamic binding of synfire chains, *J. Comp. Neurosci.* **17**, pp. 179–201.
[8] Abrams, D. and Strogatz, S. (2004). Chimera states for coupled oscillators, *Phys. Rev. Lett.* **93**, p. 174102.
[9] Ariav, G., Polsky, A. and Schiller, J. (2003). Submillisecond precision of the input-output transformation function mediated by fast sodium dendritic spikes in basal dendrites of CA1 pyramidal neurons, *J. Neurosci.* **23**, pp. 7750–7758.
[10] Ashwin, P. and Timme, M. (2005). Unstable attractors: Existence and robustness in networks of oscillators with delayed pulse coupling, *Nonlinearity* **18**, pp. 2035–2060.
[11] Ashwin, P. and Timme, M. (2005). When instability makes sense, *Nature* **436**, pp. 36–37.

[12] Aviel, Y., Mehring, C., Abeles, M. and Horn, D. (2003). On embedding synfire chains in a balanced network, *Neural Comp.* **15**, pp. 1321–1340.

[13] Bainov, D. and Simeonov, P. (1989). *Systems with Impulse Effect. Stability, Theory and Applications* (Horwood, London).

[14] Baker, S. and Lemon, R. (2000). Precise spatiotemporal repeating patterns in monkey primary and supplementary motor areas occur at chance levels, *J. Neurophysiol.* **84**, pp. 1770–1780.

[15] Bazsó, F., Zalányi, L. and Csárdi, G. (2003). Channel noise in Hodgkin-Huxley model neurons, *Phys. Lett. A* **311**, pp. 13–20.

[16] Ben-Yishai, R., Lev Bar-Or, R. and Sompolinsky, H. (1995). Theory of orientation tuning in visual cortex, *Proc. Natl. Acad. Sci.* **92**, pp. 3844–3848.

[17] Bienenstock, E. (1996). Composition, in A. Aertsen and V. Braitenberg (eds.), *Brain Theory: Biological Basis and Computational Principles* (Elsevier).

[18] Börgers, C. and Kopell, N. (2003). Synchronization in networks of excitatory and inhibitory neurons with sparse, random connectivity, *Neural Comp.* **15**, pp. 509–538.

[19] Branicky, M. (2005). Introduction to hybrid systems, in D. Hristu-Varsakelis and W. Levine (eds.), *Handbook of Networked and Embedded Control Systems* (Birkhauser).

[20] Bressloff, P. (1995). Dynamics of a compartmental model integrate-and-fire neuron with somatic potential reset, *Physica D* **4**, pp. 399–412.

[21] Bressloff, P. and Coombes, S. (1996). Traveling waves in a chain of pulse-coupled oscillators, *Phys. Rev. Lett.* **80**, pp. 4815–4818.

[22] Brette, R. and Gerstner, W. (2005). Adaptive exponential integrate-and-fire model as an effective description of neuronal activity, *J. Neurophysiol.* **94**, pp. 3637–3642.

[23] Broer, H., Estefanios, K. and Subramanian, E. (2008). Heteroclinic cycles between unstable attractors, *Nonlinearity* **21**, pp. 1385–1410.

[24] Brunel, N. (2000). Dynamics of sparsely connected networks of excitatory and inhibitory spiking neurons, *J. Comp. Neurosci.* **8**, pp. 183–208.

[25] Brunel, N. and Hakim, V. (1999). Fast global oscillations in networks of integrate-and-fire neurons with low firing rates, *Neural Comp.* **11**, pp. 1621–1671.

[26] Brunel, N. and Hansel, D. (2006). How noise affects the synchronization properties of recurrent networks of inhibitory neurons, *Neural Comp.* **18**, pp. 1066–1110.

[27] Burkitt, A. (2006). A review of the integrate-and-fire neuron model: I. Homogeneous synaptic input, *Biol. Cybern.* **95**, pp. 1–19.

[28] Burkitt, A. (2006). A review of the integrate-and-fire neuron model: II. Inhomogeneous synaptic input and network properties, *Biol. Cybern.* **95**, pp. 97–112.

[29] Cessac, B. (2008). A discrete time neural network model with spiking neurons, *J. Math. Biol.* **56**, pp. 311–345.

[30] Cessac, B. and Viéville, T. (2008). On dynamics of integrate-and-fire neural networks with conductance based synapses, *Front. Comp. Neurosci.* **2**, p. 2.

[31] Christen, M. (2006). The role of spike patterns in neuronal information processing, Doctor of Sciences Thesis, ETH Zürich.

[32] Davis, G. (2006). Homeostatic control of neural activity: From phenomenology to molecular design, *Annu. Rev. Neurosci.* **29**, pp. 307–323.

[33] Dayan, P. and Abbott, L. (2001). *Theoretical Neuroscience: Computational and Mathematical Modeling of Neural Systems* (MIT Press, Cambridge).

[34] de Solages, C., Szapiro, G., Brunel, N., Hakim, V., Isope, P., Buisseret, P., Rousseau, C., Barbour, B. and Léna, C. (2008). High-frequency organization and synchrony of activity in the Purkinje cell layer of the cerebellum, *Neuron* **58**, pp. 775–788.

[35] Denker, M., Szucs, A., Pinto, R., Abarbanel, H. and Selverston, A. (2006). A network of electronic neural oscillators reproduces the dynamics of the periodically forced pyloric pacemaker group, *IEEE Trans. Biomed. Engin.* **52**, pp. 792–798.

[36] Denker, M., Timme, M., Diesmann, M., Wolf, F. and Geisel, T. (2004). Breaking synchrony by heterogeneity in complex networks, *Phys. Rev. Lett.* **92**, p. 074103.

[37] Derrida, B., Gardner, E. and Zippelius, A. (1987). An exactly solvable asymmetric neural network model, *Europhys. Lett.* **4**, pp. 167–173.

[38] Diesmann, M., Gewaltig, M.-O. and Aertsen, A. (1999). Stable propagation of synchronous spiking in cortical neural networks, *Nature* **402**, pp. 529–533.

[39] Ermentrout, B. and Kopell, N. (1986). Parabolic bursting in an excitable system coupled with a slow oscillation, *SIAM J. Appl. Math.* **2**, pp. 233–253.

[40] Ernst, U., Pawelzik, K. and Geisel, T. (1995). Synchronization induced by temporal delays in pulse-coupled oscillators, *Phys. Rev. Lett.* **74**, pp. 1570–1573.

[41] Ernst, U., Pawelzik, K. and Geisel, T. (1998). Delay induced multistable synchronization of biological oscillators, *Phys. Rev. E* **57**, pp. 2150–2162.

[42] Fourcaud-Trocmé, N. and Brunel, N. (2005). Dynamics of the instantaneous firing rate in response to changes in input statistics, *J. Comp. Neurosci.* **18**, pp. 311–321.

[43] Fourcaud-Trocmé, N., Hansel, D., van Vreeswijk, C. and Brunel, N. (2003). How spike generation mechanisms determine the neuronal response to fluctuating inputs, *J. Neurosci.* **23**, pp. 11628–11640.

[44] Gansel, K. and Singer, W. (2005). Replay of second-order spike patterns with millisecond precision in the visual cortex, Soc. Neurosci. Abstr. 276.8.

[45] Gansel, K. and Singer, W. (2007). Repeating spatiotemporal spike patterns reflect functional network states in the visual cortex, NCCD Meeting Abstr.

[46] Gasparini, S. and Magee, J. (2006). State-dependent dendritic computation in hippocampal CA1 pyramidal neurons, *J. Neurosci.* **26**, pp. 2088–2100.

[47] Gasparini, S., Migliore, M. and Magee, J. (2004). On the initiation and propagation of dendritic spikes in CA1 pyramidal neurons, *J. Neurosci.* **24**, pp. 11046–11056.

[48] Geisler, C., Brunel, N. and Wang, X.-J. (2005). Contributions of intrinsic membrane dynamics to fast network oscillations with irregular neuronal discharges, *J. Neurophysiol.* **94**, pp. 4344–4361.

[49] Gerstein, G. and Mandelbrot, B. (1964). Random walk models for the spike activity of a single neuron, *Biophys. J.* **4**, pp. 41–68.

[50] Gerstner, W. (1996). Rapid phase locking in systems of pulse-coupled oscillators with delays, *Phys. Rev. Lett.* **76**, pp. 1755–1758.

[51] Gerstner, W. and Kistler, W. (2001). *Spiking Neuron Models: Single Neurons, Populations, Plasticity* (Cambridge Univ. Press, Cambridge).

[52] Gerstner, W., van Hemmen, L. and Cowan, J. (1996). What matters in neuronal locking? *Neural Comp.* **8**, pp. 1653–1676.

[53] Gewaltig, M.-O., Diesmann, M. and Aertsen, A. (2001). Propagation of cortical synfire activity: Survival probability in single trials and stability of the mean, *Neural Netw.* **14**, pp. 657–673.

[54] Gong, P. and van Leeuwen, C. (2007). Dynamically maintained spike timing sequences in networks of pulse-coupled oscillators with delays, *Phys. Rev. Lett.* **98**, p. 048104.

[55] Gray, C. and Singer, W. (1989). Stimulus-specific neuronal oscillations in orientation columns of cat visual cortex, *Proc. Natl. Acad. Sci.* **86**, pp. 1698–1702.

[56] Griffith, J. (1963). On the stability of brain-like structures, *Biophys. J.* **3**, pp. 299–308.

[57] Hansel, D. and coworkers (2008). In preparation.

[58] Hansel, D. and Mato, G. (2001). Existence and stability of persistent states in large neuronal networks, *Phys. Rev. Lett.* **86**, pp. 4175–4178.

[59] Hansel, D. and Mato, G. (2003). Asynchronous states and the emergence of synchrony in large networks of interacting excitatory and inhibitory neurons, *Neural Comp.* **15**, pp. 1 – 56.

[60] Hansel, D., Mato, G. and Meunier, C. (1995). Synchrony in excitatory neural networks, *Neural Comp.* **7**, pp. 307–337.

[61] Hayon, G., Abeles, M. and Lehmann, D. (2003). A model for representing the dynamics of a system of synfire chains, *J. Comp. Neurosci.* **18**, pp. 41–53.

[62] Herrmann, M., Hertz, J. and Prügel-Bennett, A. (1995). Analysis of synfire chains, *Network* **6**, pp. 403–414.

[63] Hertz, J. and Prügel-Bennett, A. (1996). Learning short synfire chains by self-organization, *Network* **7**, pp. 357–363.

[64] Hindmarsh, J. and Rose, R. (1984). A model of neuronal bursting using three coupled first order differential equations, *Proc. R. Soc. Lond. Ser. B* **221**, pp. 87–102.

[65] Hodgkin, A. and Huxley, A. (1952). A quantitative description of membrane current and its application to conduction and excitation in nerve, *J. Physiol.* **117**, pp. 500–544.

[66] Hopfield, J. (1982). Neural networks and physical systems with emergent collective computational abilities, *Proc. Natl. Acad. Sci.* **79**, pp. 2554–2558.

[67] Ikegaya, Y., Aaron, G., Cossart, R., Aronov, D., Lampl, I., Ferster, D. and Yuste, R. (2004). Synfire chains and cortical songs: Temporal modules of cortical activity, *Science* **304**, pp. 559–564.

[68] Itzhikevich, E. (2005). Polychronization: Computation with spikes, *Neural Comp.* **18**, pp. 245–282.

[69] Itzhikevich, E., Gally, J. and Edelman, G. (2004). Spike-timing dynamics of neuronal groups, *Cereb. Cortex* **14**, pp. 933–944.

[70] Izhikevich, E. (2003). Simple model of spiking neurons, *IEEE Trans. Neur. Netw.* **14**, pp. 1569–1572.

[71] Izhikevich, E. (2007). *Dynamical Systems in Neuroscience: The Geometry of Excitability and Bursting* (MIT Press, Cambridge).

[72] Jaeger, H. and Haas, H. (2004). Harnessing nonlinearity: Predicting chaotic systems and saving energy in wireless communication, *Science* **304**, pp. 78–80.

[73] Jahnke, S., Memmesheimer, R.-M. and Timme, M. (2009). How chaotic is the balanced state? *Frontiers in Comput. Neurosci.*, under review.

[74] Jahnke, S., Memmesheimer, R.-M. and Timme, M. (2008). Stable irregular dynamics in complex neural networks, *Phys. Rev. Lett.* **100**, p. 048102.

[75] Jin, D. (2002). Fast convergence of spike sequences to periodic patterns in recurrent networks, *Phys. Rev. Lett.* **89**, p. 208102.

[76] Jolivet, R., Lewis, T. and Gerstner, W. (2004). Generalized integrate-and-fire models of neuronal activity approximate spike trains of a detailed model to a high degree of accuracy, *J. Neurophysiol.* **92**, pp. 959–976.

[77] Jolivet, R., Rauch, A., Lscher, H.-R. and Gerstner, W. (2006). Integrate-and-fire models with adaptation are good enough: Predicting spike times under random current injection, in M. Taketani and M. Baudry (eds.), *Advances in network electrophysiology using multi-electrode arrays* (Springer).

[78] Jung, p. (1995). Stochastic resonance and optimal design of threshold detectors, *Phys. Lett. A* **207**, pp. 93–104.

[79] Kandel, E., Schwartz, J. and Jessell, T. (1995). *Principles of Neural Science* (Prentice Hall, London).

[80] Kestler, J. and Kinzel, W. (2006). Multifractal distribution of spike intervals for two neurons with unreliable synapses, *J. Phys. A* **39**, pp. L461–466.

[81] Kielblock, H., Kirst, C. and Timme, M. (2008). Breakdown of order preservation in networks of pulse-coupled oscillators with permutation symmetry, Under review.

[82] Kirst, C., Geisel, T. and Timme, M. (2009). Sequential desynchronization in networks of spiking neurons with partial reset, *Phys. Rev. Lett.* **102**, 068101.

[83] Kumar, A., Rotter, S. and Aertsen, A. (2008). Conditions for propagating synchronous spiking and asynchronous firing rates in a cortical network model, *J. Neurosci.* **28**, pp. 5268–5280.

[84] Kumar, A., Schrader, S., Aertsen, A. and Rotter, S. (2007). The high-conductance state of cortical networks, *Neural Comp.* **20**, pp. 1–34.

[85] Kuramoto, Y. and Battogtokh, D. (2002). Coexistence of coherence and incoherence in nonlocally coupled phase oscillators, *Nonlinear Phenom. Complex Sys.* **5**, pp. 380–385.

[86] Lapicque, L. (1907). Recherches quantitatives sur l'excitation electrique des nerfs traiteé comme une polarisation, *J. Physiol. Pathol. Gen.* **9**, p. 357.

[87] Lestienne, R. and Strehler, B. (1987). Time structure and stimulus dependence of precisely replicating patterns present in monkey cortical neuronal spike trains, *Brain Res.* **437**, pp. 214–238.

[88] Levina, A., Herrmann, J. and Geisel, T. (2007). Dynamical synapses causing self-organized criticality in neural networks, *Nat. Phys.* **3**, pp. 857–860.

[89] Levina, A., Herrmann, J. and Geisel, T. (2008). In preparation.

[90] Levy, N., Horn, D., Meilijson, I. and Ruppin, E. (2001). Distributed synchrony in a cell assembly of spiking neurons, *Neural Netw.* **14**, pp. 815–824.

[91] Little, W. (1974). The existence of persistent states in the brain, *Math. Biosci.* **39**, pp. 281–290.

[92] Little, W. and Shaw, G. (1975). A statistical theory of short and long term memory, *Behav. Biol.* **14**, pp. 115–133.

[93] Loebel, A. and Tsodyks, M. (2002). Computation by ensemble synchronization in recurrent networks with synaptic depression, *J. Comp. Neurosci.* **13**, pp. 111–124.

[94] Luczak, A., Bartho, P., Marguet, S., Buzsaki, G. and Harris, K. (2007). Sequential structure of neocortical spontaneous activity in vivo, *Proc. Natl. Acad. Sci.* **104**, pp. 347–452.

[95] Maass, W. and Natschläger, T. (2002). Real-time computing without stable states: A new framework for neural computation based on perturbations, *Neural Comp.* **14**, pp. 2531–2560.

[96] Mainen, Z. and Sejnowski, T. (1995). Reliability of spike timing in neocortical neurons, *Science* **268**, pp. 1503–1506.

[97] Makarov, V., Panetsos, F. and de Feo, O. (2005). A method for determining neural connectivity and inferring the underlying network dynamics using extracellular spike recordings, *J. Neurosci. Meth.* **144**, pp. 265–279.

[98] Matus Bloch, I. and Romero Z., C. (2002). Firing sequence storage using inhibitory synapses in networks of pulsatil nonhomogeneous integrate-and-fire neural oscillators, *Phys. Rev. E* **66**, p. 036127.

[99] McLelland, D. and Paulsen, O. (2007). Cortical songs revisited: A lesson in statistics, *Neuron* **53**, pp. 319–321.

[100] Mead, C. (1989). *Analog VLSI and Neural Systems* (Addison Wesley, Reading, MA).

[101] Mehring, C., Hehl, U., Kubo, M., Diesmann, M. and Aertsen, A. (2003). Activity dynamics and propagation of synchronous spiking in locally connected random networks, *Biol. Cybern.* **88**, pp. 395–408.

[102] Memmesheimer, R.-M. (2007). Precise spike timing in complex neural networks, Doctoral thesis, Department of Physics, Georg-August University of Göttingen.

[103] Memmesheimer, R.-M. and Timme, M. (2006). Designing complex networks, *Physica D* **224**, pp. 182–201.

[104] Memmesheimer, R.-M. and Timme, M. (2006). Designing the dynamics of spiking neural networks, *Phys. Rev. Lett.* **97**, p. 188101.

[105] Memmesheimer, R.-M. and Timme, M. (2009). *Nonlinearity*, under revision.

[106] Memmesheimer, R.-M. and Timme, M. (2008). Non-additive coupling enables propagation of synchronous spiking activity in purely random networks, Under review.

[107] Milo, R., Shen-Orr, S., Itzkovitz, S., Kashtan, N., Chklovskii, D. and Alon, U. (2002). Network motifs: Simple building blocks of complex networks, *Science* **298**, pp. 824–827.

[108] Mirollo, R. and Strogatz, S. (1990). Synchronization of pulse coupled biological oscillators, *SIAM J. Appl. Math.* **50**, pp. 1645–1662.

[109] Mokeichev, A., Okun, M., Barak, O., Katz, Y., Ben-Shahar, O. and Lampl, I. (2007). Stochastic emergence of repeating cortical motifs in spontaneous membrane potential fluctuations in vivo, *Neuron* **53**, pp. 413–425.

[110] Mongillo, G. and Amit, D. (2001). Oscillations and irregular emission in networks of linear spiking neurons, *J. Comp. Neurosci.* **11**, pp. 249–261.

[111] Morris, C. and Lecar, H. (1981). Voltage oscillations in the barnacle giant muscle fiber, *Biophys. J.* **35**, pp. 193–213.

[112] Morrison, A., Aertsen, A. and Diesmann, M. (2007). Spike-timing-dependent plasticity in balanced random networks, *Neural Comp.* **19**, pp. 1437–1467.

[113] Nadasdy, Z., Hirase, H., Czurko, A., Csicsvari, J. and Buzsaki, G. (1999). Replay and time compression of recurring spike sequences in the hippocampus, *J. Neurosci.* **19**, pp. 9479–9507.

[114] Naundorf, B., Geisel, T. and Wolf, F. (2005). Action potential onset dynamics and the response speed of neuronal populations, *J. Comp. Neurosci.* **18**, pp. 297–309.

[115] Naundorf, B., Wolf, F. and Volgushev, M. (2006). Unique features of action potential initiation in cortical neurons, *Nature* **440**, pp. 1060–1063.

[116] Nevian, T., Larkum, M., Polsky, A. and Schiller, J. (2007). Properties of basal dendrites of layer 5 pyramidal neurons: A direct patch-clamp recording study, *Nat. Neurosci.* **10**, pp. 206–214.

[117] Nowak, L., Sanchez-Vivez, M. and McCormick, D. (1997). Influence of low and high frequency inputs on spike timing in visual cortical neurons, *Cereb. Cortex* **7**, pp. 487–501.

[118] Omel'chenko, O., Maistrenko, Y. and Tass, P. (2008). Chimera states: The natural link between coherence and incoherence, *Phys. Rev. Lett.* **100**, p. 044105.

[119] Oram, M., Wiener, M., Lestienne, R. and Richmond, B. (1999). Stochastic nature of precisely timed spike patterns in visual system neuronal responses, *J. Neurophysiol.* **81**, pp. 3021–3033.

[120] Peskin, C. (1984). *Mathematical Aspects of Heart Physiology* (Courant Institute of Mathematical Sciences, New York University).

[121] Pipa, G., Riehle, A. and Grün, S. (2007). Validation of task-related excess of spike coincidences based on NeuroXidence, *Neurocomputing* **70**, pp. 2064–2068.

[122] Polsky, A., Mel, B. and Schiller, J. (2004). Computational subunits in thin dendrites of pyramidal cells, *Nat. Neurosci.* **7**, pp. 621–627.

[123] Prinz, A., Bucher, D. and Marder, E. (2004). Similar network activity from disparate circuit parameters, *Nat. Neurosci.* **7**, pp. 1345–1352.

[124] Prut, Y., Vaadia, E., Bergman, H., Haalman, I., Slovin, H. and Abeles, M. (1998). Spatio-temporal structure of cortical activity: Properties and behavioral relevance, *J. Neurophysiol.* **79**, pp. 2857–2874.

[125] Pulvermller, F. and Shtyrov, Y. (2008). Spatiotemporal signatures of large-scale synfire chains for speech processing as revealed by MEG, *Cereb. Cortex*, to be published.

[126] Purves, D., Augustine, G., Fitzpatrick, D., Katz, L., LaMantia, A.-S. and McNamara, J. (1997). *Neuroscience* (Sinauer, Sunderland).

[127] Ricciardi, L. (1976). Diffusion approximation for a multi-input model neuron, *Biol. Cybern.* **24**, pp. 237–240.

[128] Ricciardi, L. and Sacerdote, L. (1979). The Ornstein-Uhlenbeck process as a model for neuronal activity, *Biol.Cybern.* **35**, pp. 1–9.

[129] Richardson, M. (2007). Firing-rate response of linear and nonlinear integrate-and-fire neurons to modulated current-based and conductance-based synaptic drive, *Phys. Rev. E* **76**, p. 021919.

[130] Riehle, A., Grün, S., Diesmann, M. and Aertsen, A. (1997). Spike synchronization and rate modulation differentially involved in motor function, *Science* **278**, pp. 1950–1953.

[131] Rosenblatt, F. (1958). The perceptron: A probabilistic model for information storage and organization in the brain, *Psychol. Rev.* **65**, pp. 386–408.

[132] Rudolph, M. and Destexhe, A. (2006). Analytical integrate-and-fire neuron models with conductance-based dynamics for event-driven simulation, *Neural Comp.* **18**, pp. 2146–2210.

[133] Schaft, A. and Schumacher, J. (2000). *An introduction to hybrid dynamical systems* (Springer, London).

[134] Schrobsdorff, H., Herrmann, J. and Geisel, T. (2007). A feature-binding model with localized excitations, *Neurocomputing* **70**, pp. 1706–1710.

[135] Segev, I. (1992). Single neurone models: Oversimple, complex and reduced, *Trends Neurosci.* **15**, pp. 414–421.

[136] Senn, W. and Urbanczik, R. (2000). Similar nonleaky integrate and fire neurons with instantaneous couplings always synchronize, *SIAM J. Appl. Math.* **61**, pp. 1143–1155.

[137] Sethia, G., Sen, A. and Atay, F. (2008). Clustered chimera states in delay-coupled oscillator systems, *Phys. Rev. Lett.* **100**, p. 144102.

[138] Shadlen, M. and Newsome, W. (1998). The variable discharge of cortical neurons: Implications for connectivity, computation, and information coding, *J. Neurosci.* **18**, pp. 3870–3896.

[139] Shmiel, T., Drori, R., Shmiel, O., Ben-Shaul, Y., Nadasdy, Z., Shemesh, M., Teicher, M. and Abeles, M. (2005). Neurons of the cerebral cortex exhibit precise interspike timing in correspondence to behavior, *Proc. Natl. Acad. Sci.* **102**, pp. 18655–18657.

[140] Softky, W. and Koch, C. (1993). The highly irregular firing of cortical cells is insonsistent with temporal integration of random EPSPs, *J. Neurosci.* **13**, pp. 334–350.

[141] Song, S., Sjöström, P., Reigl, M., Nelson, S. and Chklovskii, D. (2005). Highly nonrandom features of synaptic connectivity in local cortical circuits, *PLoS Biology* **3**, p. 0507.

[142] Sporns, O. and Kötter, R. (2004). Motifs in brain networks, *PLoS Biology* **2**, p. e369.

[143] Svirskis, G. and Hounsgaard, J. (2003). Influence of membrane properties on spike synchronization in neurons: Theory and experiments. *Network* **14**, pp. 747–763.

[144] Tchumachenko, T. and Wolf, F. (2008). Correlations and synchrony in threshold neurons, In preparation.

[145] Tetzlaff, T., Geisel, T. and Diesmann, M. (2002). The ground state of cortical feed-forward networks, *Neurocomputing* **44-46**, pp. 673–678.

[146] Tetzlaff, T., Morrison, A., Geisel, T. and Diesmann, M. (2004). Consequences of realistic network size on the stability of embedded synfire chains, *Neurocomputing* **58-60**, pp. 117–121.

[147] Timme, M. (2007). Revealing network connectivity from response dynamics, *Phys. Rev. Lett.* **98**, p. 224101.

[148] Timme, M. and Wolf, F. (2008). The simplest problem in the collective dynamics of neural networks: Is synchrony stable? *Nonlinearity* **21**, pp. 1579–1599.

[149] Timme, M., Wolf, F. and Geisel, T. (2002). Coexistence of regular and irregular dynamics in complex networks of pulse-coupled oscillators, *Phys. Rev. Lett.* **89**, p. 258701.

[150] Timme, M., Wolf, F. and Geisel, T. (2002). Prevalence of unstable attractors in networks of pulse-coupled oscillators, *Phys. Rev. Lett.* **89**, p. 154105.

[151] Timme, M., Wolf, F. and Geisel, T. (2003). Unstable attractors induce perpetual synchronization and desynchronization, *Chaos* **13**, p. 377.

[152] Timme, M., Wolf, F. and Geisel, T. (2004). Topological speed limits to network synchronization, *Phys. Rev. Lett.* **92**, p. 074101.

[153] Tsodyks, M. and Markram, H. (1997). The neural code between neocortical pyramidal neurons depends on neurotransmitter release probability, *Proc. Natl. Acad. Sci.* **94**, pp. 719–723.

[154] Tsodyks, M., Mitkov, I. and Sompolinsky, H. (1993). Pattern of synchrony in inhomogeneous networks of oscillators with pulse interactions, *Phys. Rev. Lett.* **71**, pp. 1280–1283.

[155] Tuckwell, H. (1988). *Introduction to theoretical neurobiology: Volume 1. Linear cable theory and dendritic structure* (Cambridge Univ. Press, Cambridge).

[156] Tuckwell, H. (1988). *Introduction to theoretical neurobiology: Volume 2. Nonlinear and stochastic theories* (Cambridge Univ. Press, Cambridge).

[157] Turrigiano, G. (2007). Homeostatic signaling: The positive side of negative feedback, *Curr. Opinion. Neurobiol.* **17**, pp. 318–324.

[158] Turrigiano, G. and Nelson, S. (2004). Homeostatic plasticity in the developing nervous system, *Nat. Rev. Neurosci.* **5**, pp. 97–107.

[159] van Rossum, M., Turrigiano, G. and Nelson, S. (2002). Fast propagation of firing rates through layered networks of noisy neurons, *J. Neurosci.* **2**, pp. 1956–1966.

[160] van Vreeswijk, C. (1996). Partial synchronization in populations of pulse-coupled oscillators, *Phys. Rev. E* **54**, pp. 5522–5537.

[161] van Vreeswijk, C., Abbott, L. and Ermentrout, G. (1996). When inhibition not excitation synchronizes neural firing, *J. Comp. Neurosci.* **1**, pp. 313–321.

[162] van Vreeswijk, C. and Sompolinsky, H. (1996). Chaos in neuronal networks with balanced excitatory and inhibitory activity, *Science* **274**, pp. 1724–1726.

[163] van Vreeswijk, C. and Sompolinsky, H. (1998). Chaotic balanced state in a model of cortical circuits, *Neural Comp.* **10**, p. 1321.

[164] Vogels, T. and Abbott, L. (2005). Signal propagation and logic gating in networks of integrate-and-fire neurons, *J. Neurosci.* **25**, pp. 10786–10795.

[165] Wilson, H. and Cowan, J. (1972). Excitatory and inhibitory interactions in localized populations of model neurons, *Biophys. J.* **12**, pp. 1–24.

[166] Yu, D., Righero, M. and Kocarev, L. (2006). Estimating topology of networks, *Phys. Rev. Lett.* **97**, p. 188701.

[167] Zillmer, R., Livi, R., Politi, A. and Torcini, A. (2006). Desynchronization in diluted neural networks, *Phys. Rev. E* **74**, p. 036203.

[168] Zillmer, R., Livi, R., Politi, A. and Torcini, A. (2007). Stability of the splay state in pulse-coupled networks, *Phys. Rev. E* **76**, p. 046102.

[169] Zumdieck, A., Timme, M., Geisel, T. and Wolf, F. (2004). Long chaotic transients in complex networks, *Phys. Rev. Lett.* **93**, p. 244103.

PART 3

Networks at the Individual and Population Levels

Chapter 14

Ideas for Moving Beyond Structure to Dynamics of Ecological Networks

Daniel B. Stouffer, Miguel A. Fortuna and Jordi Bascompte

Integrative Ecology Group
Estación Biológica de Doñana, CSIC
C/Americo Vespucio s/n, E-41092 Sevilla, Spain

14.1. Introduction

There are between seven and fifty million different species of plants and animals on Earth [Pimm and Raven (2000)]. About two-thirds of these species live in the tropics, largely in the tropical forests [Pimm and Raven (2000)]. In fact, studies show that about 30–50% of plant, amphibian, reptile, mammal, and bird species occur in 25 hotspots that occupy no more than 2% of the terrestrial land mass [Myers *et al.* (2000)]. It is believed that fish and other marine organisms are similarly concentrated [McAllister *et al.* (1994)].

The concentration of natural species demands that hotspots be managed with particular attention and caution [Ceballos *et al.* (2005); Ceballos and Ehrlich (2006); Hurlbert and Jetz (2007)]. Unfortunately only about one half of the original 16 million square kilometers of tropical rain forests remain [Skole and Tucker (1993)] and clearing eliminates about 0.2 million square kilometers every year [Nepstad *et al.* (1999); Cochrane *et al.* (1999); Hansen *et al.* (2008)]. This and other factors, such as human population growth and global warming, place us in the midst of the sixth largest extinction event in natural history [Thomas *et al.* (2004)].

The impacts are far reaching as extinctions of species represent one of the most dramatic ecosystem perturbations, taking place on quicker time scales than evolution and the introduction of new species into a habitat [Thomas *et al.* (2004)]. Extinctions have the ability to greatly alter an ecosystem's biodiversity; they can affect ecosystem stability, its resilience to environmental change, or its resistance to invasion of exotic species [Chapin *et al.* (2000)].

In the ocean, the story is no different. Currently 75% of global fish stocks are fully- or over-exploited [United Nations Food and Agriculture Organization (2006)]. Amongst these stocks are a considerable number of predators, such as sharks, which occupy the highest trophic levels; it is observed that these species have been declining at an alarming pace [Myers and Worm (2003); Worm *et al.* (2005); Heithaus *et al.* (2008)]. Moreover, recent empirical and theoretical studies have demonstrated that top predator removal often induces large-scale cascading effects [Bascompte *et al.* (2005)]. Many developed nations have made progress in better fishery management [Griffith (2008)]. However, nearly 20% of fish stocks in the United States, for example, remain over-fished or are fished unsustainably [National Marine Fishery Service (2008)]. The problem is more dire in developing countries, many of which rely upon fishing as an important economic activity [United Nations Food and Agriculture Organization (2006)].

There are numerous ecological, environmental, economic, and social aspects to the problems which stand before us. In order to remedy them, and avoid recreating such problems in the future, we must develop effective environmental policies which are firmly based on current environmental and ecological research. Reflecting upon the recent past and prospective future, research in the field of ecological networks is a strong candidate to successfully lead us in this direction.

In more traditional ecology, studies are restricted to an analysis of one to a few species within an ecosystem or of one habitat patch within a landscape. In the study of food webs or mutualistic networks, in contrast, the focus is upon understanding the properties of the *entire* ecosystem [Bascompte *et al.* (2003); Pascual and Dunne (2006)]. Similarly, in the study of spatial networks, the focus is upon understanding the properties of the entire landscape by applying the network formalism to problems in spatial ecology [Urban and Keitt (2001)].

The network approach in ecology has the longest tradition in community ecology and food webs, where it has been utilized over the last thirty years [Cohen (1978); Pimm (2002)]. Even so, the static and structural properties of food webs are only recently becoming better understood partly because a recent acquisition of more resolved data [Pascual and Dunne (2006)]. This leaves open a number of important questions regarding food-web dynamics and stability. Similar conclusions are reached when reviewing the mutualistic and spatial network literature.

If we wish to transform ecology, and in particular ecological networks, from a descriptive to predictive science, we must move beyond static characterization to the topic of dynamics upon these network structures. Consider, for example, the issue of overfishing. The effect of overfishing of a single species cannot be considered in isolation; it is imperative that the dynamics be viewed as a component of a far larger and more complex system. To effectively manage fisheries, then, we must understand the dynamics of the complete ecosystem and the influences of individual species upon all others. Similarly in spatial networks, we understand that species cannot be managed from the perspective of isolated local processes but through a combination of local and regional dynamics and exchanges across the system.

We will cover the topic of ecological networks in three separate sections which focus on the most developed sub-fields: food webs, mutualistic networks, and spatial networks. In this chapter, we authors shall not provide comprehensive reviews of previous work in the field of ecological networks. We will instead first briefly discuss the state of the field, emphasizing efforts and conclusions in understanding the systems' structural properties. We then discuss examples of investigations of the networks' dynamics, followed by what we believe to be some of the most exciting unanswered questions regarding the dynamic behavior, both for the near and far term.

14.2. Food Webs

The tremendous diversity of ecosystems around the globe is apparent to even the most casual of observers. These ecosystems can differ in the population, sizes, and type of species present, the type of environment, the assembly history, and the rate of change. This diversity poses a very real challenge to the development of a general understanding of community dynamics.

Food webs are a description of who eats whom in an ecosystem [Sugihara (1984); Cohen *et al.* (1990)]. The food webs reported in the literature appear increasingly complex [Cohen *et al.* (1990); Pimm (2002); Pascual and Dunne (2006)]. Understanding the structure of food webs is of great importance because it provides insights into, for example, how ecosystems behave under perturbations [Berlow (1999); Chapin *et al.* (2000); McCann (2000)].

14.2.1. *The structure of food webs*

Recent research on the structure of food webs has lead to a solution to the problem of developing a general understanding: there exist a number of universal features that hold for a large number of empirical food webs [Camacho *et al.* (2002); Dunne *et al.* (2002b); Cohen *et al.* (2003); Bascompte and Melián (2005); Stouffer *et al.* (2005); Pascual and Dunne (2006); Stouffer *et al.* (2006); Camacho *et al.* (2007); Stouffer *et al.* (2007); Allesina *et al.* (2008); Dunne *et al.* (2008); Petchey *et al.* (2008); Williams and Martinez (2008)]. These quantitative patterns describe the most highly-resolved empirical food webs available in the literature (Fig. 14.1). The most remarkable aspect of this result is that there were no *a priori* reasons to believe that the food webs studied have anything in common; in fact, the food webs studied have different sizes (ranging from 25 to 155 species), the species themselves are different, the empirical data was collected by different investigators, and they come from a variety of environments located across the globe — lakes, streams, deserts, rain forests, estuaries, bays, and islands.

Relying upon this "universality" found in food webs, scientists have begun to reach a consensus on their static structural characteristics [Pascual and Dunne (2006); Williams and Martinez (2008); Allesina *et al.* (2008)]. These universal patterns demonstrate that there may indeed be fundamental principles which act as the determinants of food-web structure. Reinforcing this conclusion is a study by Dunne *et al.* (2008) which demonstrated that Cambrian food webs also exhibit the

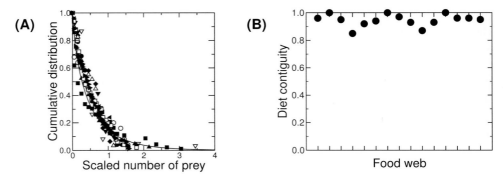

Fig. 14.1. Example of "universal" features of food web structure. (**A**): We plot the cumulative distribution of the scaled number of prey per species for 11 empirical food webs. The values for each web are scaled by twice the average number of predator-prey interactions per species in that web. The solid line is the analytical prediction for many successful static food-web models, including the niche model [Williams and Martinez (2000)]. The cumulative probability measures the fraction of events in a sample which has values larger than the set value. That the data "collapse" onto the same curve is a hallmark of universality. The same phenomenon is observed for the distributions of numbers of predators and links. (Adapted from Stouffer *et al.* (2005).) (**B**): We show the degree of diet contiguity for 14 empirical food webs, as measured by Stouffer *et al.* (2007). All 14 empirical food webs exhibit contiguity very close to 1.0, implying that species and their diets can very nearly be mapped onto a single dimension. This fact validates an important assumption of the niche model in which species and diets are explicitly one-dimensional.

same structural properties. An explanation for this universality is the principle that there are emergent properties in complex systems which arise from constraints acting upon the system [Amaral and Ottino (2004)]. While bioenergetic constraints could be considered a major factor controlling food-web structure, it actually appears that the manner in which species select their prey may be a stronger driver force [Williams and Martinez (2000); Stouffer *et al.* (2006, 2007); Petchey *et al.* (2008); Williams and Martinez (2008)]. These robust patterns have led to the development [Williams and Martinez (2000)], and subsequent validations [Stouffer *et al.* (2005, 2006, 2007); Williams and Martinez (2008)] of a simple static food web model — the so-called niche model — that reproduces the key structural features observed in empirical food webs.

To this point, researchers who study food webs have followed a maxim observed in other disciplines, such as physics: one must understand statics before attempting to understand dynamics. As noted above, however, ecologists have begun to reach a consensus regarding the structure and mechanisms which shape the network of predator-prey interactions [Pascual and Dunne (2006); Stouffer *et al.* (2006); Camacho *et al.* (2007); Stouffer *et al.* (2007)]; therefore we believe that the time to advance our understanding of community-wide dynamics is at hand. In the following sub-sections, we will outline what we believe to be some of the most exciting and promising possibilities regarding the topic of food-web dynamics.

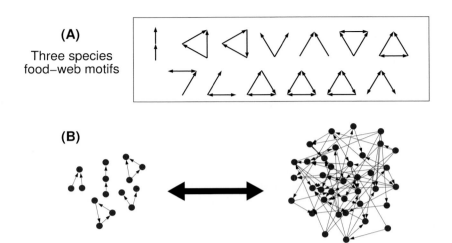

Fig. 14.2.   Food-web motifs and the scale of food-web stability. (**A**): The set of 13 *unique* food-web motifs composed of three species [Milo *et al.* (2002); Stouffer *et al.* (2007)]. Each vertex represents a species and arrows represent predator-prey interactions pointing from prey to predator. (**B**): A significant unanswered question in the study of food webs is how the dynamic properties and stability of food-web motifs relate to those of the entire food web.

14.2.2.   *The scale of food-web stability*

The complexity of empirical food webs has lead many ecologists to concentrate on dynamic ecosystem studies in terms of small sub-webs — "community modules" [Holt (1997); Holt and Hochberg (2001)] — which bridge the gap between "the baroque complexity of entire communities and the bare bones of single and pair-wise population dynamics" [Holt (1997)]. Community modules are comprised of three to five species and a set of interactions likely to have ecological relevance, such as a three-species food chain or intraguild predation.

The dynamic stability of community modules has been investigated previously; particular focus has been placed upon the implications of varying species interaction strengths and the role of weak interactions [McCann *et al.* (1998); Fussmann and Heber (2002); Emmerson and Yearsley (2004); Bascompte *et al.* (2005); Nakazawa and Yamamura (2006); Rooney *et al.* (2006)]. More recent research has also uncovered the important role of predator-prey body size ratios in stabilization of specific community modules such as a tri-trophic food chain [Otto *et al.* (2007)]. These bottom-up studies provide a theoretical foundation for the stability of individual modules, but leave open the question of which modules empirical food webs are actually composed of. To address this question, scientists have investigated "food-web motifs", the structural counter-part to community modules [Milo *et al.* (2002); Bascompte and Melián (2005); Stouffer *et al.* (2007)]. Food-web motifs consist of the complete set of unique connected subgraphs containing *n*-species [Milo *et al.* (2002)] (Fig. 14.2). These top-down studies provide an indication of the subgraphs which appear within empirical food webs more or less often than expected at random. They are unable, however, to address if there is a dynamic justification for why a particular subgraph appears with greater or less frequency.

It thus follows that a significant unanswered question in the study of food-webs is the integration of the bottom-up and top-down approaches to better understand food-web dynamics and the origins of food-web stability. Is there a direct relationship between the "local" stability of a subgraph and on whether or not that subgraph appears more or less frequently in a complete food web? If no direct relationship exists, it is worth noting that this does not preclude the existence of a relationship at a "mesoscale" level of combinations of motifs.

The identification of each modules' contribution to community stability will provide crucial information regarding the forces acting upon the ecosystem. We will be able to answer an important question regarding the mechanisms responsible for maintaining stability in the presence of external perturbations. Namely, is a community stable because it is composed of stable sub-elements, or is a community stable because of cooperative and synergistic interactions between individual, and potentially unstable, sub-elements? Do species participate in interactions which would maximize their own persistence or instead that of the entire community?

14.2.3. *Whole food-web dynamics*

An ever-present problem in the study of empirical food webs is the relative scarcity of empirical data. Much of this stems directly from the difficulties involved in collecting data of high quality [Paine (1988)]. These difficulties including the long hours required for direct observation and for data collection to conduct stomach content analysis or scatology. Additionally, it is still an issue to know when a scientist may faithfully declare any data as complete; how does one account for the brief and sole appearance of a migratory bird, for example, within the environment under investigation? Consider also that these difficulties are faced when the food web will ultimately represent an aggregate over time and space [Lawton (1989)] and therefore the result is generally absent of dynamic data.

To date, the most frequently utilized approach to quantitatively measure a dynamic food-web property is through the interaction strength [Berlow *et al.* (2004)]. Most commonly this is represented as the fraction of a species' diet or incoming biomass which comes from particular prey; however, many definitions exist Berlow *et al.* (2004). While such interaction strengths are informative and can even help parametrize dynamics models [Yodzis and Innes (1992); Brose *et al.* (2006)], they themselves often represent averages across time and space. This is because of the fact that ecosystem dynamics necessarily implies dynamics of all aspects of the food web, from abundances to species' interaction strengths.

For future characterization and understanding of community dynamics, it is essential that we support initiatives for the collection of dynamic data for entire ecosystems, potentially even to the scale of the Long Term Ecological Research

Network (LTER) or National Ecological Observatory Network (NEON) programs in the United States. Such initiatives, however, will represent decades of coordinated work with the majority of benefits realized far into the future.

A different possibility is to approach the problem from the bottom-up instead of top-down. There exist databases (e.g., NERC Centre for Population Biology (1999)) which have compiled species-specific dynamic data. However, as in the case off food-web stability discussed earlier, it is not readily apparent how laboratory or field experiments of three species, for example, might translate to dynamics of greater collections of species.

One should consider that the species-specific data often represent the dynamics of individual species not in isolation but actually embedded within larger communities. We must first develop a robust and general characterization of the dynamic behaviors exhibited in this species-specific data. This could be conducted through a mixture of data analysis and modeling. Then the dynamics of species within food-web models should be similarly characterized for comparison purposes. These two characterizations will greatly help the refining of dynamic food-web models by determining cause of differences between dynamics in isolation or within an entire food web.

14.3. Mutualistic Networks

To this point, we have defined the broad type of food web, which is the type of network with a longest tradition in ecological research. In the last few years, there has been a tendency to consider other types of interactions besides the strict who-eats-whom. The interactions between hosts and their animal parasitoids (i.e., insects that lay their eggs near, inside, or on the surface of the insect host which will provide food for the developing pupa) is a good example. A second type of interaction is that between plants and their specific herbivores, or the one between plants and the animals that pollinate their flowers or disperse their seeds. In all cases, we are now considering networks represented as bipartite networks (Fig. 14.3). In this section, we will focus on one type, that which describes the mutualistic interactions between plants and animals. There are several reasons for this. On one hand, these interactions of mutual benefit have played a major role in the generation of Earth's biodiversity [Ehrlich and Raven (1964); Thompson (2005)]. On the other hand, ecologists have compiled an impressive data set amenable to analysis by tools from the study of complex networks.

Mutualistic interactions involve dozens to hundreds of species and form complex networks (Fig. 14.3). This is a complicated object, and one may be tempted to conclude that there is no pattern beyond an ocean of links and nodes. The earliest work on mutualisms was focusing on a highly specific plant-animal interaction, or on a few interactions among a target group of species. In the last few years, however, ecologists and evolutionary biologists acknowledged the community context of these

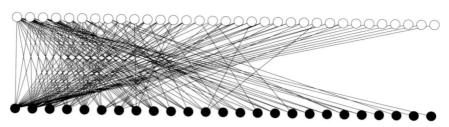

Fig. 14.3.   Example of a bipartite plant-animal mutualistic network. Solid and empty nodes represent plant species, and their seed dispersers, respectively. These mutually beneficial interactions form networks of dependence involving dozens of species.

interactions and worked on small groups of interacting species [Waser *et al.* (1996); Thompson (2005)]. More recently, these studies have been extended by scaling all the way up from small to entire networks and by providing a rational framework to characterize their complexity.

14.3.1. *The structure of mutualistic networks*

The first round of papers on mutualistic networks were, not surprisingly, descriptions of their structure, following the tradition in food webs as described in the previous section. The first pattern analyzed was the connectivity distribution, and ecologists found a common structural pattern defined by truncated power-law connectivity distributions. This put mutualistic networks in a similar context as other networks because they exhibit heterogeneous connectivity distributions. While the bulk of species had only one or a few interactions with other species, a few generalists had far more interactions than would be expected under conventional models [Jordano *et al.* (2003)]. Regardless of the type of mutualism (seed dispersal or pollination) and other biological differences, all communities had the same structure. A small difference, in relation to non-biological networks such as the Internet, was the truncation for high numbers of interactions. This truncation can be explained by several non-exclusive mechanisms such as forbidden links — some interactions between a plant and an animal can not occur due to, for example, size or phenological constraints [Jordano *et al.* (2003)].

   Mutualistic networks are, thus, heterogeneous. However, this property only refers to a statistical description of the probability to interact with a given number of species. It does not tell us anything about the identity of the interacting species. The next step in the road to disentangle the structure of mutualistic networks was describing its nested structure (Fig. 14.4). In a nested matrix, the number of interactions per species are arranged in such a way that specialists interact with proper subsets of the species with which generalists interact [Bascompte *et al.* (2003)]. This generates a network with a core of interactions, generalist plants and generalist animals interacting among themselves, and asymmetric tails, specialists interacting with the most generalist species.

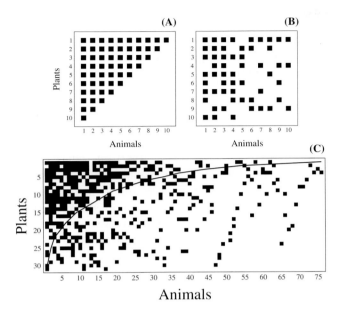

Fig. 14.4. The nested structure of mutualistic networks. The mutualistic network is now represented in matrix format, where rows represent plants, columns represent animals, and a square indicates an interaction between that plant and animal. **(A)** represents a perfectly nested matrix, where specialist species interact with well-defined subsets of the species generalists interact with, **(B)** represents a similar matrix with interactions randomly shuffled, and **(C)** shows a real mutualistic network where the continuous line indicates the isocline of perfect nestedness. (Figure modified from Bascompte *et al.* (2003).)

This asymmetry in specialization, independently reported by Vázquez and Aizen (2004), is also evident when looking at weighted networks. The weight of a link indicates the intensity of the mutualistic effect or dependence of one species on another. The bulk of dependencies between species are quite weak, but a few are strong. In the few cases in which, for example, a plant depends strongly on an animal, the animal depends on the plant very weakly, and vice versa [Bascompte *et al.* (2006)]. That is, pair-wise interactions are very asymmetric. Similarly, one can also look at the quantitative extension of species degree, species strength. The strength of a plant species, for example, is the sum of all dependencies of the animals which depend on that plant. It describes the importance of a species from the point of view of those in the other set. Species strength varies widely across species [Bascompte *et al.* (2006)]. As for binary data, mutualistic networks are also very heterogeneous in the distribution of species strength.

14.3.2. *Assembly of mutualistic networks*

So far, we have provided a descriptive account of mutualistic networks. The dynamics, as always, are more complicated to tackle from the experimental point of view. There are a few exceptions from the literature, however, that provide a link from structure to dynamics. Olesen *et al.* (2008) analyzed the day-to-day assembly of one pollination network in the Zackenberg field station, Greenland. This is a particularly convenient system since the whole area is covered by ice the bulk of the year. This implies that from the beginning of each season, one can record the entire network assembly from the appearance of the first flower, the first pollinator, and so on.

One example of a connection between network structure and dynamics was the pioneering paper by Barabasi and Albert (1999) on the preferential-attachment mechanism as a simple process leading to a scale-free network. In this case, one starts with a core of interacting nodes; at each time step, a new node enters the network and attaches to an existing node with probability proportional to the existing node's number of links. This process results in a network with a scale-free degree distribution [Barabasi and Albert (1999)]. It is one thing, however, to theoretically show that a process leads to a particular statistical pattern and quite another to demonstrate that an empirical system truly behaves in such a manner. To elaborate, upon finding a power-law connectivity distribution it may be tempting to assume a preferential-attachment process. Similar difficulties have permeated the study of other complex systems when scientists were tempted to conclude the existence of self-organized criticality from the presence of a power-law in the frequency distribution of events with a given size or energy.

The Zackenberg pollination network provided a wonderful opportunity to test the true dynamical assembly mechanism in the field. The temporal sequence for the whole season can be broken into day-to-day intervals so that, when a new interaction is formed, one can identify the precise identity of the species a new one attaches to and then test relative to the likelihood that a species would attach preferentially to the most connected species (Fig. 14.5). Olesen *et al.* (2008) found that attachment is intermediate between preferential and random. Intriguingly, this dynamic result is compatible with the previously reported pattern of a truncated power-law connectivity distribution.

Despite the previous result on actual processes of assembly, we are missing information on the dynamics of mutualistic networks, i.e., on how they change through years (see however Petanidou *et al.* (2008) and Olesen *et al.* (2008) for interesting exceptions). The most natural way to proceed due to the scarcity of dynamic data, is to use numerical simulations or mathematical models of network dynamics. A few papers have addressed models of network build-up which are directly comparable to the actual assembly in Zackenberg. For example, defining a preferential-attachment model on a bipartite network has provided an analogy to the Barabási-Albert model for unipartite networks [Guimarães *et al.* (2007)].

Interestingly enough, the nature of these bipartite networks imposes itself some changes in the resulting connectivity distribution when there is some difference in size between the two types of nodes. In this case, a truncated power-law connectiv-

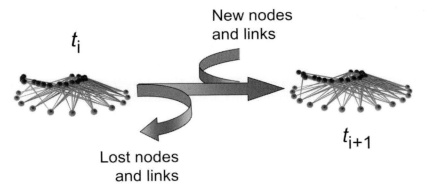

Fig. 14.5. The dynamics of network assembly. The figure illustrates the turnover in number of nodes and links between two consecutive days. Knowing whether new species tend to become attached to the already well-connected species is important in bridging structural patterns of mutualistic networks to their underlying assembly dynamics. (Figure courtesy of J. Olesen.)

ity distribution (as opposed to a power-law) naturally arises. Similarly, Santamaría and Rodríguez-Gironés (2007) have explored a suite of assembly models, concluding that the best fit to the structure of mutualistic networks is performed by a model that combined species abundance and phenotypic complementarity between morphological traits. In line with this previous result, phenotypic complementarity and the hierarchical evolutionary relationships between species have also been adduced to be a good explanation for the values of nestedness observed in nature [Rezende *et al.* (2007a)]. This is in agreement with conclusions from observations from the field [Stang *et al.* (2007)]. A combination of species abundance and niche-traits (e.g., morphological constraints) seems also to account for the levels of nestedness observed in nature [Krishna *et al.* (2008)].

We need general models that account not only for a network property such as nestedness but for a broad combination of network patterns, or, as Allesina *et al.* (2008) have recently put it, for the likelihood of reproducing an entire mutualistic network. We believe it is quite feasible to soon have such a generalized model of assembly of mutualistic networks which is able to reproduce the full range of network patterns discovered in the initial studies of mutualistic networks.

14.3.3. *Models of mutualistic-network disassembly*

A more difficult challenge — and a much more interesting question, as tends to be the case — would be to develop an experimental setting to examine the disassembly of mutualistic networks as a function of some external driver. This is quite difficult to realize in the field, and even in a mesocosm experiment, yet is probably the most relevant question from the point of view of exploring the consequences of global change on biodiversity. The most exciting possibility would be an experimental system where one could simulate the effects of species extinction or habitat alteration on the network of interactions. In the meantime, we again must rely on numerical simulations. In the most common methodology a progressive number of species goes extinct and the influence of network structure on the size of the co-extinction cascade, or the subsequent loss of evolutionary history, is investigated. The consensus result, as shown by Memmott *et al.* (2004) building on the papers by Albert *et al.* (2000) for the Internet and Solé and Montoya (2001) and Dunne *et al.* (2002a) for food webs, is that the heterogeneous, nested, structure of mutualistic networks makes them robust to the loss of specialists or to the random loss of species [Memmott *et al.* (2004)].

When one examines, however, the identities of the coextinct species and their phylogenetic positions, there is a significant phylogenetic signal on both the number of interactions per species and on whom they interact with [Rezende *et al.* (2007b)]; this is true for as many as half of the communities examined. Because of this, coextinction cascades do not involve randomly picked species but phylogenetically close species. The loss of evolutionary history then proceeds faster than expected in the absence of such phylogenetic signal, leading to a biased pruning of the evolutionary tree [Rezende *et al.* (2007b)].

Another driver of global change is habitat loss. There is only one study we are aware of that empirically studies the consequences of habitat transformation on network structure. Tylianakis *et al.* (2007) explored how the structure of host-parasitoid networks changes across an environmental gradient. They demonstrated that even without a reduction in the number of species, habitat loss changes the structure of interaction networks with important implications for their collapse. To further assess the influence of habitat transformation, requires the use of models.

An important area of research is the study of metacommunities, species interactions across discrete habitat patches maintained by a balance between local processes and regional dispersal across patches [Leibold *et al.* (2004); Holyoak *et al.* (2005)]. This important area has mainly focused on theoretical work consisting of a small number of interacting species. Melián *et al.* (2005) and Fortuna and Bascompte (2006), on the other hand, have studied real ecological networks across space from a theoretical perspective but with added realism by using the structure of the real ecological networks.

A first step towards understanding the consequences of habitat loss on mutualistic networks was the study of a spatially implicit model of metacommunities in which two real mutualistic networks were used as a skeleton for the theoretical model [Fortuna and Bascompte (2006)]. The heterogeneous, nested, structure of mutualistic networks confers a higher level of robustness to habitat loss. While species start to go extinct sooner than for expected at random, the network as a whole persists for greater values of habitat loss [Fortuna and Bascompte (2006)]. This is only a first step, however, because the metacommunity is assumed to live in an idealized spatially implicit model composed by an infinite number of patches with similar dispersal to any other patch. It maintains the structure of the mutualistic network but not the structure of the habitat landscape. We anticipate that the tendency to study full ecological networks on real ecological landscapes will be an important area in the near future. This will integrate networks of networks. The spatial component of networks has already generated important contributions which we will review in the following section. A next step will be to integrate all sections, that is, to study the spatial component of networks of ecological interactions.

14.4. Spatial Networks

We have thus far examined where we are in bridging the gap between structural understanding and system dynamics in food webs and plant-animal mutualistic networks. Such ecological networks depict relationships which principally affect the population growth rates of the interacting species. An altogether different type of ecological network, that representing spatial dynamics, has been also studied. Spatial dynamics influences both the organization and stability of communities [Tilman and Kareiva (1997); Hanski and Gilpin (1997); Bascompte and Solé (1998)] and hence, the spatial aspects in which ecological processes take place cannot be ignored. For example, in fragmented landscapes, the spatial distribution of habitat patches can influence on the dispersal movements of individuals [Wiens (2001)]. In plants, gene flow mediated by animals is a key demographic process which serves to shape the spatial pattern of intraspecific genetic variability [Barrett and Harder (1996)]. In animal societies, the relationships between individuals are often determined by the common use of resources patchily distributed across the landscape [Gibbons and Lindenmayer (2002)]. All these cases can be described and analyzed as networks in which nodes represent habitat resources and links indicate dispersal, gene flow, or social interactions. By considering spatial dynamics from a network perspective, we can shed light into problems as diverse as species persistence in spatially-extended environments and gene flow in plant populations.

14.4.1. *The structure of spatial networks*

The spatial dynamics which result from the dispersal of individuals from one population to another have been successfully studied using the metapopulation framework [Hanski and Gilpin (1997)]. In this approach, the probability for an empty habitat patch to be colonized depends, among other factors, on the distance between that patch and the rest of occupied patches [Ovaskainen and Hanski (2004)]. By considering different functional forms for the probability of an individual to reach a patch located at a particular distance (i.e., the dispersal kernel), we can build stochastic networks of connected patches in fragmented landscapes. The structure of the resulting networks of patches provides a straightforward way to quantify the robustness of a patchy population to habitat loss [Urban and Keitt (2001)]. It also provides useful information for conservation planning because it allows the identification of the most important patches — termed keystone patches — that

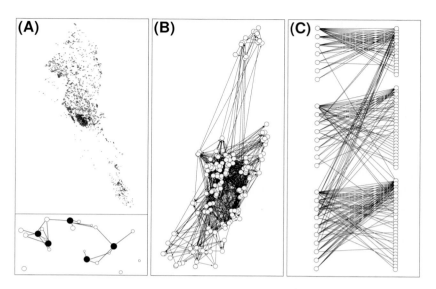

Fig. 14.6.   Examples of spatial networks. (**A**): Spatial location of more than 3000 ponds in Doñana National Park (Spain). Below, schematic representation of a subset of ponds showing links from dry (white circles) to flooded (black circles) ponds mediated by the maximum dispersal distances of an amphibian species. (**B**): The mating network of a *Prunus mahaleb* population. Nodes represent trees and arrows indicate pollination events mediated by insects. (**C**): The bipartite roosting network of a bird-predator bat. On the left, nodes represent individual bats. On the right, nodes represent individual trees. A link between a bat and a tree indicates that that particular bat used the particular tree. (Adapted from Fortuna *et al.* (2006, 2008a,b).)

are critical for landscape connectivity and hence population persistence [Urban and Keitt (2001); Minor and Urban (2008)].

There are a few key examples of how the spatial dynamics resulting from individual movements between habitat sites shapes the structure of a spatial network of patches. Fortuna *et al.* (2006) showed that the structure of a large network of temporary ponds — used as breeding sites by several amphibian species — changes with increasing levels of drought (Fig. 14.6A). When drought was very intense and the number of dry ponds increased, only a few flooded ponds were accessible from a high number of dry ones. In this way, a ranking of ponds was established as a function of their connections to dry ponds, constrained by the dispersal abilities of species. In similar systems, such as riverine networks, the fragmentation over time of network of populations of a fish species was investigated and used to propose mechanisms for ecologically successful restoration [Schick and Lindley (2007)].

The ecological and evolutionary processes shaping the spatial distribution of genetic variability include natural selection, genetic drift, mutation, non-random mating, and gene flow by migration, to name a few. The cessation of gene movement because of limited dispersal or assortative mating results in genetic isolation and allows drift and other microevolutionary processes to differentiate non-connected populations. This in turn can lead to speciation. Understanding the spatial structure of the genetic variation is therefore a major goal for conservation and evolutionary biology because it contributes to the management of threatened populations and, ultimately, can lead to evolutionary change.

To this end, molecular markers provide invaluable information about gene flow at individual and population levels. It is now possible to trace back pollen movement between the mother and the siring tree using paternity assignment techniques on collected seeds. One can then calculate the distance traveled by pollen carried by the wind or by insects. With sufficient seed collection and by knowing the precise identity and location of all the trees of a population, we can build the mating network. In the mating network, nodes represent trees and links indicate the pollen movement from donor to mother tree (Fig 14.6B). Fortuna *et al.* (2008a) have shown that there exists a non-random pattern of pollen movement in an insect-pollinated tree. The population was structured in well-defined compartments formed by groups of mother trees and their shared pollen donors. They also found that the few long-distance pollination events reduced the compartmentalized structure of the mating network, potentially increasing gene flow and hence reducing the role of genetic drift. As we will see later, there is a huge potential in the application of the complex network framework to gene-flow and population genetics.

Some species depend on resources patchily distributed across the landscape for shelter, such as hollows in trees [Gibbons and Lindenmayer (2002)]; bat colonies are a good example. Some bat species constitute fission-fusion societies whose members spread every day in multiple trees for shelter. The regular roost-switching movements of animals can be considered as channels that transport information or parasites among individuals. What structural patterns emerge from this spatial dynamics? In public parklands, the old trees used for shelter by bats are in danger of being removed by management agencies because of the potential danger to people from falling branches. In this case, the identification of the most important roosting locations would favor more efficient management solutions.

Using information on radio-tracked bats, Fortuna *et al.* (2008b) built a bipartite network establishing links between bats and trees when a particular bat used a particular tree (Fig. 14.6C). They observed the existence of well-defined colonies of bats, a compartmentalized structure similar to that found for the mating network of plants. Moreover, when the pattern of individual use of each tree inside each colony was examined, a nested network structure was detected. That is, as observed in mutualistic networks, bats using a few roosting trees are a subset of the bats that use trees used by a high number of bats. In the same way, trees that are used by a few bats are a subset of the trees used by bats that use many trees [Fortuna *et al.* (2008b)].

We have seen how the network approach can be successfully applied to characterize structural patterns emerging from spatially-distributed data in ecology. What can the topological properties of these networks tell us about the dynamics of spatial processes?

14.4.2. *Unraveling the dynamics of spatial networks*

Following on the last case study, the spatial structure of the roosting network can provide insights into the functionality of roost changes and social grouping or segregation by describing the way that information, diseases, or parasites can travel through the network. The compartmentalized structure implies that each colony of bats uses a subset of trees for shelter and bats from one colony only occasionally use trees belonging to the subset of trees used by the other colonies. This structure slows down the spread of diseases and the exchange of information through the entire network. The correlation between network structure and infection dynamics was showed by the use of a simple epidemiological model [Fortuna *et al.* (2008b)]. We can go further and ask what changes must happen to the spatial dynamics in order to maintain the structure of the roosting network? That is, if some trees disappear, will the colonies reorganize the use of trees in order to best preserve individual colonies' knowledge by minimizing the transfer of information about food resources to outsiders?

Considering instead the spatial network of temporary ponds used by amphibian species [Fortuna *et al.* (2006)]. Habitat loss experiments, analogous to the removal of species in food webs and plant-animal mutualistic networks, can tell us the importance of keystone patches on the persistence of the species. In the same way, we can estimate where would be the most suitable site to create a new pond in order to most effectively increase the connectivity during, for example, an especially intense dry season. It would also be interesting to explore how the structure of the network influences the dispersal movements of amphibian species. Could particular network topologies connecting the temporary ponds induce changes in the dispersal kernels of amphibians?

The most challenging and promising field for the application of networks of spatial dynamics is within population genetics. We have seen how network approaches can shed light into the characterization of gene flow patterns inside a tree population [Fortuna *et al.* (2008a)]. We can also investigate the dynamical implications of the patterns of intraspecific genetic variability among populations. These patterns can be well-characterized with spatial network approaches, as demonstrated by Dyer and Nason (2004). They described the statistical relationships between all populations simultaneously defining a "population graph" that can also be viewed as a "genetic landscape". In a genetic landscape, nodes represent populations and a link indicates a high degree of genetic similarity between them [Dyer and Nason (2004)]. What are the dynamical implications of the topology of the genetic landscape on microevolutionary processes, as these ultimately can lead to a significant evolutionary change? Do the dynamics of communities generate a universal spatial pattern in the distribution of genetic variability between species?

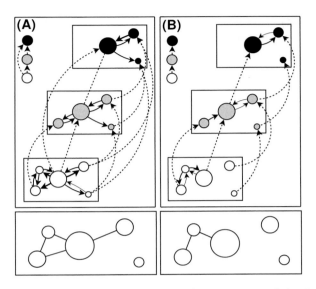

Fig. 14.7. Hypothetical diagram showing the effect of the structure of simple trophic modules such as **(A)**, omnivory and **(B)**, the simple trophic chain on the gene landscape of their constituent species in a spatial network of habitat patches. Each circle represents a patch holding a population. Black, gray, and white patches are occupied by the top, intermediate, and basal species, respectively. Solid arrows represent dispersal movements among populations (i.e., gene flow). Dashed arrows indicate trophic links between species occupying the same patches. Below, the gene landscape of the basal species is represented for both cases. In here, links among patches indicate dependent genetic covariance among them, that is, the holding populations are genetically similar. Isolated patches represent independent evolutionary units (i.e., their genetic covariance is independent of the other populations). If this hypothetical situation is observed in nature, it could imply that more complex networks (omnivory versus simple trophic chain) reduce the genetic structure of populations by maintaining the genetic diversity.

We can go further by coupling the networks of interacting species with the complex spatial context in which they live. The spatial distribution of species is heterogeneous. Different species are therefore found in different habitat patches. We can think about networks of interacting species embedded into the spatial networks of these habitat patches. These represent "meta-webs" which operate on multiple scales. One component of biodiversity depends on local processes, such as interactions among species, and the other component depends on regional processes, such as dispersal between habitat patches. By studying the dynamics of these networks in spatially explicit landscapes, we could integrate population genetics to explore how specific interactions among species shape the structure of gene landscapes (Fig. 14.7).

14.5. Concluding Remarks

Throughout this chapter, we hoped to leave the reader with at least one indelible idea. The time is now in ecology to stop considering species or patches as independent units but instead as parts of a larger, more complex, network of interacting pieces. The structural properties which describe these interactions — food webs, mutualistic networks, and spatial networks — have begun to take shape, but the understanding of these networks' dynamics is only just beginning. We have provided but a few examples of where work on ecological network dynamics can lead us. Wherever this happens to be, the future is bright, the potential results critical, and the possibilities endless.

References

Albert, R., Jeong, H. and Barabási, A. (2000). Error and attack tolerance of complex networks, *Nature* **406**, 6794, pp. 378–382.

Allesina, S., Alonso, D. and Pascual, M. (2008). A general model for food web structure. *Science* **320**, 5876, pp. 658–661.

Amaral, L. A. N. and Ottino, J. (2004). Complex networks: Augmenting the framework for the study of complex systems, *Eur. Phys. J. B* **38**, 2, pp. 147–162.

Barabasi, A. and Albert, R. (1999). Emergence of scaling in random networks, *Science* **286**, 5439, pp. 509–512.

Barrett, S. C. H. and Harder, L. D. (1996). Ecology and evolution of plant-mating, *Trends Ecol. Evol.* **11**, pp. 73–79.

Bascompte, J., Jordano, P., Melian, C. J. and Olesen, J. M. (2003). The nested assembly of plant-animal mutualistic networks, *Proc. Natl. Acad. Sci. USA* **100**, 16, pp. 9383–9387.

Bascompte, J., Jordano, P. and Olesen, J. M. (2006). Asymmetric coevolutionary networks facilitate biodiversity maintenance, *Science* **312**, 5772, pp. 431–433.

Bascompte, J. and Melián, C. J. (2005). Simple trophic modules for complex food webs, *Ecology* **86**, 11, pp. 2868–2873.

Bascompte, J., Melián, C. J. and Sala, E. (2005). Interaction strength combinations and the overfishing of a marine food web, *Proc. Natl. Acad. Sci. USA* **102**, 15, pp. 5443–5447.

Bascompte, J. and Solé, R. (eds.) (1998). *Modeling Spatiotemporal Dynamics in Ecology* (Springer-Verlag, Berlin).

Berlow, E. L. (1999). Strong effects of weak interactions in ecological communities, *Nature* **398**, pp. 330–334.

Berlow, E. L., Neutel, A.-M., Cohen, J. E., de Ruiter, P. C., Ebenman, B., Emmerson, M., Fox, J. W., Jansen, V. A. A., Iwan Jones, J., Kokkoris, G. D., Logofet, D. O., McKane, A. J., Montoya, J. M. and Petchey, O. (2004). Interaction strengths in food webs: issues and opportunities, *J. Anim. Ecol.* **73**, 3, pp. 585–598.

Brose, U., Jonsson, T., Berlow, E. L., Warren, P., Banasek-Richter, C., Bersier, L.-F., Blanchard, J. L., Brey, T., Carpenter, S. R., Blandenier, M.-F. C., Cushing, L., Dawah, H. A., Dell, T., Edwards, F., Harper-Smith, S., Jacob, U., Ledger, M. E., Martinez, N. D., Memmott, J., Mintenbeck, K., Pinnegar, J. K., Rall, B. C., Rayner, T. S., Reuman, D. C., Ruess, L., Ulrich, W., Williams, R. J., Woodward, G. and Cohen, J. E. (2006). Consumer-resource body-size relationships in natural food webs. *Ecology* **87**, 10, pp. 2411–2417.

Camacho, J., Guimerà, R. and Amaral, L. A. N. (2002). Robust patterns in food web structure, *Phys. Rev. Lett.* **88**, p. art. no. 228102.

Camacho, J., Stouffer, D. B. and Amaral, L. A. N. (2007). Quantitative analysis of the local structure of food webs, *J. Theor. Biol.* **246**, 2, pp. 260–268.

Ceballos, G. and Ehrlich, P. R. (2006). Global mammal distributions, biodiversity hotspots, and conservation, *Proc. Natl. Acad. Sci. USA* **103**, 51, pp. 19374–19379.

Ceballos, G., Ehrlich, P. R., Soberón, J., Salazar, I. and Fay, J. P. (2005). Global mammal conservation: What must we manage? *Science* **309**, 5734, pp. 603–607.

Chapin, F. S., Zavaleta, E. S., Eviner, V. T., Naylor, R. L., Vitousek, P. M., Reynolds, H. L., Hooper, D. U., Lavorel, S., Sala, O. E., Hobbie, S. E., Mack, M. C. and Díaz, S. (2000). Consequences of changing biodiversity, *Nature* **405**, pp. 234–242.

Cochrane, M. A., Alencar, A., Schulze, M. D., Sourza, C. M., Jr., Depstad, D. C., Lefebvre, P. and Davidson, E. A. (1999). Positive feedbacks in the fire dynamic and closed canopy tropical forests, *Science* **284**, pp. 1832–1835.

Cohen, J. E. (1978). *Food Webs and Niche Space* (Princeton University Press, Princeton, NJ).

Cohen, J. E., Briand, F. and Newman, C. M. (1990). *Community Food Webs: Data and Theory* (Springer-Verlag, Berlin).

Cohen, J. E., Jonsson, T. and Carpenter, S. R. (2003). Ecological community description using the food web, species abundance, and body size, *Proc. Natl. Acad. Sci. USA* **100**, pp. 1781–1786.

Dunne, J., Williams, R. and Martinez, N. (2002a). Network structure and biodiversity loss in food webs: robustness increases with connectance, *Ecol. Lett.* **5**, pp. 558–567.

Dunne, J. A., Williams, R. J. and Martinez, N. D. (2002b). Food-web structure and network theory: The role of connectance and size, *Proc. Natl. Acad. Sci. USA* **99**, pp. 12917–12922.

Dunne, J. A., Williams, R. J., Martinez, N. D., Wood, R. A. and Erwin, D. H. (2008). Compilation and network analyses of cambrian food webs. *PLoS Biol.* **6**, 4, p. e102.

Dyer, R. J. and Nason, J. D. (2004). Population graphs: the graph theoretic shape of genetic structure, *Mol. Ecol.* **13**, pp. 1713–1727.

Ehrlich, P. and Raven, P. (1964). Butterflies and plants: a study in coevolution, *Evolution* **18**, pp. 586–608.

Emmerson, M. C. and Yearsley, J. M. (2004). Weak interactions, omnivory and emergent food-web properties, *Proc. R. Soc. Lond. B* **271**, pp. 397–405.

Fortuna, M. A. and Bascompte, J. (2006). Habitat loss and the structure of plant-animal mutualistic networks, *Ecol. Lett.* **9**, pp. 281–286.

Fortuna, M. A., García, C., Guimaraes, P. and Bascompte, J. (2008a). Spatial mating networks in insect-pollinated plants, *Ecol. Lett.* **11**, pp. 490–498.

Fortuna, M. A., Gomez-Rodríguez, C. and Bascompte, J. (2006). Spatial network structure and amphibian persistence in stochastic environments, *Proc. R. Soc. B* **273**, pp. 1429–1434.

Fortuna, M. A., Popa-Lisseanu, A., Ibañez, C. and Bascompte, J. (2008b). The roosting spatial network of a bird-predator bat, *Ecology* **In press**.

Fussmann, G. F. and Heber, G. (2002). Food web complexity and chaotic population dynamics, *Ecol. Lett.* **5**, pp. 394–401.

Gibbons, P. and Lindenmayer, D. (2002). *The hollows and wildlife conservation in Australia* (Csiro Publishing, Australia).

Griffith, D. R. (2008). The ecological implications of individual fishing quotas and harvest cooperatives, *Front. Ecol. Environ.* **6**, 4, pp. 191–198.

Guimarães, P., Jr, Machado, G., de Aguiar, M., Jordano, P., Bascompte, J., Pinheiro, A. and dos Reis, S. (2007). Build-up mechanisms determining the topology of mutualistic networks, *J. Theor. Biol.* **249**, pp. 181–189.

Hansen, M. C., Stehman, S. V., Potapov, P. V., Loveland, T. R., Townshend, J. R. G., Defries, R. S., Pittman, K. W., Arunarwati, B., Stolle, F., Steininger, M. K., Carroll, M. and Dimiceli, C. (2008). Humid tropical forest clearing from 2000 to 2005 quantified by using multitemporal and multiresolution remotely sensed data. *Proc. Natl. Acad. Sci. USA* **105**, 27, pp. 9439–9444.

Hanski, I. and Gilpin, M. E. (1997). *Metapopulation biology: ecology, genetics and evolution* (Academic Press).

Heithaus, M. R., Frid, A., Wirsing, A. J. and Worm, B. (2008). Predicting ecological consequences of marine top predator declines, *Trends Ecol. Evol.* **23**, 4, pp. 202–210.

Holt, R. D. (1997). Community modules, in A. C. Gange and V. K. Brown (eds.), *Multitrophic Interactions in Terrestrial Ecosystems, 36th Symposium of the British Ecological Society* (Blackwell Science, Oxford), pp. 333–349.

Holt, R. D. and Hochberg, M. E. (2001). Indirect interations, community modules and biological control: a theoretical perspective, in E. Waijnberg, J. K. Scott and P. C. Quimby (eds.), *Evaluation of Indirect Ecological Effects of Biological Control* (CAB International), pp. 13–37.

Holyoak, M., Leibold, M. and Holt, R. (eds.) (2005). *Metacommunities: spatial dynamics and ecological communities* (Chicago University Press).

Hurlbert, A. H. and Jetz, W. (2007). Species richness, hotspots, and the scale dependence of range maps in ecology and conservation, *Proc. Natl. Acad. Sci. USA* **104**, 33, pp. 13384–13389.

Jordano, P., Bascompte, J. and Olesen, J. M. (2003). Invariant properties in coevolutionary networks of plant-animal interactions, *Ecol. Lett.* **6**, 1, pp. 69–81.

Krishna, A., Guimarães, P., Jr, Jordano, P. and Bascompte, J. (2008). A neutral-niche theory of nestedness in mutualistic networks, *Oikos* **In press**.

Lawton, J. H. (1989). Food webs, in J. Cherrett (ed.), *Ecological Concepts* (Blackwell Scientific, Oxford, UK), pp. 43–78.

Leibold, M., Holyoak, M., Mouquet, N., Amarasekare, P., Chase, J., Hoopes, M., Holt, R., Shurin, J., Law, R., Tilman, D., Loreau, M. and Gonzalez, A. (2004). The metacommunity concept: a framework for multi-scale community ecology, *Ecol. Lett.* **7**, pp. 601–613.

McAllister, D., Scheuler, F. W., Roberts, C. M. and Hawkins, J. P. (1994). Mapping and GIS analysis of the global distribution of coral reef fishes on an equal-area grid, in R. I. Miller (ed.), *Mapping the Diversity of Nature* (Chapman & Hall, London), pp. 155–175.

McCann, K. S. (2000). The diversity-stability debate, *Nature* **405**, pp. 228–233.

McCann, K. S., Hastings, A. and Huxel, G. R. (1998). Weak trophic interactions and the balance of nature, *Nature* **395**, pp. 794–798.

Melián, C., Bascompte, J. and Jordano, P. (2005). Spatial structure and dynamics in a marine food web, in A. Belgrano, U. Scharler, J. Dunne and R. Ulanowicz (eds.), *Aquatic Food Webs*, chap. Spatial structure and dynamics in a marine food web (Oxford Univ Press, Oxford), pp. 19–24.

Memmott, J., Waser, N. and Price, M. (2004). Tolerance of pollination networks to species extinctions. *Proc. Biol. Sci.* **271**, 1557, pp. 2605–2611.

Milo, R., Shen-Orr, S., Itzkovitz, S., Kashtan, N., Chklovskii, D. and Alon, U. (2002). Network motifs: Simple building blocks of complex networks, *Science* **298**, 5594, pp. 824–827.

Minor, E. S. and Urban, D. L. (2008). A graph-theory framework for evaluating landscape connectivity and conservation planning, *Conserv. Biol.* **22**, pp. 297–307.

Myers, N., Mittermeier, R. A., Mittermeier, C. G., da Fonseca, G. A. B. and Kent, J. (2000). Biodiversity hotspots for conservation priorities, *Nature* **403**, pp. 853–858.

Myers, R. A. and Worm, B. (2003). Rapid worldwide depletion of predatory fish communities. *Nature* **423**, 6937, pp. 280–283.

Nakazawa, T. and Yamamura, N. (2006). Community structure and stability analysis for intraguild interactions among host, parasitoid, and predator, *Popul. Ecol.* **48**, pp. 139–149.

National Marine Fishery Service (2008). Annual report to Congress on the Status of US Fisheries-2007, Tech. rep., U.S. Department of Commerce, NOAA, Natl., Mar. Fish. Serv., Silver Spring, Maryland.

Nepstad, D. C., Verissimo, A., Alencar, A., Nobre, C., Lima, E., Lefebvre, P., Schlesinger, P., Potter, C., Moutinho, P., Mendoza, E., Cochrane, M. and Brooks, V. (1999). Large-scale impoverishment of Amazonian forests by logging and fire, *Nature* **398**, pp. 505–508.

NERC Centre for Population Biology, I. C. (1999). The Global Population Dynamics Database. http://www.sw.ic.ac.uk/cpb/cpb/gpdd.html.

Olesen, J., Bascompte, J., Elberling, H. and Jordano, P. (2008). Temporal dynamics in a pollination network, *Ecology* **89**, pp. 1573–1582.

Otto, S. B., Rall, B. C. and Brose, U. (2007). Allometric degree distributions facilitate food-web stability. *Nature* **450**, 7173, pp. 1226–1229.

Ovaskainen, O. and Hanski, I. (2004). *Ecology, genetics and evolution of metapopulations,* chap. Metapopulation dynamics in highly fragmented landscapes (Academic Press), pp. 73–104.

Paine, R. T. (1988). Food webs: Road maps of interactions or the grist for theoretical development? *Ecology* **69**, 6, pp. 1648–1654.

Pascual, M. and Dunne, J. A. (eds.) (2006). *Ecological Networks: Linking Structure to Dynamics in Food Webs* (Oxford University Press, Oxford, UK).

Petanidou, T., Kallimanis, A., Tzanopoulos, J., Sgardelis, S. and Pantis, J. (2008). Long-term observation of a pollination network: fluctuation in species and interactions, relative invariance of network structure and implications for estimates of specialization, *Ecol. Lett.* **11**, pp. 564–575.

Petchey, O. L., Beckerman, A. P., Riede, J. O. and Warren, P. H. (2008). Size, foraging, and food web structure. *Proc. Natl. Acad. Sci. USA* **105**, 11, pp. 4191–4196.

Pimm, S. L. (2002). *Food webs,* 1st edn. (University of Chicago Press, Chicago, IL).

Pimm, S. L. and Raven, P. (2000). Extinction by numbers, *Nature* **403**, pp. 843–845.

Rezende, E., Jordano, P. and Bascompte, J. (2007a). Effects of phenotypic complementarity and phylogeny on the nested structure of mutualistic networks, *Oikos* **116**, pp. 1919–1929.

Rezende, E., Lavabre, J., Guimarães, P., Jr, Jordano, P. and Bascompte, J. (2007b). Non-random coextinctions in phylogenetically structured mutualistic networks, *Nature* **448**, pp. 925–928.

Rooney, N., McCann, K., Gellner, G. and Moore, J. C. (2006). Structural assymetry and the stability of diverse food webs, *Nature* **442**, pp. 265–269.

Santamaría, L. and Rodríguez-Gironés, M. (2007). Linkage rules for plant-pollinator networks: trait complementarity or exploitation barriers? *PLoS Biol.* **5(2)**, p. e31.d0i:10.

Schick, R. S. and Lindley, S. (2007). Directed connectivity among fish populations in a riverine network, *J. Appl. Ecol.* **44**, pp. 1116–1126.

Skole, D. and Tucker, C. J. (1993). Evidence for tropical deforestation, fragmented habitat, and adversely affected habitat in the Brazilian Amazon:1978–1988, *Science* **260**, pp. 1905–1910.

Solé, R. and Montoya, J. (2001). Complexity and fragility in ecological networks. *Proc. Biol. Sci.* **268**, 1480, pp. 2039–2045.

Stang, M., Klinkhamer, G., Peter and van der Meijden, E. (2007). Asymmetric specialization and extinction risk in plant-flower visitor webs: a matter of morphology or abundance? *Oecologia* **151**, pp. 442–453.

Stouffer, D. B., Camacho, J. and Amaral, L. A. N. (2006). A robust measure of food web intervality, *Proc. Natl. Acad. Sci. USA* **103**, 50, pp. 19015–19020.

Stouffer, D. B., Camacho, J., Guimerà, R., Ng, C. A. and Amaral, L. A. N. (2005). Quantitative patterns in the structure of model and empirical food webs, *Ecology* **86**, pp. 1301–1311.

Stouffer, D. B., Camacho, J., Jiang, W. and Amaral, L. A. N. (2007). Evidence for the existence of a robust pattern of prey selection in food webs, *Proc. R. Soc. B* **274**, 1621, pp. 1931–1940.

Sugihara, G. (1984). Graph theory, homology, and food webs, in S. A. Levin (ed.), *Population Biology, Proceedings of Symposia in Applied Mathematics*, Vol. 30 (American Mathematical Society), pp. 83–101.

Thomas, J. A., Telfer, M. G., Roy, D. B., Preston, C. D., Greenwood, J. J. D., Asher, J., Fox, R., Clarke, R. T. and Lawton, J. H. (2004). Comparative losses of British butterflies, birds, and plants and the global extinction crisis, *Science* **303**, pp. 1879–1881.

Thompson, J. N. (2005). *The Geographic Mosaic of Coevolution* (University of Chicago Press, Chicago, IL, USA).

Tilman, D. and Kareiva, P. (1997). *Spatial ecology: the role of space in population dynamics and interespecific interactions* (Princeton University Press).

Tylianakis, J. M., Tacharntke, T. and Lewis, O. T. (2007). Habitat modification alters the structure of tropical host-parasitoid food webs, *Nature* **445**, pp. 202–205.

United Nations Food and Agriculture Organization (2006). The State of World Fisheries and Aquaculture, Tech. rep., United Nations, Rome, Italy.

Urban, D. L. and Keitt, T. (2001). Landscape connectivity: a graph-theoretic perspective, *Ecology* **82**, pp. 1205–1218.

Vázquez, D. P. and Aizen, M. A. (2004). Asymmetric specialization: a pervasive feature of plant-pollinator interactions, *Ecology* **85**, pp. 1251–1257.

Waser, N., Chittka, L., Price, M., Williams, N. and Ollerton, J. (1996). Generalization in pollination systems, and why it matters, *Ecology* **77**, 4, pp. 1043–1060.

Wiens, J. A. (2001). *Dispersal*, chap. The landscape context of dispersal (Oxford University Press), pp. 96–109.

Williams, R. J. and Martinez, N. D. (2000). Simple rules yield complex food webs, *Nature* **404**, pp. 180–183.

Williams, R. J. and Martinez, N. D. (2008). Success and its limits among structural models of complex food webs, *J. Anim. Ecol.* **77**, 3, pp. 512–519.

Worm, B., Sandow, M., Oschlies, A., Lotze, H. K. and Myers, R. A. (2005). Global patterns of predator diversity in the open oceans. *Science* **309**, 5739, pp. 1365–1369.

Yodzis, P. and Innes, S. (1992). Body size and consumer-resource dynamics, *Am. Nat.* **139**, pp. 1151–1175.

Chapter 15

Evolutionary Models for Simple Biosystems

Franco Bagnoli

Department of Energy and CSDC, University of Florence
Via S. Marta, 350139 Firenze, Italy
Also INFN, sez. Firenze
franco.bagnoli@unifi.it

The concept of evolutionary development of structures constituted a *real* revolution in biology: it was possible to understand how the very complex structures of life can arise in an out-of-equilibrium system. The investigation of such systems has shown that indeed, systems under a flux of energy or matter can self-organize into complex patterns, think for instance to Rayleigh-Bernard convection, Liesegang rings, patterns formed by granular systems under shear. Following this line, one could characterize life as a state of matter, characterized by the slow, continuous process that we call evolution. In this paper we try to identify the organizational level of life, that spans several orders of magnitude from the elementary constituents to whole ecosystems.

Although similar structures can be found in other contexts like ideas (memes) in neural systems and self-replicating elements (computer viruses, worms, etc.) in computer systems, we shall concentrate on biological evolutionary structure, and try to put into evidence the role and the emergence of network structure in such systems.

15.1. Introduction

The study of evolution has been largely aided by theoretical and computer models. First, it is hard to perform experiments in evolution, although some of them have been carried out. As a consequence, one cannot observe the dynamics of an evolutionary phenomenon, but only a (partial) snapshot of some phases. In general, there are many dynamics compatible with these observations, and therefore it is not easy to decide among the possible "sources" of an observed behavior. The decision is often quantitative: many hypotheses can in principle originate an observed evolutionary feature, and one has not only to decide which one is compatible with other characteristics, but also which is the more *robust, i.e.*, which requires less parameters (that in general are poorly known) and less tuning.

An unique characteristic of evolution is its historical character. The genotypic space is so large and fiddled with "holes" (corresponding to non-viable phenotypes), there is no chance to observe twice the same genotype in evolution. Similar phenotypes may emerge (convergent evolution). Due to the autocatalytic (either you eat or you are eaten) character of ecosystems, once a branch of a bifurcation is taken, the other alternative is forever forbidden (this generally corresponds to extinctions).

It is difficult to carry out a completely theoretical approach. Most of people interested in theoretical results want to apply their results to the interpretation of real biological systems. Secondly, in some 3 billion years, evolution has developed such an intricate variety of "case studies" of applied evolution that we still have a lot of poorly understood examples to be studied. Finally, the difficulty in performing experiments, discourages the study of effects that *could* be compatible with evolution but have never been realized on the earth.

The theoretical approach allows to recognize that evolutionary dynamics applies not only to biological system. There are at least other two environments that support (or may support) an evolutionary dynamics. The first one is the human mind, and in this case the replicators are concepts, ideas and myths that can be tracked in human culture. Indeed, ideas may propagate from individual to individual, can mutate and are selected for their "infectivity" and their fitness to the cultural corpus. Many cultural traits like religions proved quite capable of propagating, in spite of the load and the detrimental effects they have on their hosts. The term *meme* (sounding like gene and resembling memory) has been coined for this replicator [1]. A particularly important aspect of cultural evolution concerns languages, and in particular language competition [2]. We have not enough space to exanime these aspects here.

The second example is given by the giant computer network, Internet. Computer viruses and worms are examples of replicating objects that can mutate and are selected by their ability of escaping antiviral software and infecting other computers. With the increasing power of computers and their pervasivity, it is expected that such "life" forms will be much more common. Computer life forms and memes may cooperate, think for instance to hoaxes, e-mail viruses, etc. We shall concentrate here on "standard", "biological" life.

Some parts of this paper were previously published in Ref. [3]. See also Ref. [4].

15.2. The Structure of an Evolving System

The three basic blocks of evolutionary dynamics are replication, mutation and selection. Evolution works on a population of reproducing individuals, often called replicators.

15.2.1. *Modeling living individuals*

In order to visualize the problem, let us consider a pool of bacteria, growing in a liquid medium in a well-stirred container, so that we do not have to be concerned with spatial structures. We have chosen bacteria because they are able to synthesize most of the compounds they need from simple chemical sources, and appear to have a simpler structure than eukaryotes (cells with nucleus) and archea. The ultimate reason for this simplicity is evolution: bacteria have "chosen" to exploit the speed of replication in spite of complexity, and optimized accordingly their working machinery. Viruses have progressed a lot in the direction of speed, but they need a much more structured background: the presence of a chemical machinery assembled by other living cells.

Bacteria absorb nutrients (energy), amino acids and other chemicals (building blocks) from the medium, and uses them to increase their size, up to a point in which they divide in two (or sprout some buddies). These tasks are carried out by a biochemical network of proteins. Proteins act as enzymes, transforming chemical elements, as structural elements and have also a regulatory functions either by directly activating or inactivating other proteins, or by promoting or blocking the production of other or same proteins. In fact, in a growing bacteria, there is continuously the need of producing new proteins and all the other constituents of the body.

A protein is synthesized as an one-dimensional chain of amino acid. The tri-dimensional shape of a protein, and thus its chemical function, is defined (at least in many simpler cases) by its one-dimensional sequence, that folds and assumes its working conformation by itself.[a] There are only 20 amino acids used by living beings. The one-dimensional sequence of a protein is *coded* into the one-dimensional sequence of basis (gene) in a chromosome. Many bacteria have one large chromosome, plus a variable number of smaller ones (plasmids). A chromosome is a double helix of DNA, that can be view (in computer terms) as a sequence of symbols from a four-letter alphabet (ATCG). The translation of a gene into a protein takes place in several steps. First the gene is transcribed into an intermediate form, using RNA. In the RNA alphabet, thymine (T) is replaced by uracil (U). This messenger RNA (mRNA) in eukaryotes are further processed by splicing pieces (introns), but the basic working is similar.[b]

Messenger RNA is then translated into a protein by a large complex of proteins and RNA, called ribosome. A ribosome acts as a catalyst, by allowing other small RNA-amino acid complexes called tRNA to bind to mRNA. This binding is rather specific. A tRNA present a triplet of basis that must complement (with some tolerance) to the triplet (called codon) in the specific region of mRNA that is processed by the ribosome. The tRNA carries an amino acid that is specific to its anti-codon (this pairing is performed by other enzymes). Thus, by this large biochemical network, we have the production of a protein following the information stored in a gene.

---

[a]Many larger complexes are formed by more that just one protein, and some protein need the help of other enzymes to stabilize the three-dimensional shape with covalent bonds or to add metals ions, sugar chains and other elements.

[b]Actually, in archea, too, there are introns, and also in bacteria self-splicing portions of RNA are present.

When a cell has grown sufficiently, it divides. In order to perform this step, also the DNA has to be replicated, using again other proteins and RNA fragments. Actually, in growing bacteria, both processes (production of body material and DNA copying) are performed in parallel, so that some part of the chromosome can be present in multiple copies. The splitting of a cell into two can also be seen as balance between the exponential growth of the body and the (almost) linear copying of the DNA. All life long, but especially during the copy phase, *mutations* may occur, changing some basis, inserting or deleting others, etc. Mutations are essential to generate diversity, but, as we shall see, they need to occur quite rarely. Clearly, in multi-cellular organisms, only mutations that occur in the germinal line are transmitted to the progeny.

The translation of a gene depends by other proteins: they can block the translation by binding themselves to DNA, or by altering the DNA itself (methylation). Other proteins are also needed to promote the transcription of a gene. Thus, from the proteins' point of view, a gene is just a way of producing other proteins, under the influence of the specific pool of proteins that are working together into the organism.

Not all genes are translated into proteins. For instance, tRNA genes are only transcribed into RNA. Other portions of DNA are never transcribed, but can alter the biochemistry of the cell. For instance, specific sequences may inhibit the expression of genes by bending the DNA, or preventing the binding of promotor proteins. Other DNA sequences may interfere with the process of DNA copying, for instance by increasing the number of repeats of themselves (microsatellites, ALU). Other sequences may *jump* from a position of the chromosome to another (transposons), eventually carrying some portion of DNA with itself. By this mechanism they can inactivate, activate or merge genes.

Bacteria can also alter their genetic contents by capturing or exchanging pieces of DNA (plasmids), a process that resemble the sexual reproduction of some eukaryotes.

Finally, retroviruses can alter the genome of the host by inserting their own code (or a precursor of it in RNA retroviruses). This process, that *reverts* the usual information path from DNA to RNA and proteins (the so-called "central dogma" of biology), is reputed to be one of the most powerful mechanisms of genetic transformations. Transposons are probably the relics of viruses, and a substantial part of some eukaryotic genome can be recognised as inactivated viruses.

The last important element is selection. In our example, bacteria living in a test tube will proliferate, duplicate and probably differentiate. Sooner or later, they will consume all the nutrients in the solution. A few of them can survive sometime by "eating" others, but clearly, in a closed environment, the last fate is the extinction of all living forms.[c] Life (replicators) needs an open environment, with fluxes of

---

[c]Actually, this can take a long time, since many bacteria are able to "freeze" themselves in form of spores, that can survive for a long time in rather harsh environments.

energy and matter. We can simulate this in a lab, by picking a drop of the culture, and putting it into a fresh medium. In this way, the survival from a test tube to another is a random process, with a vanishing probability from the point of view of individuals. However, when viewed by genes, there are many individuals that share the same genetic information, while other ones may have a mutated version. The genetic selection favors the genes that are present in a larger number of copies, and have therefore more chances to survive. The growth of bacteria just after being inoculated into a fresh medium is exponential, so those genomes that allow a faster replication of individuals will be present in more copies. This is the winning strategy of most bacteria. The speed is not the only quantity that can be *optimized*. One could replicate more slowly, but produce some toxic chemical to which it is immune. Or, if the refreshing occurs lately, one could prefer to develop ways of surviving starvation. Finally, if there are oscillations of temperature, acidity, or in the presence of predators, a better strategy could be that of developing ways to overcome these accidents. All these additional instruments need more proteins, more DNA coding, and therefore a slower replications speed. How can the optimum compromise between speed and complexity be reached?

The main idea of evolution is that this is a self-organized process. Mutations produce variety, and selection prunes it. We have to stress that selection is not an absolute limit, it depends on the rest of the environment. The probability of a given genome to pass to next "flask" generation (the *fitness*) does not depend on how fast in absolute it replicates, but on how fast it replicates in comparisons with the other genotypes present in the environment. And, as soon as the "best" genotype, possibly recently arisen by a mutation, is picked up and colonizes a new flask, its comparative advantage vanishes. Due to selection, copies of it become more numerous, and the competition becomes harder. This effect, named "red queen" is the main motor of evolution.

But let us dig more on this subject. The way in which an individual if perceived by others, or influences the environment is called its "phenotype". It is a product of proteins (mainly) and therefore of the genes in its genome. It may depend also on the age of the individual, on past experiences, and on the interactions with other genomes (often very similar, as we shall see). A genome is a "bag of genes", extending the "definition" (not very mathematical) of gene to all genetic elements that have a recognizable persistence along generations, and that influence in some way their phenotype.

A gene is successful and tends to spread among the population if it confers a selective advantage to the individuals that carry it, and this happens through its expression, *i.e.*, the phenotype. The phenotype is in general a rather complex functions of the genes that constitute it. Genes very rarely control directly some aspect of the phenotype. What a gene does, is to produce a protein, that participates to the intricate biochemical machinery of the cell, and in multicellular organisms to the development of the organism. Variants of the gene (alleles) produces similar proteins in other bodies. The mechanism of blind production of variant plus selection is able to "optimize" this assembly of organisms. This optimization is always at short term: the competition (red queen) forbids the development or maintenance of "accessories" that may be useful in the future, or that may lead to the development of new functions. Selection tends to prefer quick-and-dirty solutions.[d]

We have to consider that the phenotype is often best understood as the result of a pool of genotypes, not as a characteristic of an individual. For instance, an ant or termite nest is the result of the cooperation of several individuals, and of many generations. A multicellular body is the result of the cooperation of many cells. What these cells have in common is the sharing of large portions of their genome (all for multicellular bodies). In other cases (symbionts, larger communities like human societies), cooperation tends to favor assemblies of rather unrelated individuals. How cooperation can arise from the "selfishness" of genes, is one of the most interesting outcome of evolution.

We have spoken mainly of bacteria, and of asexual reproduction. However, many eukaryotes reproduce sexually. Sex poses a puzzling problem [5, 6]. In true sexual reproduction (not just exchange of genetic material like in some bacteria), the male does offer only genetic information, in form of sperm. The female adds her half of genetic material, plus all the body. The production rate of offspring is therefore proportional to the abundance of females. The sex rate at birth is generally about 50%, and in comparisons with an asexual replicator or a parthenogenic female, the efficiency is just the half. In spite of this, many organisms that are able to reproduce both sexually and parthenogenycally (many plants, some insects and amphibians) maintain the sexual habits.

Sex induces also dangerous habits. Especially in large animals, a kind of runoff is observed: one sex develops a phenotypic characteristic, like the tail of peacock, and the other sex finds that the partners with that characteristic are more interesting, and therefore reproduce more. The following generations enhance those characteristics, until the limit (trade-off between survival and exaggeration of the character) is reached.

15.2.2. *The sequence space*

Let us start our investigation by considering "standard" living forms, based on nucleic acid. The storage of information is an one-dimensional structure, possibly fragmented into many chromosomes. We shall call this information the *genome*, and in usual living beings it is formed by one or more filaments of DNA or RNA. Such filaments are formed by a sequence of symbols (basis), and therefore are digital pieces of information.

---

[d]This effect is particularly evident in the fact that birds living in islands with no predator tend to lose the capacity of flying: a non-flying relative that can invest more energy in egg production tends to be selected, even if when predators reappear it would be useful to have working wings.

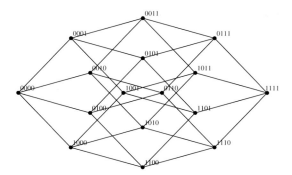

Fig. 15.1.   The two-dimensional projection of the Boolean hypercube for $L = 4$.

Any base can assume one of four values (ATCG in DNA, AUCG in RNA). A sequence of L basis can therefore assume one out of 4^L values. The space of all possible sequences is called the sequence space. It is a high-dimensional, discrete space.

Due to the extremely large number of possible sequences, there has not been enough time during evolution to "try" all possible sequences nor a relevant fraction of it. Therefore, we cannot apply "equilibrium" concepts to evolution.

The sequence space is well defined for a fixed length L, but the length of the genome in real organisms is not fixed. In order to allow for variable L, one can consider an extended space with an extra symbol "*", which stands for an *empty* space, that can occasionally be replaced by a real basis. This is the same representation used to align sequences, to be illustrated in the following. We can assume that between any subsequent couple of basis there are empty symbols.

For simple modeling, we limit our investigation to fixed-length genomes $g = (g_1, \ldots, g_L)$, and assume that each component g_i is a Boolean quantity, taking values 0 and 1. The quantity g_i should represent "genes" in a rather general meaning. Sometimes, we use g_i to represent two allelic form of a gene, say $g_i = 0$ for the wild type, and $g_i = 1$ for the less dangerous mutation, even if the quantity is not a "gene" in the standard assumption (*e.g.*, information to produce a protein). In other contexts, g_i might represent the presence or the absence of a certain gene, so that we can theoretically identify of all possible genomes with different sequences.

Every possible sequence of length L corresponds to a corner in the Boolean, L-dimensional hypercube (see Fig. 15.1). Each link in the hypercube correspond to the change of one element g_i, *i.e.*, to a single-element mutation. Other mutations correspond to larger jumps.

In the following, we shall denote by $d(x, y)$ the *genetic distance*, *i.e.*, the minimum number of mutations needed to pass from x to y. The quantity $d(x, y)$ is a *topological* concept, in real life one has to consider the probability of mutations in order to estimate if a longer path could occur with higher probability. In our digital approach, $d(x, y)$ is just the Hamming distance between the two sequences.

15.2.3. *Mutations*

During lifetime, or in the genome replication phase (when an organism reproduces), some symbol in the genome can be changed: a mutation. Many different types of mutations has been observed. The most common (either because they are actually more common than others, or because they have less chance to cause harmful consequences) are point mutations (an exchange of a basis with another. Other common mutations are deletions (the replacement of a symbol with a "*") and insertions (the replacement of a "*" with a basis). These mutations correspond to *jumps* is the sequence space. Therefore, a given lineage can be seen as a path in sequence space. By considering all descendants of a given individual (assuming asexual reproduction), we have a set of branching walks, where an "arm" stops when an individual dies [7, 8].

The probability that a path comes back to an already visited sequence (disregarding the empty symbol) vanishes with growing L. Moreover, as we shall see, competition will prevent the merging of different species.

In our model, a mutation is just a change in the sequence g, that can occur with a probability that depends only on the sub-sequence that changes, or that may depend on other portion of the genotype. For instance, *point mutations* are the replacement of a zero with an one and vice versa. The probability of this occurrence ($g_i \rightarrow g_i'$) may depend on the position i of the "gene". A translocation, on the other hand, involves correlated changes in different positions. Considering the interpretation of g_i as the presence or absence of a certain genetic characteristic, it could be represented as the simultaneous change of two locations g_i and g_j in the genome, that for instance may happen only of $g_i = 1$ (presence of the transposable element). Similar representations may be interpreted as gene duplications, etc.

Point mutations correspond to base substitutions. If this happens in a coding region, it causes the change of one codon. Due to the redundancy of genetic code, this mutation may not affect at all the resulting protein, either because the altered codon is read by the same tRNA as the original one, or because it is read by a different tRNA carrying the same amino acid. The modifications in the protein may also be marginal if a amino acid is replaced by another one with similar characteristics.

A deletion or insertion in a coding region causes a "frame shift", completely altering the resulting protein. Transpositions also causes similar damages, and the protein is normally not working.[e] If the protein has a crucial role, this modification lead to the death of the individual. However, it may happen that the same protein is produced by more that one copy of the gene. This generally occurs after a gene duplication, and is vastly more common in eukaryotes (due to the larger genome size and less pressure towards "efficiency") than in prokaryotes. In this case, the mutation only reduces the amount of protein produced, a mutation that may be harmful or not. In the latter case, the duplicated gene is "allowed" to mutate,

---

[e]Some cases of partially-working proteins correspond to genetic diseases, like Huntington's corea, that allow the survival of affected individuals for some time.

generating varied proteins that, inserted in the biochemical network of the cell, may induce new functions (although in general it simply increases the amount of "junk" genetic material). This is supposed to be one of the main "motors" of evolution, together with the *horizontal* transfer of DNA among individuals (possibly belonging to different species) by retroviruses.[f]

In non-coding regions (essentially eukaryotes), the substitutions, insertions and deletions may have "smoother" effects, unless they destroy the promoter region of a gene.

In general, we may say that genetic modifications are either lethal, or almost neutral, at least at the level of proteins and biochemical networks. How these almost-neutral modifications (like for instance a change of pigmentation) correspond to the alteration of survival characteristics, depends on their influence on the phenotype, and in how phenotypes of different individuals (and species) interact.

Thus, in many cases one can get rid of the genotypic space: for large scale evolution (formation of species) one can assume that mutations induces a diffusion in the phenotypic space, while for small scale evolution (asymptotic distribution of intra-specific traits) one can assume that mutations are able to generate all possible phenotypic traits.

For simplicity we assume that all point mutations are equally likely, while in reality they depend on the identity of the symbol and on its positions on the genome. For real organisms, the probability of observing a mutation is quite small. We assume that at most one mutation is possible per generation. We denote with μ_s the probability of having one point mutation per generation.

The probability to have a point mutation from genotype y to genotype x is given by the short-range mutation matrix $\boldsymbol{M}_s(x, y)$ which is

$$\boldsymbol{M}_s(x|y) = \begin{cases} 1 - \mu_s & \text{if } x = y, \\ \dfrac{\mu_s}{L} & \text{if } d(x, y) = 1, \\ 0 & \text{otherwise}. \end{cases} \tag{15.1}$$

Other mutations correspond to long-range jumps in the genotypic space. A very rough approximation consists in assuming all mutations are equally probable. Let us denote with μ_ℓ the probability per generation of this kind of mutations. The long-range mutation matrix, $\boldsymbol{M}_\ell$, is defined as

$$\boldsymbol{M}_\ell(x|y) = \begin{cases} 1 - \mu_\ell & \text{if } x = y, \\ \dfrac{\mu_\ell}{2^L - 1} & \text{otherwise}. \end{cases} \tag{15.2}$$

---

[f]Transposable elements and gene duplications are often the effects of viral elements, still able to manipulate the genome, but "trapped" into the genome by mutations that have destroyed their capability of producing the protective cap.

In the real world, only a certain kind of mutations are possible, and in this case $\boldsymbol{M}_\ell$ becomes a sparse matrix $\hat{\boldsymbol{M}}_\ell$. We introduce a sparseness index s which is the average number of nonzero off-diagonal elements of $\hat{\boldsymbol{M}}_\ell$. The sum of these off-diagonal elements still gives μ_ℓ. In this case $\hat{\boldsymbol{M}}_\ell$ is a quenched sparse matrix, and $\boldsymbol{M}_\ell$ can be considered the average of the annealed version.

Both $\boldsymbol{M}_s$ and $\boldsymbol{M}_\ell$ are Markov matrices. Moreover, they are circular matrices, since the value of a given element does not depend on its absolute position but only on the distance from the diagonal. This means that their spectrum is real, and that the largest eigenvalue is $\lambda_0 = 1$. Since the matrices are irreducible, the corresponding eigenvector ξ_0 is non-degenerate, and corresponds to the flat distribution $\xi_0(x) = 1/2^L$.

15.2.4. The phenotype

The phenotype of an individual is how it appears to others, or, better, how it may affect others. Clearly, the phenotype concerns the characteristics of one's body, but may also include other "extended" traits, like the dam for beavers, termite nests, or the production of oxygen that finally leaded ancient reductive organisms to extinction.

This function is in general rather complex: genes interact among themselves in an intricate way (epistatic interactions), and while it is easy to "build" a bad gene (for instance, a gene that produces a misfolded (incorrectly folded) protein, or a working protein that interferes with the biochemistry of the cell), a "good" gene is good only as long there is cooperation with other genes. This is the main reason of the "general" failure of genetic engineering: it is difficult to design a gene that produces the desired result (and indeed this is never attempted: one generally tries to move a gene with some effect from an organism to another), but it is yet more difficult to avoid this gene to interfere with the rest of machinery. Only some genetic modifications do not lead to drastic lowering of the output, and most of "successful stories" of transgenic plants concerns genes coding for proteins that inhibit some specific poison.

Genes that have additive effects on the phenotype are called "non-epistatic"; this in general holds only for a given phenotypic trait.

Whithin our modeling, we represent the phenotype of an individual as a set of quantitative characters u, that depend on the genotype, the age and on past experiences. A simplification that ease the treatment of the subject consists in assuming that the phenotype is just a function of the genotype:

$$u = u(g).$$

With this assumption, the fitness can be considered a function of the genotype, without introducing the phenotype [9–11].

As an example of phenotypes that depend on accumulated characteristics (age), we shall briefly discuss the Penna model in Section 15.4.

15.2.5. *Fitness and the mutation-selection equation*

Selection acts on phenotypes. The fact that an individual survives and reproduces (so perpetuating its genes) may depend on chance. We may however compute (and sometime measure) the "propensity" of an individual to survive and produce viable offspring. This quantity is termed "fitness", and for our assumptions we may assume that it is proportional to the average number of sons reaching the reproductive age for a given phenotype u, and for a given time interval (generation).

We are now in the position of simulating the evolution of a population. We may represent it as a set of N individuals $x^{(j)}$, $j = 1, N$, characterized by their genotype (that unambiguosly characterize the phenotype within our assumptions). The spatial structure of the system may be that of a regular graph (or continuous space with an interaction range), or a social network that may evolve with the system (and in this case may be related to the genotype of the organisms), or a "well stirred" environment where all interactions are equally probable.

In each generation, each individual interact with one or more other individuals, accumulating a "score" (fitness), that may determine its survival or, equivalently, the probability of producing offspring.

In the simplest arrangement, the time evolution (discrete generations) of an asexual population is given by the following phases:

1. Scoring phase: each individual is allowed to interact with others, according with its spatial connectivity (and possibly to displacement like evasion, pursuit, etc.). Their interactions depend on the relative phenotypes, and contribute to the accumulation of a score (generally reset to a default value at the beginning of the phase). The score may be negative or positive.
2. Survival phase: individuals can survive with a probability that is a monotonic function of the score.
3. Reproduction phase: empty locations may be colonized by neighboring ones. Colonization implies copying the genome with errors (mutations).

The previous model represents a whole ecosystem, but one is often interested only on some subpart of an ecosystem, like one or few species. If the consistency of a species does not affect others (for instance, one can assume that a species of herbivores does not modify the abundance of grass), then one can neglect to simulate the invariable species, by changing the default value of the score.

Let me be more explicit. Assume that a positive score increases survival, and a negative one decreases it. The presence (interaction) of grass, increases the survival of herbivores, while the encounter with a predator decreases it. If one is interested in the simulation of the interplay among grass, herbivores and carnivores, then all three species should be simulated. But if one assumes that grass is abundant and not modified by herbivores, one can neglect to include the vegetable substrate, and let herbivore start with a positive score at each step (while carnivores has to meet food to survive) [12].

Sexual reproduction implies some modification in the reproductive phase. One can simulate haploid or diploid individuals, with explicit sex determination of just with a recombination among the genomes of the two parents.

Also the displacement may depend on the phenotypic distribution near a given location: preys may try to escape predators, who are pursuing the first. Different sexes may find convenient to stay nearby, and so on.

With the assumption of "well mixed" population, *i.e.*, disregarding spatial correlation, and in the limit of very large population the dynamics is described by the mutation-selection equation [13] (mean-field approach)

$$n(x, t+1) = \left(1 - \frac{N}{K}\right) A(x, \boldsymbol{n}(t)) \sum_y M(x|y) n(y, t+1), \quad (15.3)$$

where $n(x, t)$ is the number of individuals with genome x at time t, M is the mutation matrix ($\sum_x M(x|y) = 1$) and $A(x, \boldsymbol{n}(t))$ if the fitness of genotype x (or better: of the relative phenotype), given the rest of the population $\boldsymbol{n}(t)$. In the following, we shall neglect to indicate the time dependence, and use a prime to indicate quantities computed one time step further.

The quantity K denotes the carrying capacity, and $N = \sum_x n(x)$ is the total number of individuals. The logistic term $1 - N/K$ implies that there is no reproduction in the absence of free space, and therefore models the competition among all individual for space. In the limit of populations whose size is artificially kept constant, this term may be included into A.

In vector terms, Eq. (15.3) can be written as

$$\boldsymbol{n}' = \left(1 - \frac{N}{K}\right) \boldsymbol{AMn},$$

where $\boldsymbol{A}$ is a diagonal matrix.

By summing over x, we obtain

$$N' = \langle A \rangle \left(1 - \frac{N}{K}\right) N, \quad (15.4)$$

i.e., the logistic equation, where the reproduction rate $\langle A \rangle = \sum_x A(x, \boldsymbol{n}) n(x)/N$ depends on the population and therefore on time.

By dividing Eq. (15.3) by Eq. (15.4), and introducing the frequencies $p(x) = n(x)/N$, we get

$$p'(x) = \frac{A(x, \boldsymbol{n}(t))}{\langle A \rangle} \sum_y M(x|y) p(y). \quad (15.5)$$

or

$$\boldsymbol{p}' = \frac{\boldsymbol{A}}{\langle A \rangle} \boldsymbol{Mp}$$

in the limit of vanishing mutation probability (per generation) and weak selection, a continuous approach (overlapping generations) is preferred. The mutation-selection equation becomes

$$\dot{p} = (A - \langle A \rangle)Mp \simeq (A - \langle A \rangle)p + \Delta p\,,$$

where $\Delta = M - 1$, and 1 is the identity. Considering only symmetric point mutations, Δ is the L-dimensional diffusion operator. One can recognize here the structure of a reaction-diffusion equation: evolution can be considered as a reaction-diffusion process in sequence space. The typical patterns are called *species*.

This simple model allows to recover some known results.

15.3. Evolution on a Fitness Landscape

The analysis of the replication equation, Eq. (15.5) is much simpler if the fitness A does not depend on the population structure, *i.e.*, $A = A(x)$. The lineage of an individual corresponds to a walk on a static landscape, called the *fitness landscape*. A generic fitness function may be considered a landscape for short times, in which the species, other than the ones under investigations, can be considered constant.

In this case the evolution really corresponds to an optimization process: in the case of infinite population, and for mutations that are able to "connect" any two genomes (in an arbitrary number of passes – M is an irreducible matrix), the system may reach an equilibrium state (in the limit of infinite time). The mutation-selection equation may be linearized by considering that the relative probability $q(x)$ of a path x (composed by a time sequence of genotypes $x = x^{(1)}, x^{(2)}, \ldots, x^{(t)}$) is

$$q(x) = \frac{\prod_{t=1}^{T+1} M(g^{(t+1)}|x^{(t)})A(x^{(t)})}{\prod_{t=1}^{T+1} \langle A \rangle(t)}\,, \tag{15.6}$$

where the denominator is common to all paths. Thus, the unnormalized probabilities of genotypes at time t, $z(x) \equiv z(x,t)$, evolve as

$$z'(x) = \sum_y M(x|y)A(y)z(y) \qquad \text{or} \qquad z' = MAz\,. \tag{15.7}$$

Notice that $z(x,t)$ can be considered as a restricted partition function, see Section 15.3.2.

The structure of probabilities in Eq. (15.6) suggests to use an exponential form for A:

$$A(x, p) = \exp(H(x, p))\,,$$

that makes evident the analogy with statistical mechanics: the fitness landscape behaves as a sort of energy, while mutations act like temperature [14–16]. See also Section 15.3.2.

There are a few landscapes that have been studied in details [17]. The simplest one is the flat landscape, where selection plays no role. It is connected to the concept of *neutral evolution*, that we shall examine in Section 15.3.6. In this landscape, evolution is just a random walk in sequence space, and one is interested in the probability of fixation of mutations in a finite populations, which corresponds to the divergence of an isolated bunch of individuals (allopatric speciation).

A species is in general a rather stable entity, in spite of mutations. One can assume that it occupies a *niche*, which corresponds to a maximum of the fitness in the phenotypic space. The presence of this niche generally depends on other species, like for instance insects specialized in feeding on a particular flower have their niche tied to the presence of their preferred food. An extremal case is that of a *sharp* fitness landscape constituted by an isolated peak. This landscape is particular since it is not possible to guess where the top of the fitness is, unless one is exactly on it. A lineage in a flat genotypic space is similar to a random walk, due to mutations. The only possibility to find the top is though casual encounter, but is a high-dimensional space, this is an extremely unfavorable event.

For small mutation probability, the asymptotic distribution is a *quasispecies* grouped around the peak. By increasing the mutation probability (or equivalently the genome length), this cloud spreads. It may happen that in finite populations no one has the right phenotype, so that the peak is lost. This is the *error threshold* transition, studied in Section 15.3.3. Clearly, this transition poses the problem of how this peak has been populated for the first time. The idea (see Section 15.5.6), is that the absence of population fitness space is actually similar to a Swiss cheese: paths of flat fitness and "holes" of unviable phenotypes (corresponding essentially to proteins that are unable to fold). The "corrugation" of the fitness is due to the presence of other species. So for instance a highly specialized predator may become so tied to its specific prey, that its effective fitness landscape (for constant prey population) is extremely sharp.

Another consequences of an increased mutation rate, more effective for individuals with accurate replication machinery like multicellular ones, is the extinction of the species without losing their "shape", the so-called Muller's ratchet [18–20] or stochastic escape [21, 22], which, for finite populations, causes the loosing of fitter strains by stochastic fluctuations. Muller's ratchet is studied in Section 15.3.3.

Genes that have an additive effect (non-epistatic) lead to *quantitative traits*, and to smoother landscape, shaped like the Fujiyama mount. It is possible to obtain a good approximation for the asymptotic distribution near such a maximum, as shown in Section 15.3.4.1. Such a results will come at hand in dealing with competition, Section 15.5.1. On a Fujiyama landscape, no error threshold transition is present [17].

Finally, one can study the problem of the evolution in a landscape with variable degree of roughness, a problem similar to that of disordered media in statistical mechanics [11, 23, 24].

A rather generic formulation follows lines similar to that of disordered statistical problems [17]. The phenotype $u(x)$ of a genotype $x = (x_1, x_2, \ldots, x_L)$ is given by

$$u(x) = \sum_{\{i_1 \ldots i_K\}} J_{\{i_1 \ldots i_K\}} x_{i_1} \cdots x_{i_K} , \qquad (15.8)$$

where $J_{\{i_1 \ldots i_K\}}$, for each different set of indexes $\{i_1 \ldots i_K\}$, are independent, identically distributed, random variables, so that for every K we are dealing with a random ensemble of phenotypes (and of fitness). The larger K, the faster the fitness correlations decay in sequence space, so that the fitness landscape is less and less correlated, *i.e.*, as it is usually said, more and more rugged. In the limit $k \to \infty$, one has a complete disordered fitness landscape, where the values of the fitness at different positions in sequence space are independent random variables (random energy model [25]). For $K = 2$ this model is similar to the one introduced by Kauffman [26] for modeling the genetic network of a cells.

On an extremely rugged landscape one observes an error threshold transition [17, 27] between a state in which the population distribution is concentrated on local maxima (localized phase), and a wide distribution. Due to the disorder, the distribution in the localized phase depends on the initial conditions, *i.e.*, on the past history of the population.

15.3.1. *Evolution and optimization, replicator equation*

The evolution may appear as an optimization process. Indeed, if we neglect mutations, Eq. (15.5) becomes

$$\boldsymbol{p}' = \frac{\boldsymbol{A}}{\langle A \rangle} \boldsymbol{p} ,$$

called the *replicator equation* (discrete time).

Assume that we start from a uniform distribution over the whole phenotypic space, and that the fitness shows a single, smooth maximum. The phenotypes u with $A(u) < \langle A \rangle$ tend to decrease in frequency, while those with $A(u) > \langle A \rangle$ tend to increase their frequency. Because of this, the average fitness $\langle A \rangle$ increases with time (Fisher theorem [28]). Notice that this results may be compatible with the extinction of the population, either due to chaotic oscillations for high reproductive rates, oscillations that may substantially reduce the population to a level in which random sampling is important, or because $\langle A \rangle$ is below one in Eq. (15.4). An example of this is given by genes that alter the sex ratio. These genes are able to "kill" male embryos, and thus increase their own population (since they reside on the female chromosome) although the final fate of the populatiion is extinction. This effect may be the origin of the smallness of male Y chromosome in mammals: a smaller chromosome implies less genes that may be targeted by the proteins produced by the killer genes.

The structure of Eq. (15.5) says that the fitness does not have absolute values, except that the average fitness (the basic reproductive rate) has to be greater than one. For a given genome, the survival of individuals depends on their *relative* fitness: those that happen to have a fitness larger than average tend to survive and reproduce, the others tend to disappear. By doing so, the average fitness in general increases, so that a genome that is good at a certain time will become more common and the relative fitness more similar to the average one. The effects of this generic tendency depend on the form of the fitness. For instance, for offspring production, it is more convenient to invest in females than in males. But as soon as males become rare, those that carry a gene that increases their frequency in the population will be more successful, in that the females fertilized with this gene will produce more males, that in average will fertilize more females and so on.

The presence of sexual reproduction itself is difficult to be interpreted as an optimization process. As reported in the Introduction, the offspring production of sexual species is about a half of that of asexual ones. The main advantage of sexual reproduction (with diploidicity) is that of maintaining a genetic (and therefore genotypic) diversity by shuffling paternal and maternal genes, without the risk associated with a high mutation rate: the generation of unviable offspring, that lead to the error threshold or to the mutation meltdown, Section 15.3.3. This genetic diversity is useful in colonizing new environments or for variable ones [29], but is essential to escape the exploitation from parasites, that reproduce (and therefore evolve) faster than large animals and plants [30].

Therefore, sex can be considered an optimization strategy only for variable environments. The fact that an optimization technique inspired by sexual reproduction, genetic algorithms [31], is so widely used is somewhat surprising.

15.3.2. *Evolution and statistical mechanics*

When one takes into consideration only point mutations ($M \equiv M_s$), Eq. (15.7) can be read as the transfer matrix of a two-dimensional Ising model [14, 15, 32], for which the genotypic element $x_i^{(t)}$ corresponds to the spin in row t and column i, and $z(x,t)$ is the restricted partition function of row t. The effective Hamiltonian (up to constant terms) of a possible lineage $x = (x^{(t)})_t = 1^T$ from time $1 \leq t \leq T$ is

$$\mathcal{H} = \sum_{t=1}^{T-1} \left(\gamma \sum_{i=1}^{L} x_i^{(t)} x_i^{(t+1)} + H(x^{(t)}) \right) , \tag{15.9}$$

where $\gamma = -\ln(\mu_s/(1-\mu_s))$.

This peculiar two-dimensional Ising model has a long-range coupling along the row (depending on the choice of the fitness function) and a ferromagnetic coupling along the time direction (for small short range mutation probability). In order to obtain the statistical properties of the system one has to sum over all possible configurations (stories), eventually selecting the right boundary conditions at time $t = 1$.

The bulk properties of Eq. (15.9) cannot be reduced in general to the equilibrium distribution of an one-dimensional system, since the transition probabilities among rows do not obey detailed balance. Moreover, the temperature-dependent Hamilto-

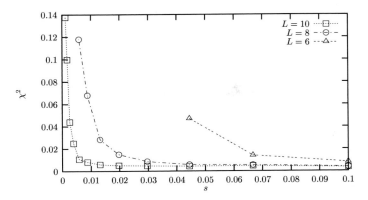

Fig. 15.2.   Scaling of χ^2 with the sparseness factor s, for three values of genome length L. $\mu_\ell = 0.1$ and $\mu_s = 0.1$.

nian (15.9) does not allow an easy identification between energy and selection, and temperature and mutation, what is naively expected by the biological analogy with an adaptive walk.

A Ising configuration of Eq. (15.9) corresponds to a possible genealogical story, *i.e.*, as a directed polymer in the genotypic space [16], where mutations play the role of elasticity. It is natural to try to rewrite the model in terms of the sum over all possible paths in genotypic space.

As shown in Ref. [33], in the case of long-range mutations, this scenario simplifies, and the asymptotic probability distribution $\tilde{p}$ is proportional to the diagonal of A^{1/μ_ℓ}:

$$\tilde{p}(x) = C \exp\left(\frac{H(x)}{\mu_\ell}\right), \qquad (15.10)$$

i.e., a Boltzmann distribution with Hamiltonian $H(x)$ and temperature μ_ℓ. This corresponds to the naive analogy between evolution and equilibrium statistical mechanics. In other words, the genotypic distribution is equally populated if the phenotype is the same, regardless of the genetic distance since we used long-range mutations. The convergence to equilibrium is more rapid for rough landscapes.

For pure short-range mutations, this correspondence holds only approximately for very weak selection or very smooth landscapes. The reason is that in this case the coupling in the time direction are strong, and mutations couple genetically-related strains, that may differ phenotypically.

Since short and long-range mutations have an opposite effect, it is interesting to study the more realistic case in which one has more probable short-range mutations, with some long-range ones that occur sporadically, always between the same genomes (quenched disorder). This scenario is very reminiscent of the small-world phenomenon [34], in which a small percentage s (sparseness) of long-range links added to a locally connected lattice is able to change its diffusional properties. Even in the limit of $s \to 0$ (after long times), one observes a mean-field (well-stirred) distribution. We computed the variance χ^2 of the deviations of the asymptotic distribution and that of Eq. (15.10) for a rough landscape. As can be seen in Fig. 15.2, for large genomes the transition to the mean-field limit occurs for $s \to 0$.

15.3.3. *Quasispecies, error threshold and Muller's ratchet*

Before going in deep studying a general model of an evolving ecosystem that includes the effect of competition (co-evolution), let us discuss a simple model [35] that presents two possible mechanisms of escaping from a local optimum, *i.e.* the error threshold and the Muller's ratchet.

We consider a *sharp peak landscape*: the phenotype $u_0 = 0$, corresponding to the master sequence genotype $x = 0 \equiv (0,0,\dots)$ has higher fitness $A_0 = A(0)$, and all other genotypes have the same, lower, fitness A_*. Due to the form of the fitness function, the dynamics of the population is fundamentally determined by the fittest strains.

Let us indicate with $n_0 = n(0)$ the number of individuals sharing the master sequence, with $n_1 = n(1)$ the number of individuals with phenotype $u = 1$ (only one bad gene, i.e. a binary string with all zero, except a single 1), and with n_* all other individuals. We assume also non-overlapping generations,

During reproduction, individuals with phenotype u_0 can mutate, contributing to n_1, and those with phenotype u_1 can mutate, increasing n_*. We disregard the possibility of back mutations from u_* to u_1 and from u_1 to u_0. This last assumption is equivalent to the limit $L \to \infty$, which is the case for existing organisms. We consider only short-range mutation with probability μ_s. Due to the assumption of large L, the multiplicity factor of mutations from u_1 to u_* (i.e. $L-1$) is almost the same of that from u_0 to u_1 (i.e. L).

The evolution equation Eq (15.3) of the population becomes

$$n_0' = \left(1 - \frac{N}{K}\right)(1 - \mu_s)A_0 n_0 \,,$$

$$n_1' = \left(1 - \frac{N}{K}\right)((1 - \mu_s)A_* n_1 + \mu_s A_0 n_0) \,, \tag{15.11}$$

$$n_*' = \left(1 - \frac{N}{K}\right)A_*(n_* + \mu_s n_1) \,.$$

and

$$\langle A \rangle = \frac{A_0 n_0 + A_*(n_1 + n_*)}{N}$$

is the average fitness of the population.

The steady state of Eq. (15.11) is given by $\boldsymbol{n}' = \boldsymbol{n}$. There are three possible fixed points $\boldsymbol{n}^{(i)} = \left(n_0^{(i)}, n_1^{(i)}, n_*^{(i)}\right)$: $\boldsymbol{n}^{(1)} = (0,0,0)$ $(N^{(1)} = 0)$, $\boldsymbol{n}_2 = (0, 0, K(1 - 1/\langle A_* \rangle))$

$(N^{(2)} = n_*^{(2)})$ and

$$
\boldsymbol{n}^{(3)} = \begin{cases}
n_0^{(3)} = N^{(3)} \dfrac{(1-\mu_s)A_0 - A_*}{A_0 - A_*}, \\[2mm]
n_1^{(3)} = N^{(3)} \dfrac{\mu_s}{1-\mu_s} \dfrac{A_0(qA_0 - A_*)}{(A_0 - A_*)^2}, \\[2mm]
n_*^{(3)} = N^{(3)} \dfrac{\mu_s^2}{1-\mu_s} \dfrac{A_0 A_*}{(A_0 - A_*)^2}, \\[2mm]
N^{(3)} = 1 - \dfrac{1}{A_0(1-\mu_s)}.
\end{cases}
$$

The fixed point $\boldsymbol{n}^{(1)}$ corresponds to extinction of the whole population, i.e. to mutational meltdown (MM). It is trivially stable if $A_0 < 1$, but it can become stable also if $A_0 > 1$, $A_* < 1$ and

$$
\mu_s > 1 - \frac{1}{A_0}. \tag{15.12}
$$

The fixed point $\boldsymbol{n}^{(2)}$ corresponds to a distribution in which the master sequence has disappeared even if it has larger fitness than other phenotypes. This effect is usually called Muller's ratchet (MR). The point P_2 is stable for $A_0 > 1$, $A_* > 1$ and

$$
\mu_s > \frac{A_0/A_* - 1}{A_0/A_*}. \tag{15.13}
$$

The fixed point $\boldsymbol{n}^{(3)}$ corresponds to a coexistence of all phenotypes. It is stable in the rest of cases, with $A_0 > 1$. The asymptotic distribution, however, can assume two very different shapes. In the quasi-species (QS) distribution, the master sequence phenotype is more abundant than other phenotypes; after increasing the mutation rate, however, the numeric predominance of the master sequence is lost, an effect that can be denoted error threshold (ET). The transition between these two regimes is given by $n_0 = n_1$, *i.e.*,

$$
\mu_s = \frac{A_0/A_* - 1}{2A_0/A_* - 1}. \tag{15.14}
$$

Our definition of the error threshold transition needs some remarks: in Eigen's original work [36, 37] the error threshold is located at the maximum mean Hamming distance, which corresponds to the maximum spread of population. In the limit of very large genomes these two definitions agree, since the transition becomes very sharp [16]. See also Refs. [38, 39].

In Fig. 15.3 we reported the phase diagram of model (15.11) for $A_* > 1$ (the population always survives). There are three regions: for a low mutation probability μ_s and high selective advantage A_0/A_* of the master sequence, the distribution has the quasi-species form (QS); increasing μ_s the distribution undergoes the error threshold (ET) effect; finally, for very high mutation probabilities, the master sequence disappears and we enter the Muller's ratchet (MR) region [32, 40]. The error threshold phase transition is not present for smooth landscapes (for an example of a study of evolution on a smooth landscape, see Ref. [41]).

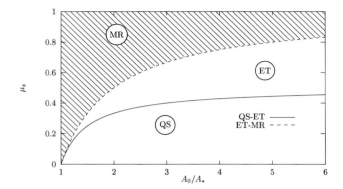

Fig. 15.3.   Phase diagram for the error threshold and Muller's ratchet transitions ($A_* > 1$). MR refers to the Muller's ratchet phase ET to the error threshold distribution and QS to quasi-species distribution.   The phase boundary between the Muller's ratchet effect and the error threshold distribution (Eq. (15.12)) is marked ET-MR; the phase boundary between the error threshold and the quasi-species distribution (Eq. (15.14)) is marked QS-ET.

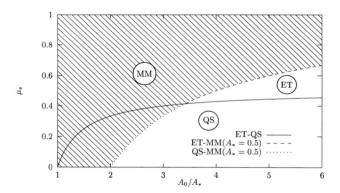

Fig. 15.4.   Phase diagram for the mutational meltdown extinction, the error threshold and the quasi-species distributions ($A_* < 1$). MM refers to the mutational meltdown phase, ET to the error threshold distribution and QS to quasi-species distribution. The phase boundary between the Mutational meltdown effect and the error threshold distribution (Eqs. (15.13) and (15.16)) is marked ET-MM; the phase boundary between the mutational meltdown and the quasi-species distribution (Eqs. (15.13) and (15.15)) is marked QS-MM.

In Fig. 15.4 we illustrate the phase diagram in the case $A_* = 0.5$. For a low mutation probability μ_s and high selective advantage A_0/A_* of the master sequence, again one observes a quasi-species distribution (QS), while for sufficiently large μ_s there is the extinction of the whole population due to the mutational meltdown (MM) effect. The transition between the QS and MM phases can occur directly, for

$$A_0/A_* < \frac{1 - \sqrt{1 - A_*}}{A_*} \tag{15.15}$$

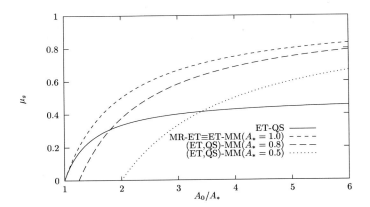

Fig. 15.5. Phase diagram for the error threshold and mutational meltdown transitions for some values of A_*. ET-QS refers to the Error threshold transition, Eq. (15.14), QS-MM to the mutational meltdown extinction without the error threshold transition, Eqs. (15.13) and (15.15), ET-MM to the mutational meltdown extinction after the error threshold transition, Eqs. (15.13) and (15.16). The line MR-ET marks the Muller's ratchet boundary, Eq. (15.12), which coincides with the mutational meltdown (MM) boundary for $A_* = 1$.

(dotted QS-MM line in Figure): during the transient before extinction the distribution keeps the QS form. For

$$A_0/A_* > \frac{1 - \sqrt{1 - A_*}}{A_*} \tag{15.16}$$

one has first the error threshold transition (QS-ET line in Figure), and then one observes extinction due to the mutational meltdown effect (dashed ET-MM line in Figure). This mutation-induced extinction has been investigated numerically in Ref. [40].

We finally report in Fig. 15.5 the phase diagram of the model in the $A_* < 1$ case, for some values of A_*. Notice that for $A_* = 1$ the mutational meltdown effect coincides with the Muller's ratchet one.

Here the mutation probability μ_s is defined on a per-genome basis. If one considers a fixed mutation probability μ per genome element, one has $\mu_s \simeq L\mu$, where L is the genome length. Thus, it is possible to trigger these phase transitions by increasing the genome length.

Numerical simulations on a related model [40] does not show a phase in which, for $A_* < 1$, the population survives after the loss of the quasispecies distribution, *i.e.*, the error theshold - mutational meltdown (ET-MM) transition was not observed.

The Error threshold for finite populations has been studied in Refs. [32, 42, 43].

15.3.4. *Evolution in a phenotypic space*

In the previous section we have obtained some information about the stability of a quasi-species distribution. In the following we want to study the stability of a distribution formed by more than one quasi-species, i.e. the speciation phenomenon. Before doing that we need to know the shape of a quasi-species given a static fitness landscape. Some analytical results can be obtained by considering the dynamics only in the phenotypic space [41].

We assume that the phenotypic index u ranges between $-\infty$ and ∞ in unit steps (the fitness landscape provides that only a finite range of the phenotypic space is viable), and that mutations connect phenotypes at unit distance; the probability of observing a mutation per unit of time is μ. The mutational matrix $M(u,v)$ has the form:

$$M(u,v) = \begin{cases} \mu & \text{if } |u,v| = 1\,, \\ 1 - 2\mu & \text{if } u = v\,, \\ 0 & \text{otherwise}\,. \end{cases}$$

Let us consider as before the evolution of phenotypic distribution $p(u)$, that gives the probability of observing the phenotype u. As before the whole distribution is denoted by $\boldsymbol{p}$.

Considering a phenotypic linear space and non-overlapping generations, we get from Eq. (15.5)

$$p'(u) = \frac{(1 - 2\mu)A(u,\boldsymbol{p})p(u) + \mu(A(u+1,\boldsymbol{p})p(u+1) + A(u-1,\boldsymbol{p})p(u-1)}{\langle A \rangle}\,.$$

In the limit of continuous phenotypic space, u becomes a real number and

$$p'(u) = \frac{1}{\langle A \rangle}\left(A(u,\boldsymbol{p})p(u) + \mu\frac{\partial^2 A(u,\boldsymbol{p})p(u)}{\partial u^2}\right)\,, \tag{15.17}$$

with

$$\int_{-\infty}^{\infty} p(u)du = 1\,, \qquad \int_{-\infty}^{\infty} A(u,\boldsymbol{p})p(u)du = \langle A \rangle\,. \tag{15.18}$$

Equation (15.17) has the typical form of a nonlinear reaction-diffusion equation. The numerical solution of this equation shows that a stable asymptotic distribution exists for almost all initial conditions.

The fitness $A(u,\boldsymbol{p}) = \exp(H(u,\boldsymbol{p}))$ can be written as before, with

$$H(u,\boldsymbol{p}) = V(u) + \int_{-\infty}^{\infty} J(u,v)p(v)dv\,.$$

Before studying the effect of competition and the speciation transition let us derive the exact form of $p(u)$ in case of a smooth and sharp static fitness landscape.

15.3.4.1. *Evolution near a smooth and sharp maximum*

In the presence of a single maximum the asymptotic distribution is given by one quasi-species centers around the global maximum of the static landscape. The effect of a finite mutation rate is simply that of broadening the distribution from a delta peak to a bell-shaped curve.

We are interested in deriving the exact form of the asymptotic distribution near the maximum. We take a static fitness $A(u)$ with a smooth, isolated maximum for $u = 0$ (*smooth maximum* approximation). Let us assume that

$$A(u) \simeq A_0(1 - au^2),\tag{15.19}$$

where $A_0 = A(0)$. Substituting $\exp(w) = Ap$ in Eq. (15.17) we have (neglecting to indicate the phenotype u, and using primes to denote differentiation with respect to it):

$$\frac{\langle A \rangle}{A} = 1 + \mu(w'^2 + w''),$$

and approximating $A^{-1} = A_0^{-1}(1 + au^2)$, we have

$$\frac{\langle A \rangle}{A_0}(1 + au^2) = 1 + \mu(w'^2 + w'').\tag{15.20}$$

A possible solution is

$$w(u) = -\frac{u^2}{2\sigma^2}.$$

Substituting into Eq. (15.20) we finally get

$$\frac{\langle A \rangle}{A_0} = \frac{2 + a\mu - \sqrt{4a\mu + a^2\mu^2}}{2}.\tag{15.21}$$

Since $\langle A \rangle/A_0$ is less than one we have chosen the minus sign. In the limit $a\mu \to 0$ (small mutation rate and smooth maximum), we have

$$\frac{\langle A \rangle}{A_0} \simeq 1 - \sqrt{a\mu}\tag{15.22}$$

and

$$\sigma^2 \simeq \sqrt{\frac{\mu}{a}}.\tag{15.23}$$

The asymptotic solution is

$$p(u) = \frac{1 + au^2}{\sqrt{2\pi}\sigma(1 + a\sigma^2)} \exp\left(-\frac{u^2}{2\sigma^2}\right),\tag{15.24}$$

so that $\int p(u)du = 1$. The solution is then a bell-shaped curve, its width σ being determined by the combined effects of the curvature a of maximum and the mutation rate μ.

For completeness, we study here also the case of a *sharp maximum*, for which $A(u)$ varies considerably with u. In this case the growth rate of less fit strains has a large contribution from the mutations of fittest strains, while the reverse flow is negligible, thus

$$p(u - 1)A(u - 1) \gg p(u)A(u) \gg p(u + 1)A(u + 1)$$

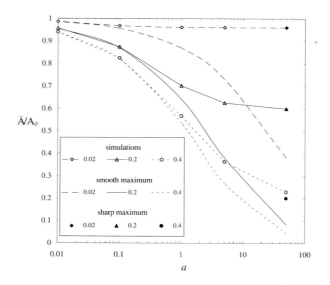

Fig. 15.6. Average fitness $\langle A \rangle / A_0$ versus the coefficient a, of the fitness function, Eq. (15.19), for some values of the mutation rate μ. Legend: _numerical solution_ corresponds to the numerical solution of Eq. (15.17), _smooth maximum_ refers to Eq. (15.21) and _sharp maximum_ to Eq. (15.25).

neglecting last term, and substituting $q(u) = A(u)p(u)$ in Eq. (15.17) we get:

$$\frac{\langle A \rangle}{A_0} = 1 - 2\mu \qquad \text{for } u = 0 \tag{15.25}$$

and

$$q(u) = \frac{\mu}{(\langle A \rangle A(u) - 1 + 2\mu)} q(u-1) \qquad \text{for } u > 0. \tag{15.26}$$

Near $u = 0$, combining Eq. (15.25), Eq. (15.26) and Eq. (15.19), we have

$$q(u) = \frac{\mu}{(1 - 2\mu)au^2} q(u-1).$$

In this approximation the solution is

$$q(u) = \left(\frac{\mu}{1 - 2\mu a} \right)^u \frac{1}{(u!)^2},$$

and

$$y(u) = A(u)q(u) \simeq \frac{1}{A_0}(1 + au^2) \left(\frac{\mu A_0}{\langle A \rangle a} \right)^u \frac{1}{u!^2}.$$

We have checked the validity of these approximations by solving numerically Eq. (15.17); the comparisons are shown in Fig. (15.6). We observe that the _smooth maximum_ approximation agrees with the numerics for small values of a, when $A(u)$ varies slowly with u, while the _sharp maximum_ approximation agrees with the numerical results for large values of a, when small variations of u correspond to large variations of $A(u)$.

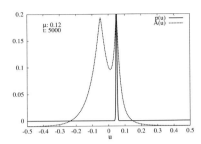 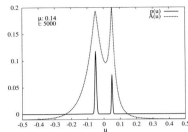

Fig. 15.7. Mutation-induced speciation. A two peaks static fitness landscape, increasing the mutation rate we pass from a single quasi-species population (left, $\mu = 0.12$) to the coexistence of two quasi-species (right, $\mu = 0.14$).

15.3.5. *Coexistence on a static fitness landscape*

We investigate here the conditions for which more than one quasi-species can coexist on a static fitness landscape without competition.

Let us assume that the fitness landscape has several distinct peaks, and that any peak can be approximated by a quadratic function near its maximum. For small but finite mutation rates, as shown by Eq. (15.24), the distribution around an isolated maxima is a bell curve, whose width is given by Eq. (15.23) and average fitness by Eq. (15.22). Let us call thus distribution a quasi-species, and the peak a niche.

If the niches are separated by a distance greater than σ, a superposition of quasi-species (15.24) is a solution of Eq. (15.5). Let us number the quasi-species with the index k:

$$p(u) = \sum_k p_k(u) \,;$$

each $p_k(u)$ is centered around u_k and has average fitness $\langle A \rangle_k$. The condition for the coexistence of two quasi-species h and k is $\langle A \rangle_h = \langle A \rangle_k$ (this condition can be extended to any number of quasi-species). In other terms one can say that in a stable environment the fitness of all co-existing individuals is the same, independently on the species.

Since the average fitness (15.22) of a quasi-species depends on the height A_0 and the curvature a of the niche, one can have coexistence of a sharper niche with larger fitness together with a broader niche with lower fitness, as shown in Fig. 15.7. This coexistence depends crucially on the mutation rate μ. If μ is too small, the quasi-species occupying the broader niche disappears; if the mutation rate is too high the reverse happens. In this case, the difference of fitness establishes the time scale, which can be quite long. In presence of a fluctuating environment, these time scales can be long enough that the extinction due to global competition is not relevant. A transient coexistence is illustrated in Fig. 15.8. One can design a special form of the landscape that allows the coexistence for a finite interval of values of μ, but

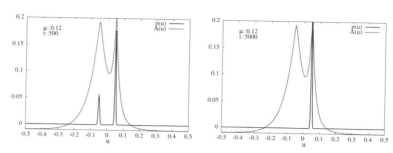

Fig. 15.8.   Evolution on a two-peaks static fitness landscape, after 500 (left) and 5000 (right) time steps. For a transient period of time the two species co-exist, but in the asymptotic limit only the fittest one survives.

certainly this is not a generic case. This condition is known in biology under the name of Gause or *competitive exclusion* principle [44].

We show in the following that the existence of a degenerate effective fitness is a generic case in the presence of competition, if the two species can co-adapt before going extinct.

15.3.6. *Flat landscapes and neutral evolution*

The lineage is the sequence of connections from a son to the father at birth. Going backward in time, from an existing population, we have a tree, that converges to the last universal common ancestor (LUCA). In the opposite time direction, we observe a spanning tree, with lot of branches that go extinct. The lineage of an existing individual is therefore a path in genetic space. Let us draw this path by using the vertical axes for time, and the horizontal axes for the genetic space. Such paths are hardly experimentally observable (with the exception of computer-generated evolutionary populations), and has to be reconstructed from the differences in the existing populations.

The probability of observing such a path, is given by the probability of survival, for the vertical steps, and the probability of mutation for the horizontal jumps. While the latter may often be considered not to depend on the population and the genome itself,[g] the survival is in general a function of the population. Only in the case of neutral evolution, one can disregard this factor. Within this assumption, it is possible to compute the divergence time between two genomes, by computing their differences, and reconstruct phylogenetic trees.

15.3.7. *A simple example of fitness landscape*

The main problem in theoretical evolutionary theory is that of forecasting the consequences of mutations, *i.e.*, the fitness landscape. If a mutation occurs in a coding region, the resulting protein changes. The problem of relating the sequence of a

[g]the inverse of this probability gives the "molecular clock"

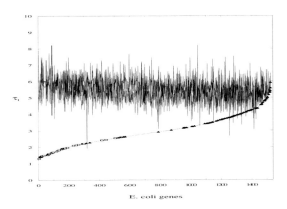

Fig. 15.9. Mean translation time τ_1 for 1530 *Escherichia coli* coding regions calculated using data on tRNA abundances published by Ikemura [46]. The sequences are ordered according to the values of τ_1. The lower line shows τ_1 for the correct reading frame. The upper lines show the value of τ_1 for the +1 and +2 reading frames. The open squares correspond to ribosomal genes, the filled triangles correspond to genes carried by plasmids.

protein to its tri-dimensional shape is still unsolved. Even more difficult is the problem of computing the effects of this modified shape in the chemical network of a cell. Therefore, the only landscape that is considered is the neutral (flat) one.

One of the few cases in which it is possible to guess the consequences of mutations is the case of synonymous substitutions in bacteria [45].

The genetic code (correspondence between codons and amino acids) is highly redundant: there are 64 different codons, and only 20 amino acids, plus a stop command that signals the end of the protein chain. This redundancy occurs in two ways: the third codon is often uninfluenced for the binding of the tRNA, and many different tRNA (cognate to different codons) correspond to the same amino acid (synonymous tRNA). Therefore, all these substitutions are neutral with respect to the protein produced. Synonymous tRNA occurs in different abundances.

The time needed for a ribosome to attach a new amino acid to the growing chain depends mainly on the waiting time for the right charged tRNA to arrive. This time is proportional to the abundance of tRNAs, and supposing that all of them are always charged, it is proportional to the copies of the tRNAs in the genome. Therefore, even if synonymous substitutions are neutral for the functionality of proteins, they imply different time intervals for their production.

Bacteria in an abundant medium grow exponentially, and a small difference in the time needed for a duplication can make a big difference in fitness. Proteins are needed in various abundances for duplications: structural ones, like ribosomal proteins, are essential (and consequently they are well optimized and very conserved), while those carried by plasmids, being transmittable among different strains, are presumably not very optimized. This assumption is confirmed by Fig. 15.9, in which it is reported the translation time τ_1 (computed using the abundances of

cognate tRNAs) for various proteins. One can see that the time obtained shifting
the frame of $+1$ or $+2$ basis, that produces a meaningless protein, is substantially
higher for ribosomal protein, and similar to the correct one for proteins carried by
plasmids [47].

We can simplify the problem by grouping "abundant" tRNAs by the symbol 0,
and "rare" tRNAs by 1. Disregarding "intermediate" tRNAs, we get a sequence x
of zeros and ones, and a fitness, assumed monotonic with the replication time τ_1,
that has the form

$$A(x) = \exp(\tau_1(x)) = \exp\left(A + B\sum_i x_i\right),$$

i.e., a monotonic function of the "magnetization" of an Ising chain. Here the codons
does not interact, and their effect is additive. If one considers that the discharging
of a tRNA may slower the translation of cognate codons, one may introduce other
couplings among codons [48, 49].

15.3.8. *Finite populations and random drift*

In finite population, the fluctuations plays a fundamental role. Let us just consider a simple system x composed by N individuals belonging either to variant (or
species) A $(x = 0)$ or B $(x = 1)$. At each time step an individual x_i is chosen
at random, replicates with a probability $p(x_i)$ and substitutes another randomly
chosen individual (Moran process [50]). We have a birth-death process of the quantity $n = \sum_i x_i$ with two absorbing states, $n = 0$ and $n = N$ corresponding to the
extinction of B and A, respectively. It can be shown [51] that the probability $S(1)$
that a single mutant B will take over the whole population is

$$S(1) = \frac{1 - r}{1 - r^N} \quad \xrightarrow[r \to 1]{} \quad \frac{1}{N},$$

where $r = p(1)/p(0)$ is the selective advantage (or disadvantage) of variant B with
respect to A. If $\mu \simeq 0$ is the "mutation probability" from A to B, the probability
of appearance of a mutant in a population of N A's is μN, and therefore the
probability of fixation of a mutant is $Q = \mu N S(1)$, for neutral selection $(r = 1)$ we
get $Q = \mu$, irrespective of the population size [52]. This result by Kimura provides
the *molecular clock* for estimating evolution times from the measure of the distance
between two genomes (with the assumption of neutral evolution).

15.3.9. *Speciation in a flat landscape*

The reasons for the existence of species is quite controversial [53]. The definition
itself of a species is not a trivial task. One definition (a) is based simply on the
phenotypic differentiation among the phenotypic clusters, and can be easily applied
also to asexual organisms, such as bacteria. Another definition (b), that applies
only to sexual organism, is based on the inter-breeding possibility. Finally (c), one

can define the species on the basis of the genotypic distance of individuals, taking into consideration their genealogical story [54].

We shall see in Section 15.5.3 a model of how such a segregation, preparatory to speciation, can arise. The role of genetic segregation in maintaining the structure of the species is particularly clear in the model of Refs. [55–57]. In this model, there is no fitness selection, and one can have asexual reproduction, or sexual (recombination) one with or without a segregation based on genetic distance. Their conclusion is that species (defined using the reproductive isolation, definition (b)) can appear in flat static landscapes provided with sexual reproduction and discrimination of mating. In some sense these authors have identified definitions (b) and (c).

In the first model reproduction is asexual: each offspring chooses its parent at random in the previous generation, with mutations. In this model there are no species, and the distribution fluctuates greatly in time, even for very large populations. The addition of competition, Section 15.5.1, may stabilize the distribution.

In the second model reproduction is recombinant with random mating allowed between any pair of individuals. In this case, the population becomes homogeneous and the genetic distance between pairs of individuals has small fluctuations which vanish in the limit of an infinitely large population. Without segregation (due to competition and/or to genetic incompatibilities) one cannot have the formation of stable, isolated species.

In the third model reproduction is still recombinant, but instead of random mating, mating only occurs between individuals which are genetically similar to each other. In that case, the population splits spontaneously into species which are in reproductive isolation from one another and one observes a steady state with a continual appearance and extinction of species in the population.

15.4. Ageing

A simple example of age-dependent (pleyotropic) effects of genes is given by the Penna model [58, 59]. In this model, each individual is characterized by a genome of length L, and an age counter. The genome is arranged in such a way that gene x_i is activated at age i. There are two alleles, a good one $x_i = 0$ and a bad one $x_i = 1$. If a given number m of bad genes are activated, the individual dies.

The phenotype $u(x, a)$ depends here of the genotype x and of the age a, and can be written as

$$u(x, a) = \sum_{i=1}^{a} x_i \, ,$$

while the fitness is sharp:

$$A(u) = \begin{cases} 1 & \text{if } u < m \, , \\ 0 & \text{otherwise} \, . \end{cases}$$

This model can explain the shape of Gompertz's mortality law [60]. If one imposes that reproduction occurs only after a certain age, one can easily explain the accumulation of bad genes after that age, as exhibited by the catastrophic senescence of many semelparous animals (like the pacific salmon) [61]. By inserting maternal cares, one can also give a motivation for the appearance of menopause in women [62]: after a certain age, the fitness may be more increased by stopping giving birth to more sons, with the risk of loosing own life and that of not-yet-independent offspring.

15.5. Dynamic Ecosystems

15.5.1. *Speciation in the phenotypic space*

We are here referring to the formation of species in a spatially homogeneous environment, i.e. to *sympatric* speciation. In this frame of reference, a niche is a phenotypic realization of relatively high fitness. Species have obviously to do with niches, but one cannot assume that the coexistence of species simply reflects the presence of "pre-existing" niches; on the contrary, what appears as a niche to a given individual is co-determined by the presence of other individuals (of the same or of different species). In other words, niches are the product of co-evolution.

In this section we introduce a new factor in our model ecosystem: a short-range (in phenotypic space) competition among individuals. As usual, we start the study of its consequences by considering the evolution in phenotypic space [63, 64].

We assume that the static fitness $V(u)$, when not flat, is a linear decreasing function of the phenotype u except in the vicinity of $u = 0$, where it has a quadratic maximum:

$$V(u) = V_0 + b\left(1 - \frac{u}{r} - \frac{1}{1 + u/r}\right) \tag{15.27}$$

so that close to $u = 0$ one has $V(u) \simeq V_0 - bu^2/r^2$ and for $u \to \infty$, $V(u) \simeq V_0 + b(1 - u/r)$. The master sequence is located at $u = 0$.

We have checked numerically that the results are qualitatively independent on the exact form of the static fitness, providing that it is a smooth decreasing function. We have introduced this particular form because it is suitable for analytical computation, but a more classic Gaussian form can be used.

For the interaction matrix W we have chosen the following kernel K

$$K\left(\frac{u - v}{R}\right) = \exp\left(-\frac{1}{\alpha}\left|\frac{u - v}{R}\right|^\alpha\right).$$

The parameter J and α control the intensity and the steepness of the intra-species competition, respectively. We use a Gaussian ($\alpha = 2$) kernel, for the motivations illustrated in Section 15.5.1.

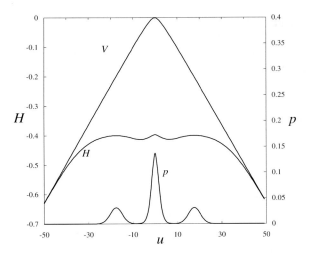

Fig. 15.10. Static fitness V, effective fitness H, and asymptotic distribution p numerically computed for the following values of parameters: $\alpha = 2$, $\mu = 0.01$, $V_0 = 1.0$, $b = 0.04$, $J = 0.6$, $R = 10$, $r = 3$ and $N = 100$.

For illustration, we report in Fig. 15.10 the numerical solution of Eq. (15.5), showing a possible evolutionary scenario that leads to the coexistence of three quasi-species. We have chosen the smooth static fitness $V(u)$ of Eq. (15.27) and a Gaussian ($\alpha = 2$) competition kernel. One can realize that the effective fitness H is almost degenerate (here $\mu > 0$ and the competition effect extends on the neighborhood of the maxima), i.e. that the average fitness of all coexisting quasi-species is the same.

We now derive the conditions for the coexistence of multiple species. We are interested in its asymptotic behavior in the limit $\mu \to 0$, which is the case for actual organisms. Actually, the mutation mechanism is needed only to define the genotypic distance and to populate all available niches. Let us assume that the asymptotic distribution is formed by $\mathcal{L}$ quasi-species. Since $\mu \to 0$ they are approximated by delta functions $p_k(u) = \gamma_k \delta_{u,u_k}$, $k = 0, \ldots, \mathcal{L} - 1$, centered at u_k. The weight of each quasi species is γ_k, i.e.

$$\int p_k(u)du = \gamma_k, \qquad \sum_{k=0}^{\mathcal{L}-1} \gamma_k = 1.$$

The quasi-species are ordered such as $\gamma_0 \geq \gamma_1, \ldots, \geq \gamma_{\mathcal{L}-1}$.

The evolution equations for the p_k are

$$p_k'(u) = \frac{A(u_k)}{\langle A \rangle} p_k(u),$$

where $A(u) = \exp(H(u))$ and

$$H(u) = V(u) - J \sum_{j=0}^{\mathcal{L}-1} K\left(\frac{u - u_j}{R}\right) \gamma_j.$$

The stability condition of the asymptotic distribution is $(A(u_k) - \langle A \rangle)p_k(u) = 0$, i.e. either $A(y_k) = \langle A \rangle = \text{const}$ (degeneracy of maxima) or $p_k(u) = 0$ (all other points). This supports our assumption of delta functions for the p_k.

The position u_k and the weight γ_k of the quasi-species are given by $A(u_k) = \langle A \rangle = \text{const}$ and $\partial A(u)/\partial u|_{u_k} = 0$, or, in terms of the fitness H, by

$$V(u_k) - J \sum_{j=0}^{\mathcal{L}-1} K\left(\frac{u_k - u_j}{R}\right) \gamma_j = \text{const},$$

$$V'(u_k) - \frac{J}{R} \sum_{j=0}^{\mathcal{L}-1} K'\left(\frac{u_k - u_j}{R}\right) \gamma_j = 0,$$

where the prime in the last equation denotes differentiation with respect to u.

Let us compute the phase boundary for coexistence of three species for two kinds of kernels: the exponential ($\alpha = 1$) and the Gaussian ($\alpha = 2$) one. The diffusion kernel can be derived by a simple reaction-diffusion model, see Ref. [63].

We assume that the static fitness $V(u)$ of Eq. (15.27). Due to the symmetries of the problem, we have the master quasi-species at $u_0 = 0$ and, symmetrically, two satellite quasi-species at $u = \pm u_1$. Neglecting the mutual influence of the two marginal quasi-species, and considering that $V'(u_0) = K'(u_0/R) = 0$, $K'(u_1/R) = -K'(-u_1/R)$, $K(0) = J$ and that the three-species threshold is given by $\gamma_0 = 1$ and $\gamma_1 = 0$, we have

$$\tilde{b}\left(1 - \frac{\tilde{u}_1}{\tilde{r}}\right) - K(\tilde{u}_1) = -1,$$

$$\frac{\tilde{b}}{\tilde{r}} + K'(\tilde{u}_1) = 0,$$

where $\tilde{u} = u/R$, $\tilde{r} = r/R$ and $\tilde{b} = b/J$. We introduce the parameter $G = \tilde{r}/\tilde{b} = (J/R)/(b/r)$, that is the ratio of two quantities, the first one related to the strength of inter-species interactions (J/R) and the second to intra-species ones (b/r).

In the following we drop the tildes for convenience. Thus

$$r - z - G \exp\left(-\frac{z^\alpha}{\alpha}\right) = -G,$$

$$G z^{\alpha-1} \exp\left(-\frac{z^\alpha}{\alpha}\right) = 1.$$

For $\alpha = 1$ we have the coexistence condition

$$\ln(G) = r - 1 + G.$$

The only parameters that satisfy these equations are $G = 1$ and $r = 0$, i.e. a flat landscape ($b = 0$) with infinite range interaction ($R = \infty$). Since the coexistence region reduces to a single point, it is suggested that $\alpha = 1$ is a marginal case. Thus

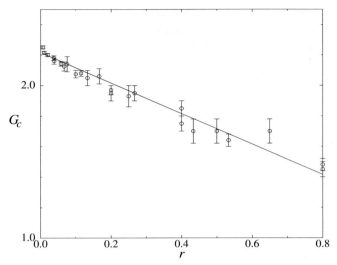

Fig. 15.11. Three-species coexistence boundary G_c for $\alpha = 2$. The continuous line represents the analytical approximation, Eq. (15.28), the circles are obtained from numerical simulations. The error bars represent the maximum error (see text for details).

for less steep potentials, such as power law decrease, the coexistence condition is supposed not to be fulfilled.

For $\alpha = 2$ the coexistence condition is given by

$$z^2 - (G + r)z + 1 = 0\,,$$

$$Gz \exp\left(-\frac{z^2}{2}\right) = 1\,.$$

One can solve numerically this system and obtain the boundary $G_c(r)$ for the coexistence. In the limit $r \to 0$ (almost flat static fitness) one has

$$G_c(r) \simeq G_c(0) - r \tag{15.28}$$

with $G_c(0) = 2.216\ldots$. Thus for $G > G_c(r)$ we have coexistence of three or more quasi-species, while for $G < G_c(r)$ only the fittest one survives.

We have solved numerically Eq. (15.17) for several different values of the parameter G. We have considered a discrete phenotypic space, with N points, and a simple Euler algorithm. The results, presented in Fig. 15.11, are not strongly affected by the integration step. The error bars are due to the discreteness of the changing parameter G. The boundary of the multi-species phase is well approximated by Eq. (15.28); in particular, we have checked that this boundary does not depend on the mutation rate μ, at least for $\mu < 0.1$, which can be considered a very high mutation rate for real organisms. The most important effect of μ is the broadening of quasi-species curves, which can eventually merge as described in Section 15.3.4.1.

This approximate theory to derive the condition of coexistence of multiple quasi-species still holds for the hyper-cubic genotypic space. The different structure of genotypic space does not change the results in the limit $\mu \to 0$. Moreover, the threshold between one and multiple quasi-species is defined as the value of parameters for which the satellite quasi-species vanish. In this case the multiplicity factor of satellite quasi-species does not influence the competition, and thus we believe that the threshold G_c of Eq. (15.28) still holds in the genotypic hyper-cubic space.

For a variable population, the theory still works substituting G with

$$G_a = NG = N\frac{J/R}{b/r} , \tag{15.29}$$

(for a detailed analysis see Ref. [3]).

This result is in a good agreement with numerical simulations, as shown in the following Section.

15.5.2. *Speciation and mutational meltdown in the hyper-cubic genotypic space*

Let us now study the consequences of evolution in presence of competition in the more complex genotypic space. We were not able to obtain analytical results, so we resort to numerical simulations. Some details about the computer code we used can be found in Appendix B.

In the following we refer always to rule (a), that allows us to study both speciation and mutational meltdown. Rule (b) has a similar behavior for speciation transition, while, of course, it does not present any mutational meltdown transition.

We considered the same static fitness landscape of Eq. (15.27), (non-epistatic interactions among genes).

We observe, in good agreement with the analytical approximation Eq. (15.28), that if G_a (Eq. (15.29)) is larger than the threshold G_c (Eq. (15.28)), several quasi-species coexist, otherwise only the master sequence quasi-species survives. In Fig. 15.12 a distribution with multiple quasi-species is shown.

We can characterize the speciation transition by means of the entropy S of the asymptotic phenotypic distribution $p(u)$ as function of G_a,

$$S = -\sum_u p(u) \ln p(u)$$

which increases in correspondence of the appearance of multiple quasi-species (see Fig. 15.13).

We locate the transition at a value $G_a \simeq 2.25$, while analytical approximation predicts $G_c(0.1) \simeq 2.116$. The entropy, however, is quite sensible to fluctuations of the quasi-species centered at the master sequence (which embraces the largest part of distribution), and it was necessary to average over several runs in order to obtain a clear curve; for larger values of μ it was impossible to characterize this transition. A quantity which is much less sensitive of fluctuations is the average

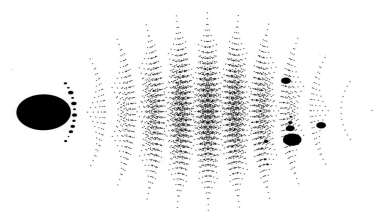

Fig. 15.12.   Easter egg representation of quasi-species in hyper-cubic space for $L = 12$. The smallest points represent placeholder of strains (whose population is less than 0.02), only the larger dots correspond to effectively populated quasi-species; the area of the dot is proportional to the square root of population. Parameters: $\mu = 10^{-3}$, $V_0 = 2$, $b = 10^{-2}$, $R = 5$, $r = 0.5$, $J = 0.28$, $N = 10000$, $L = 12$.

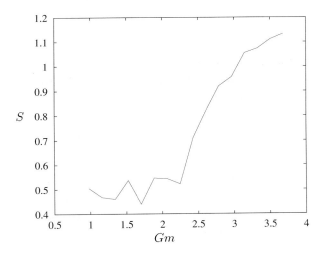

Fig. 15.13.   The speciation transition characterized by the entropy S as a function of the control parameter $G\,m$. Each point is an average over 15 runs. Same parameters as in Fig. 15.12, varying J. Errors are of the order of fluctuations.

square phenotypic distance from the master sequence $\langle g(u)^2 \rangle$

$$\langle g(u)^2 \rangle = \sum_u g(u)^2 p(u)\,.$$

In Fig. 15.14 (left) we characterize the speciation transition by means of $\langle g(u)^2 \rangle$, and indeed a single run was sufficient, for $\mu = 10^{-3}$. For much higher mutation rates ($\mu = 5\,10^{-2}$) the transition is less clear, as shown in Fig. 15.14 (right), but

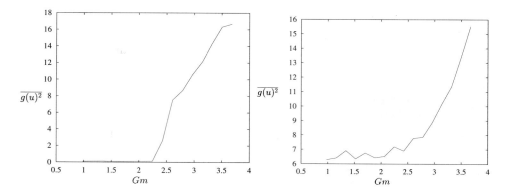

Fig. 15.14. Independence of the speciation transition by the mutation rate. The transition is characterized by the average square phenotypic distance $\langle g(u)^2 \rangle$ of phenotypic distribution $p(u)$, as a function of the control parameter $G\, m$. Each point is a single run. Same parameters as in Fig. 15.12, varying J with $\mu = 10^{-3}$ (left) and $\mu = 5\,10^{-2}$ (right).

one can see that the transition point is substantially independent of μ, as predicted by the analysis of speciation in the phenotypic space, Eq. (15.28).

15.5.3. *Sex*

By inserting sexual reproduction (recombination) and possibly horizontal transfer of genetic material (conjugation in bacteria, viruses and – in plants, also tumor-inducing bacteria) we should rather consider a different phylogenetic path for each gene. The history of a given gene (more appropriately, a locus) is in general different from that of another one, still present in the same genome. The survival probability of a gene depends on how good the gene performs in conjunction with the other genes. Even if only "egoistic" genes (those that maximize their survival and reproductive probability) passes from a generation to the other, they are nevertheless forced to collaborate. Moreover, genes are identical, in the sense that two individuals or cells may carry the *same* gene. The survival of a gene may thus profit of the death of another individual with the same gene, if this death increases sufficiently the survival and reproductive efficiency of the former. This is the origin of multicellular beings (in which reproduction is reserved to the germ line), of insect colonies, familiar structure and many other evolutionary patterns. More in this subject in Section 15.5.4.

Actually, forced cooperation is true only for those genes that may pass to the following generation through the reproduction of the individual. Genes may not follow the fate of the individual by escaping as viruses, or being horizontally transferred (plasmids). Especially in case of viruses, the direct advantage of viruses may correspond to a great damage for the host. If for some mutation the virus content is "trapped" inside the host and forced to follow the usual germ line to propagate, that automatically it is forced to collaborate.

In case of sexual reproduction, the evolution may lead to other non-optimal stable patterns, represented by the run-off development of sexual characters. This phenomenon if particularly evident in a model by Dieckmann and Doebeli [65].

In this model, the space plays no role (no spatial niches, sympatric speciation). The fitness in phenotypic space presents a smooth maximum, so in the absence of other ingredients, the asymptotic distribution is a quasispecies peaked around this maximum. The introduction of a finite-range competition broadens the distribution, but in the absence of assortativity (preference in mating), no speciation occurs [29].

In the model, the assortativity is coded in a portion of the genotype, and is related to a phenotypic character either related to the fitness or acting just like an arbitrary marker. In both cases, speciation may occur (depending on the parameters), even if in the second case the differentiation takes longer (since it is due to genetic drift).

Sex and recombination have important consequences also in the evolution over simple landscapes, like the sharp peak one. In Ref. [66] it is shown that while for asexual reproduction the error threshold is driven by the average mutation probability per unit of genome, μ, in recombinant organisms it depends on the total mutation rate μL, and the transition is near $\mu L = 1$. This observation poses a limit to the maximum length of genomes for a given accuracy of replication.

15.5.4. *Game theory*

Up to now, we have supposed that there is an instantaneous response to the environment. Considering only binary interactions, the contribution to the score of phenotype x given by an encounter with phenotype y, $W(x|y)$, can schematically be grouped as

$$
\begin{aligned}
W(x|y) < 0 \ \& \ W(y|x) < 0 : & \quad \text{competition}, \\
W(x|y) > 0 \ \& \ W(y|x) < 0 : & \quad \text{predation or parasitism of } y \text{ on } x, \\
W(x|y) < 0 \ \& \ W(y|x) > 0 : & \quad \text{predation or parasitism of } x \text{ on } y, \\
W(x|y) > 0 \ \& \ W(y|x) > 0 : & \quad \text{cooperation}.
\end{aligned}
\tag{15.30}
$$

However, the actual interactions among individuals with memory and strategy are more complex, may depend on the past encounters and on the environment status.

A more sophisticated modeling follows the ideas of game theory: the genotype of an individual is read as a small program, and the score of an encounter depends on the running of the programs of the participants [67].

In order to reduce this complexity, let us assume that there are only binary encounters, and that the participants can assume only two *external* status: zero and one, that traditionally are called cooperation or defection. We assume also that the encounter is divided into rounds (hands), and that the number of hands depends on external factors (one could alternatively have a three-letter alphabet like cooperate, defect or escape, and so on). In other words, we consider the participants as programmed automata.

A generic program therefore says what to play at hand i, given the results of previous hands, the identity of the two participants and the internal state (memory). The choice of the output state can be stochastic or deterministic, but in general one has to determine, for each hand i, the probability $p(i)$ of outputting 0 (that of outputting 1 is $1 - p(i)$), and this can be done using a look-up table that constitutes the genetic information of the individual.

To make things simpler, let us assume that the choice i does depend only on the past hand, so that a program has to say just how to start (hand 0) and the four probabilities $p(a_x, a_y)$: probability of playing 0 if in the last hand player x played a_x and player y played a_y. More complex strategies can be used [68].

The score of the hand for player x is given by the payoff matrix $Q(a_x|a_y)$. The score for player y is the transpose of Q' if the game is symmetric. The interesting situation is the *prisoner's dilemma*: $Q(1|0) > Q((0|0) > Q(1|1) > Q(0|1)$.[h] For a single-shot play, standard game theory applies. It can be shown that it is evolutionary convenient for both player to defect, since mutual cooperation ($Q(0|0)$) gives less payoff that exploitation ($Q(1|0)$); which from the point of view of the opposite player (transpose) gives the minimal payoff. The game is interesting if cooperation gives the higher *average* payoff ($2Q(0|0) > Q(0|1) + Q(1|0)$).

Axelrod [68] noted that if the the game is iterated, cooperation can arise: rule with memory, like TIT-FOR-TAT (TFT: cooperate at first hand, and then copy your opponent's last hand) may stabilize the mutual cooperation state, and resist invasion by other simpler rules (like ALL-DEFECT – ALLD, always play 1). Other more complex rules may win against TFT (and possibly loose against ALLD).

Nowak [70] gives at least five ways for which mutual cooperation can arise in an evolutionary environment:

1. Kin selection. Cooperation is preferred because individuals share a large fraction of the genome, so a gene for cooperation that is common in the population shares the average payoff, even in a single-shot game.
2. Direct reciprocity. This is just Axelrod's iterated game. If the expected length of the game is large enough, it is worth to try to cooperate for gaining the higher share.
3. Reputation. Information about past games can be useful in deciding if own opponent is inclined towards cooperation or towards exploitation.
4. Network reciprocity. In this case one considers the spatial structure of the group, represented by a graph with local connectivity k. A player plays against all neighbors, and gather the related score. If the advantage in cooperating in a cluster of k neighbors is greater than the gain I would have by switching to defect, the strategy is stable.

---

[h]The asymptotic state of the iterated Prisoner's dilemma game is not given by a "simple" optimization procedure, see Ref. [69]

5. Group selection. In this case, the system is assumed to be split in groups, each of which can become extinct, or give origin to another one. A group can be invaded by defectors, but in this case it will succumb in competition with all-cooperator groups, that have larger payoffs.

15.5.5. *Effective treatment of strategies*

The analysis of strategies with memory is quite complex, due to the large number of variables. As shown by Nowak [70], it is often possible to study the evolutionary competition among strategies by using an effective payoff matrix W, that specifies the average gain of a strategy against the others, given the parameters of the problem. In this way the evolution of the system is just given by the interaction weights of Eq. (15.30), and the analysis is quite simpler.

For the competition between two strategies A and B with a parameter p, three conditions can be found [70], according with the position of the saddle point p_c that separates the basins of all-A $(p = 0)$ and all-B $(p = 1)$ distributions:

1. If $p_c = 0$, only the distribution all-B is present. This generally corresponds to all-defectors.
2. If p_c is near zero, the strategy all-A is *evolutionary stable*: in an infinite population, a single (or small but dispersed) mutant B cannot invade, while a bunch of B (in neighboring sites) can.
3. If $p_c > 1/2$, the A startegy is *risk-dominant*: by mixing at random the population, in a larger number of sites A is dominant.
4. If $p_c > 2/3$ then the A strategy is *advantageous*. The analysis of finite population of size N shows that the fixation probability of a mutant by random drift for flat or weak selection is $1/N$. For $p_c > 2/3$, the fixation probability of a cooperator mutant in a finite population of defectors is greater than that given by random drift.

15.5.6. *Self-organization of ecosystems*

The final goal of a theoretical evolutionary theory is to explain the macro-patterns of evolution (speciation, extinctions, emergence of new characteristics like cooperation, formation of complex ecosystems) as the result of self-organization of individuals [71]. The main idea is that at the molecular level evolution is either neutral, or lethal. The latter corresponds to unviable phenotypes, for instances, mutations that lead to proteins that do not fold correctly. Within this assumption, the fitness space for a single "replicator" (an ancient bio-molecule) is like a slice of a Swiss cheese. We assume that there is one big connected component. By considering only the paths on this component, the origin of life should correspond to one largely connected "hub", where entropy tends to concentrate random walks (induced to mutations).

However, competition (red queen), due to the exhaustion of resources, soon changes the fitness of different genotypes. In particular, the location corresponding to the origin of life becomes the one where competition is maximal, due to the large connectivity. The competition induces speciation, so that a hypothetical "movie" of the ancient lineages will show a diverging phylogenetic tree, in which the locations corresponding to ancestors become minima of fitness (extinction) after a speciation event. As we have seen, competition also promotes the grouping of phenotypes and the formation of "species", even in an asexual world.

In experimental and simulated evolutionary processes, mutants soon arises trying to exploit others by "stealing" already formed products, like for instance the capsid of viruses. This is the equivalent of a predator/prey (or parasite/host) relationship. Predators induce complex networks of relationship, for which even distant (in phenotypic space) "species" coevolve synchronously. When such an intricate "ecological" network has established, it may happen that the extinction of a species may affect many others, triggering an avalanche of extinctions (mass extinctions).

The opposite point of view is assuming that chance and incidents dominate the evolution. This could be a perfect motivation for punctuated equilibrium [72] (intermittent bursts of activity followed by long quiescent periods): random catastrophic non-biological events (collision of asteroids, changes in sun activity, etc.) suddenly alter the equilibrium on earth, triggering mass extinction and the subsequent rearrangements. These events certainly happens, the problem is that of establishing their importance.

It is quite difficult to verify the first scenario computationally. A good starting point is the *food web* [73]. We can reformulate the model in our language as follows. A species i is defined by a set of L phenotypic traits $i_1, \ldots, i_k$, chosen among K possibilities. The relationship between two species is given by the match of these characteristics, according with an antisymmetric $K \times K$ matrix M. The score $W(i|j)$ accumulated by species i after an encounter with species j is given by

$$W(i|j) = \frac{1}{L} \sum_{n=1}^{k} \sum_{m=1}^{k} M_{i_n, j_m} .$$

A species accumulates the score by computing all possible encounters. A monotonic function of the score determines the growth (or decreasing) of a species. When the system has reached a stationary state, the size of a species determines its probability of survival (*i.e.*, the survival probability is determined by the accumulated score). When a species disappears, another one is chosen for cloning, with mutations in the phenotypic traits.

A special "species" 0 represents the source of energy or food, for instance solar light flux. This special species is only "predated" and it is not affected by selection.

This model is able to reproduce a feature observed in real ecosystems: in spite of many coexisting species, the number of tropic level is extremely low, and increases only logarithmically whith the number of species.

The model can be further complicated by adding spatial structure and migration, ageing, etc. [74].

Simpler, phenomenological models may be useful in exploring how coherent phenomena like mass extinction may arise from the self-organization of ecosystems. Most of these models consider species (or niches) fixed, disregarding mutation. Possible, the simplest one is the Bak-Sneppen (BS) model [75]. In the BS model, species are simply assigned a location on a graph, representing the ecological relation between two species (in the simplest case, the graph is a regular one-dimensional lattice with nearest-neighbor links). A species is assigned a random number, interpreted as its relative abundance.[i] At each time step, the smallest species is removed, and its ecological place is taken by another one, with a random consistence. With this change, the size of species connected to the replaced one are also randomly changed. The smallest species at the following time step may be one of those recently changed, or an unrelated one. The first case is interpreted as an *evolutionary avalanche* (mass extinction).

This *extremal dynamics* (choice of the smallest species) is able to self-organize the ecosystem so that the size distribution of species is not trivial (a critical value appears so that no smaller species are present), and the distribution of avalanche size follows a power-law, similar (but with a different exponent) to that observed in paleontological data by Raup [76]. This model is able to illustrate how self-organization and punctuated equilibrium can arise internally in a dynamical system with long-range coupling (due to the extremal dynamics).

The Bak-Sneppen model can be though as an extremal simplification of a food web, which in turn can be considered a simplification of the microscopic dynamics (individual-based) of a modeled ecosystem.

15.6. Conclusions

We have shown some aspects of a theoretical approach to self-organization in evolutionary population dynamics. The ideal goals would be that of showing how macro-evolutionary patterns may arise from a simplified individual-based dynamics. However, evolutionary systems tend to develop highly correlated structure, so that is it difficult to operate the scale separation typical of simple physical system (say, gases). One can develop simple models of many aspects, in order to test the robustness of many hypotheses, but a comprehensive approach is still missing.

References

[1] R. Dawkins *The Selfish Gene* (Oxford University Press, Oxford UK 1989), p. 352; S. Blackmore, *The Meme Machine* (Oxford University Press, Oxford UK 1999).

[2] There are many books about the evolution of languages. For instance: L.L. Cavalli-

---

[i]In the original paper this number is termed "fitness", but in an almost-stable ecosystem all individuals should have the same fitness.

Sforza, Proc. Natl. Acad. Sci. USA **94**, 7719–7724 (1997). L.L. Cavalli-Sforza and M.W. Feldman,*Cultural Transmission and Evolution: A quantitative approach* (Princeton University Press, Princeton 1981). M. A. Nowak, Phil. Trans.: Biol. Sc., **355**, 1615–1622 (2000). For what concerns simulations: C. Schulze, D. Stauffer and S. Wichmann, Comm. Comp. Phys. **3**, 271–294 (2008).

[3] F. Bagnoli and M. Bezzi, *An evolutionary model for simple ecosystems*, in: Annual Reviewies of Computational Physics VII, ed. by D. Stauffer (World Scientific, Singapore, 2000) p. 265.

[4] E. Baake and W. Gabriel: *Biological evolution through mutation, selection, and drift: An introductory review.* In: Annual Reviews of Computational Physics VII, ed. by D. Stauffer (World Scientific, Singapore, 0000) pp. 203–264.

[5] J. Maynard Smith, *The Evolution of Sex* (Cambridge University Press, Cambridge 1978).

[6] S. C. Stearns, *The Evolution of Sex and its Consequences*, (Birkäuser Verlag, Basel 1987).

[7] S. A. Kauffman and S. Levine, J. Theor. Biol., **128**, 11 (1987).

[8] S. A. Kauffman. *The Origins of Order.* (Oxford University Press, Oxford 1993).

[9] S. Wright, *The Roles of Mutation, Inbreeding, Crossbreeding, and Selection in Evolution, Proc. 6th Int. Cong. Genetics, Ithaca*, **1**, 356 (1932).

[10] D. L. Hartle, *A Primer of Population Genetics*, 2nd ed. (Sinauer, Sunderland, Massachusetts, 1988).

[11] L. Peliti, *Fitness Landscapes and Evolution*, in *Physics of Biomaterials: Fluctuations, Self-Assembly and Evolution*, Edited by T. Riste and D. Sherrington (Kluwer, Dordrecht 1996) pp. 287-308, and cond-mat/9505003.

[12] N. Boccara, O. Roblin, and M. Roger, Phys. Rev. E **50**, 4531 (1994).

[13] J. Hofbauer, J. Math. Biol. **23**, 41 (1985).

[14] I. Leuthäusser, J. Stat. Phys. **48**, 343 (1987).

[15] P. Tarazona, Phys. Rev. A **45**, 6038 (1992).

[16] S. Galluccio, *Phys. Rev. E* **56**, 4526 (1997).

[17] L. Peliti. Lectures at the Summer College on Frustrated System, Trieste, August 1997, cond-mat/9712027.

[18] M. Lynch and W. Gabriel, Evolution **44**, 1725 (1990).

[19] M. Lynch, R. Bürger, D. Butcher and W. Gabriel, J. Hered., **84**, 339 (1993).

[20] A.T. Bernardes, J. Physique **I5**, 1501 (1995).

[21] P.G. Higgs and G. Woodcock, J. Math. Biol. **33**, 677 (1995).

[22] G. Woodcock and P.G. Higgs, J. theor. Biol. **179** 61 (1996).

[23] C. Amitrano, L. Peliti and M. Saber, J. Mol. Evol. **29**, 513 (1989).

[24] L. Peliti, *Introduction to the statistical theory of Darwinian evolution*, Lectures at the Summer College on Frustrated System, Trieste, August 1997, and cond-mat/9712027.

[25] B. Derrida (1981) Random Energy Model: An exactly solvable model of disordered systems, Phys. Rev. B 24 2613.

[26] S.A. Kauffman, S. Levine, J. theor. Biol. **128**, 11 (1987); S.A. Kauffman, *The Origins of Order* (Oxford University Press, Oxford 1993).

[27] S. Franz, L. Peliti, M. Sellitto J. Phys. A: Math. Gen. **26**, L1195 (1993).

[28] R.A. Fisher, *The genetical theory of natural selection* (Dover, New York 1930).

[29] F. Bagnoli and C. Guardiani, Physica A**347**, 489–533 (2005); F. Bagnoli and C. Guardiani, Physica A**347**, 534–574 (2005).

[30] W.D. Hamilton, Proc. Natl. Acad. Sc. (PNAS) **87** 3566-3573 (1990).

[31] J.H. Holland, *Adaptation in Natural and Artificial Systems* (University of Michigan Press, Ann Arbor 1975).

[32] T. Wiehe, E. Baake and P. Schuster, J. Theor. Biol. **177**, 1 (1995).

[33] F. Bagnoli and M. Bezzi, Phys. Rev. E **64**, 021914 (2001).

[34] D.J. Watts and S.H. Strogatz, Nature **393**, 440 (1998).

[35] F. Bagnoli and M. Bezzi, Int. J. Mod. Phys. C **9**, 999 (1998), and cond-mat/9807398.

[36] W. Eigen, Naturwissenshaften **58**, 465 (1971).

[37] W. Eigen and P. Schuster, Naturwissenshaften **64**, 541 (1977).

[38] E. Baake and T. Wiehe, J. Math. Biol. **35**, 321 (1997).

[39] H. Wagner, E. Baake and T. Gerisch, J. Stat. Phys **92**, 1017 (1998).

[40] K. Malarz and D. Tiggemann, Int. J. Mod. Phys. C **9**, 481 (1998).

[41] L.S. Tsimring, H. Levine and D.A. Kessler, Phys. Rev. Lett. **76**, 4440 (1996); D. A. Kessler, H. Levine, D. Ridgway and L. Tsmiring, J. Stat. Phys. **87**, 519 (1997).

[42] M. Nowak and P. Schuster, L. Theor. Biol. **137**, 375 (1989).

[43] D. Alves and J. F. Fontanari, Phys. Rev. E **54** 4048 (1996).

[44] G. F. Gause, *The struggle for existence* (Williams & Wilkins, Baltimore, MD 1934). G. Hardin, Science **131**, 1292-1297 (1960).

[45] F. Bagnoli and P. Lió, J. Theor. Biol. **173**, 271 (1995), and cond-mat/9808317.

[46] T. Ikemura, J.Mol. Biol. **146**, 1-21 (1981).

[47] F. Bagnoli and P. Li, J. theor. Biol. **173**, 271-281 (1995).

[48] F. Bagnoli, G. Guasti and P. Li, *Translation optimization in bacteria: statistical models*, in *Nonlinear excitation in biomolecules*, edited by M. Peyrard (Springer-Verlag, Berlin 1995) p. 405.

[49] G. Guasti *Bacterial Translation Modeling and Statistical Mechanics*, in *Dynamical Modeling in Biotechnology*, edited by F. Bagnoli and S. Ruffo, (World Scientific, Singapore 2001) p. 59–72.

[50] P.A.P. Moran, Proc. Camb. Philos. Soc. **54**, 60–71 (1958); P.A.P. Moran *The statistical processes of evolutionary theory* (Clarendon Press, Oxford 1962).

[51] M.A. Nowak, **Evolutionary Dynamics** (Harward University Press, Cambridge Mass. 2006).

[52] M. Kimura, *The neutral theory of molecular evolution*, (Cambridge University Press, Cambridge 1983).

[53] J. Maynard Smith, *Evolutionary Genetics* (Oxford University Press, Oxford 1998).

[54] M. Ridley, *Evolution* (Blackwell Scientific Publications, Inc. Cambridge, Mass. 1993).

[55] B. Derrida and L. Peliti, J. Phys. A. **20**, 5273 (1991).

[56] P.G. Higgs and B. Derrida, J. Phys. A. **24**, L985 (1991).

[57] P.G. Higgs and B. Derrida, J. Mol. Evol. **35**, 454 (1992).

[58] T.J.P. Penna, J. Stat. Phys. **78**, 1629 (1995).

[59] D. Stauffer, S. Moss de Oliveira, P.M.C. de Oliveira and J. S. Sá Martins, *Biology, Sociology, Geology by Computational Physicists* (Elsevier, Amsterdam 2006).

[60] M. Ya. Azbel, Proc. Roy. Soc. London B **263**, 1449 (1996), Phys. Repts. **288**, 545 (1997) and Physica A **273**, 75 (1999).

[61] T.J.P. Penna, S. Moss de Oliveira and D. Stauffer, Phys. Rev. E **52**, R3309 (1995).

[62] A. O. Sousa, Physica A **326** 233-240 (2003).

[63] F. Bagnoli and M. Bezzi, Phys. Rev. Lett. **79**, 3302 (1997), and cond-mat/9708101.

[64] F. Bagnoli and M. Bezzi, *Competition in a Fitness Landscape*, Fourth European Conference on Artificial Life, P. Husbands and I. Harvey (eds.), The MIT Press (Cambridge, Massachussets, 1997) p. 101 and cond-mat/9702134 (1997).

[65] M. Doebeli, J. Evol. Biol. **9**, 893 (1999); U. Dieckmann and M. Doebeli, Nature (London) **400**, 354 (1999).

[66] P.M.C. de Oliveira, S. Moss de Oliveira, D. Stauffer and A. Pekalski, Eur. Phys. J. B **63**, 245 (2008).

[67] J. Hofbauer and K. Sigmund, *Evolutionary Games and Population Dynamics* (Cambridge University Press, Cambridge (UK) 1998).

[68] R. Axelrod, *The evolution of Cooperation* (Basic Book, NY 1984); reprinted (Penguin, UK 1989); R. Axelrod and W.D. Hamilton, Science **211**, 1390–1396 (1981).

[69] M. Lssig, L. Peliti and F. Tria, Europhys. Lett. **62**, 446-451 (2003).

[70] M.A. Nowak, Science **314**, 1560–1563 (2006)

[71] J. Maynard Smith, Proc. Royal Soc. London B **205** 475.

[72] S.J. Gould, Paleobiology **3**, 135 (1977).

[73] G. Caldarelli, P. G. Higgs and A. J. McKane, J. Theoret. Biol. **193**, 345–358 (1998); B. Drossel and A.J. McKane, Modelling food webs. In: S. Bornholdt and H. G. Schuster, Editors, *Handbook of Graphs and Networks* (Springer, Berlin 2003), pp. 218–247; A. J. McKane and B. Drossel. Models of food web evolution. In 'Food Webs', Mercedes Pascual and Jennifer Dunne (Hg.), Oxford University Press 2005.

[74] D. Stauffer, A. Kunwar and D. Chowdhury, Physica A **352**, 202–215 (z2005).

[75] P. Bak and K. Sneppen, Phys. Rev. Lett. **71**, 4083 (1993). P. Bak, *How Nature Works: The Science of Self-Organized Criticality.* (Copernicus, New York 1996).

[76] M.D. Raup, Science **251**, 1530 (1986).

Chapter 16

Evolution of Cooperation in Adaptive Social Networks

Sven Van Segbroeck

COMO, Vrije Universiteit Brussel (VUB), B-1050 Brussels, Belgium

Francisco C. Santos

IRIDIA, Université Libre de Bruxelles (ULB), B-1050 Brussels, Belgium

Arne Traulsen

Max-Planck-Institute for Evolutionary Biology, D-24306 Plön, Germany

Tom Lenaerts

MLG, Département d'Informatique, Université Libre de Bruxelles (ULB) and Computer Science Department, Vrije Universiteit Brussel (VUB), B-1050, Belgium

Jorge M. Pacheco

ATP-Group and CFTC, Departamento de Física da Faculdade de Ciências, P-1649-003 Lisboa Codex, Portugal

Humans are organized in societies, a phenomenon that would never have been possible without the evolution of cooperative behavior. Several mechanisms that foster this evolution have been unraveled over the years, with population structure as a prominent promoter of cooperation. Modern networks of exchange and cooperation are, however, becoming increasingly volatile, and less and less based on long-term stable structure. Here, we address how this change of paradigm affects the evolution of cooperation. We discuss analytical and numerical models in which individuals can break social ties and create new ones. Interactions are modeled as two-player dilemmas of cooperation. Once a link between two individuals has formed, the productivity of this link is evaluated. Links can be broken off at different rates. This individual capacity of forming new links or severing inconvenient ones can effectively change the nature of the game. We address random formation of new links and local linking rules as well as different individual capacities to maintain social interactions. We conclude by discussing how adaptive social networks can become an important step towards more realistic models of cultural dynamics.

16.1. Introduction

From human societies to the simplest biological systems, cooperative interactions thrive at all levels of organization. A cooperative act typically involves a cost (c) to the provider while conferring a benefit (b) to the recipient (with $b > c$) [Hamilton (1996); Trivers (1985); Wilson (1975); Axelrod and Hamilton (1981)]. Individuals try to maximize their own resources and are therefore expected to avoid paying any costs while gladly accepting all the benefits offered by others. This ubiquitous paradox is often analyzed in the framework of (evolutionary) game theory. Game theory describes systems in which the success of an individual depends on the action of others. The classical approach focused on the determination of optimal strategic behavior of rational individuals in such a static setting [von Neumann and Morgenstern (1944)]. Evolutionary game theory places this framework into a dynamical context by looking at the evolutionary dynamics in populations of players [Maynard Smith (1982)]. The expected payoff from the game is a function of the frequencies of all strategies. Successful behaviors spread in such a population. There are two interpretations of evolutionary game theory: In the conventional setting, the payoff is interpreted as biological fitness. Individuals reproduce proportional to their fitness and successful strategies spread by genetic reproduction. A second interpretation is the basis for cultural evolution in social systems: Successful behaviors are imitated with a higher probability. They spread by social learning instead of genetic reproduction. Both frameworks are captured by the same mathematical approach: The generic mathematical description of evolutionary game dynamics is the replicator equation [Taylor and Jonker (1978); Hofbauer and Sigmund (1998); Zeeman (1980)]. This system of nonlinear ordinary differential equations describes how the relative abundances (frequencies) of strategies change over time.

The assumption underlying the replicator equation is that individuals meet each other at random in infinitely large, well-mixed populations. But it also emerges in other cases, e.g. if the interaction rates between individuals are not random [Taylor and Nowak (2006)] or from a large-population approximation of evolutionary game dynamics in finite populations [Traulsen *et al.* (2005)].

However, in reality the probability to interact with someone else is not the same across a population or social community. Interactions occur on social networks which define the underlying topology of such cooperation dynamics. Initially, this line of research has focused on regular lattices [Nowak and May (1992); Herz (1994); Lindgren and Nordahl (1994); Szabó and Tőke (1998); Hauert (2002)]. More recently, more complex topologies and general networks have been considered in great detail [Vainstein and Arenzon (2001); Abramson and Kuperman (2001); Ebel and Bornholdt (2002a); Holme *et al.* (2003); Szabó and Vukov (2004); Santos and Pacheco (2005); Ohtsuki *et al.* (2006); Santos *et al.* (2006b, 2008)]. While the theoretical advances in this field are tremendous, there is so far a lack of experimental data. Designing and implementing such experiments has proven difficult and, so far, only general statements as "the probability to be generous is correlated with the number of social links of an individual" can be made [Branas-Garza *et al.* (2007)].

One important property of social networks that is seldom addressed in theoretical studies is that real world social networks are not static. Instead, we make new friends and lose touch with old ones, depending on the kind of interaction we have with them. This makes social networks an example of an adaptive network [Gross and Blasius (2008)]. The basic idea is that interactions which benefit both partners last longer than interactions where one partner is exploited by the other. Here, we discuss such an approach, which leads to analytical results in certain limits. These serve as important starting points for further developments.

16.2. Active Linking: Random Link Formation

We break down the model into two parts: Evolutionary dynamics of strategies (or behaviors) of the individuals associated with nodes in a network whose links describe social interactions. The adaptive nature of the social interactions leads to a network linking dynamics. We consider two-player games of cooperation in which individuals can choose to give help to the opponent (to cooperate, C), or to refuse to do so (to defect, D). The network is of constant size with N nodes. The number of links, however, is not constant and changes over time. There are N_C individuals that cooperate and $N_D = N - N_C$ individuals that defect.

16.2.1. *Linking dynamics*

An interaction between two players occurs if there is a link between these players. Links are formed at certain rates and have specific life-times. We denote by $X(t)$ the number of CC links at time t. Similarly, $Y(t)$ and $Z(t)$ are the number of CD and DD links at time t. The maximum possible number of CC, CD and DD links is given by $X_m = N_C(N_C - 1)/2$, $Y_m = N_C N_D$, and $Z_m = N_D(N_D - 1)/2$, respectively. Suppose cooperators form new links at rate α_C and defectors form new links at rate α_D. Thus, CC links are formed at a rate α_C^2, CD links are formed at a rate $\alpha_C \alpha_D$ and DD links are formed at a rate α_D^2. The death rates of CC, CD and DD links are given by γ_{CC}, γ_{CD} and γ_{DD}, respectively. If the number of links is large, we can model the dynamics of links by differential equations. We obtain a system of three ordinary differential equations for the number of links

$$\dot{X} = \alpha_C^2(X_m - X) - \gamma_{CC}X\,,$$
$$\dot{Y} = \alpha_C \alpha_D(Y_m - Y) - \gamma_{CD}Y\,, \qquad (16.1)$$
$$\dot{Z} = \alpha_D^2(Z_m - Z) - \gamma_{DD}Z\,.$$

For $\alpha^2 \gg \gamma$, the network is almost complete, which recovers the results for well-mixed populations. For $\alpha^2 \ll \gamma$, the network is sparse with few links. The most

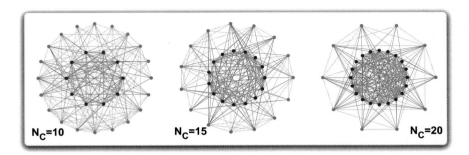

Fig. 16.1. Frequency dependent steady state dynamics. Results of active linking dynamics for a population size of $N = 30$ individuals. Cooperators are located in the "inner-rim", and are represented by blue circles, whereas defectors are located in the "outer-rim", and are represented by red circles. In this way, CC-links (solid cyan lines) live only within the "inner-rim", whereas CD-links (solid red lines) occupy the space between the rims while DD-links (solid grey lines) cross the entire region of the figure. Each panel depicts a snapshot in the steady state of the active-linking dynamics, associated with a different (and fixed) frequency of C and D players. The parameters determining the active linking dynamics are: $\alpha_C = \alpha_D = 0.5$, $\gamma_{CC} = 0.5$, $\gamma_{CD} = 0.25$ and $\gamma_{DD} = 0.5$.

interesting case we discuss below is $\alpha^2 \approx \gamma$, where the system has fixed points with intermediate ranges of X, Y and Z. Rescaling α and γ in an appropriate way (note that the equation contains squares of α and linear terms of γ) does not change the fixed points of the system, but affects the overall timescale of active linking. When this process is coupled with strategy dynamics, such changes can be crucial.

While the above is probably the simplest possibility to model linking dynamics, more sophisticated choices are possible, taking for example the number of existing links of a node into account. However, to address some general properties of the coevolution between links and strategies, we concentrate on the simplest choice first. In the steady state, the number of links of the three different types is given by

$$X^* = X_m \frac{\alpha_C^2}{\alpha_C^2 + \gamma_{CC}} = X_m \phi_{CC} \,,$$

$$Y^* = Y_m \frac{\alpha_C \alpha_D}{\alpha_C \alpha_D + \gamma_{CD}} = Y_m \phi_{CD} \,,$$

$$Z^* = Z_m \frac{\alpha_D^2}{\alpha_D^2 + \gamma_{DD}} = Z_m \phi_{DD} \,.$$

(16.2)

Here, ϕ_{CC}, ϕ_{CD}, and ϕ_{DD} are the fractions of active CC, CD and DD links in the steady states. Examples of population structures attained under steady-state dynamics for three different combinations of (N_C, N_D) are shown in Fig. 16.1.

16.2.2. *Strategy dynamics*

Next, we address the dynamics of the strategies at the nodes. We consider the stochastic dynamics of a finite population, i.e. we restrict ourselves to finite networks. We consider general two-player games of cooperation given by the payoff matrix

$$
\begin{array}{cc}
 & \begin{array}{cc} C & D \end{array} \\
\begin{array}{c} C \\ D \end{array} & \begin{pmatrix} R & S \\ T & P \end{pmatrix}.
\end{array}
\tag{16.3}
$$

Thus, a cooperator interacting with another cooperator obtains the *reward* from mutual cooperation R. Cooperating against a defector leads to the *sucker's payoff* S, whereas the defector obtains the *temptation to defect* T in such an interaction. Finally, defectors receive the *punishment* P from interactions with other defectors. A social dilemma arises when individuals are tempted to defect, although mutual cooperation would be the social optimum $(R > P)$. We distinguish three generic cases of 2-player social dilemmas:

- Dominance: When $T > R > P > S$, we enter the realm of the *Prisoner's Dilemma* (PD) [Rapoport and Chammah (1965)], where cooperation is dominated by defection. The opposite scenario, when $R > T$ and $S > P$, poses no social dilemma and is referred to as a *Harmony Game* (HG) [Posch *et al.* (1999)].
- Coordination: $R > T$ and $S < P$ leads to what is called coordination or *Stag Hunt* games (SH) [Skyrms (2003)], in which it is always good to follow the strategy of the majority in the population. Except for $R + S = T + P$, one strategy has a larger basin of attraction. This strategy is called a risk dominant strategy. For $R + S > T + P$, cooperation is risk dominant.
- Coexistence: In the case of $R < T$ and $S > P$, known as a Hawk-Dove [Maynard Smith (1982)] or *Snowdrift* game (SG) [Sugden (1986); Doebeli and Hauert (2005); Hauert and Doebeli (2004)], a small minority is favored. This means that the ultimate outcome in a population of players is a mixture of strategies C and D.

From the payoff matrix, we can calculate the payoffs of the individuals, depending on the number of interactions they have with cooperators and defectors. On a complete network, the payoffs are

$$
\pi_C = R(N_C - 1) + SN_D
\tag{16.4}
$$

and

$$
\pi_D = TN_C + P(N_D - 1).
\tag{16.5}
$$

Often, the payoffs are scaled by $1/(N-1)$, such that the payoffs do not increase with the population size. For the strategy update process defined below, this corresponds simply to a rescaling of the intensity of selection, i.e. changing the noise intensity, if all individuals have the same number of interactions. If the number of interactions is not the same for all players, the heterogeneity between players can lead to new effects [Santos *et al.* (2006b); Santos and Pacheco (2006)].

For strategy dynamics, we adopt the pairwise comparison rule [Szabó and Tőke (1998); Blume (1993)], which has been recently shown to provide a convenient framework of game dynamics at all intensities of selection [Traulsen et al. (2007, 2006)]. According to this rule, two individuals from the population, A and B are randomly selected for update (only the selection of mixed pairs can change the composition of the population). The strategy of A will replace that of B with a probability given by the Fermi function (from statistical physics)

$$p = \frac{1}{1 + e^{-\beta(\pi_A - \pi_B)}} .$$ (16.6)

The reverse will happen with probability $1 - p$. The quantity β, which in physics corresponds to an inverse temperature, controls the intensity of selection. For $\beta \ll 1$, we can expand the Fermi function in a Taylor series and recover weak selection, which can be viewed as a high temperature expansion of the dynamics [Nowak et al. (2004); Traulsen et al. (2007)]. For $\beta \gg 1$, the intensity of selection is high. In the limit $\beta \to \infty$, the Fermi function reduces to a step function: In this case the individual with the lower payoff will adopt the strategy of the other individual regardless of the payoff difference.

The quantity of interest in finite population dynamics is the fixation probability ρ, which is the probability that a single mutant individual of one type takes over a resident population with $N - 1$ individuals of another type.

16.2.3. Separation of timescales

The system of coevolving strategies and links is characterized by two timescales: One describing the linking dynamics (τ_a), the second one describing strategy dynamics (τ_e). We can obtain analytical results in two limits, where both timescales are separated. Defining the ratio $W = \tau_e/\tau_a$, separation of time scales will occur for $W \ll 1$ and $W \gg 1$.

16.2.3.1. Fast strategy dynamics

If strategies change fast compared to changes of the network structure, active linking does not affect strategy dynamics. Thus, the dynamics is identical to the evolutionary game dynamics on a fixed network. Such systems have been tackled by many authors for a long time [Nowak and May (1992); Herz (1994); Lindgren and Nordahl (1994); Szabó and Tőke (1998); Hauert (2002); Vainstein and Arenzon (2001); Szabó and Vukov (2004); Abramson and Kuperman (2001); Ebel and Bornholdt (2002a); Holme et al. (2003); Santos and Pacheco (2005); Ohtsuki et al. (2006); Santos et al. (2006b, 2008)]. The difficulty of an analytical solution for such systems is determined by the topology of the network, which corresponds to an initial condition in our case. Analytical solutions are feasible only for few topologies. One important limiting case leading to analytical solutions are complete networks corresponding to well-mixed systems. In this case, the fixation probability can be approximated by

$$\rho_C = \frac{\text{erf}[\xi_1] - \text{erf}[\xi_0]}{\text{erf}[\xi_N] - \text{erf}[\xi_0]}, \tag{16.7}$$

where erf(x) is the error function and $\xi_k = \sqrt{\frac{\beta}{u}}(ku + v)$ [Traulsen *et al.* (2006)]. We have $2u = R - S - T + P$ and $2v = -R + SN - TN + T$.

For $u \to 0$, we have $\rho_C = (1 - e^{-2\beta v})/(1 - e^{-2\beta vN})$.

If strategy dynamics is fast, the linking dynamics only becomes relevant in states where the system can no longer evolve from strategy dynamics alone, but changing the topology allows to escape from these states.

16.2.3.2. *Fast linking dynamics*

Whenever $W \gg 1$ linking dynamics is fast enough to ensure that the network will reach a steady state before the next strategy update takes place. At the steady state of the linking dynamics, the average payoffs of C and D individuals are given by

$$\pi_C = R\phi_{CC}(N_C - 1) + S\phi_{CD}N_D \tag{16.8}$$

and

$$\pi_D = T\phi_{CD}N_C + P\phi_{DD}(N_D - 1). \tag{16.9}$$

Note that the effective number of interactions of cooperators and defectors can become very different if $\phi_{CC} \gg \phi_{DD}$ or vice versa. Comparing Eqs. (16.8) and (16.9) to Eqs. (16.4) and (16.5) suggests that the linking dynamics introduces a simple transformation of the payoff matrix. We can study standard evolutionary game dynamics using the modified payoff matrix

$$\begin{array}{c} \\ C \\ D \end{array} \begin{array}{cc} C & D \\ \left(\begin{array}{cc} R\phi_{CC} & S\phi_{CD} \\ T\phi_{CD} & d\phi_{DD} \end{array} \right) \end{array} = \begin{array}{c} \\ C \\ D \end{array} \begin{array}{cc} C & D \\ \left(\begin{array}{cc} R' & S' \\ T' & P' \end{array} \right) \end{array}. \tag{16.10}$$

Consequently, linking dynamics can change the nature of the game [Pacheco *et al.* (2006b)]. So far, we have only shown this in the limit where linking dynamics is much faster than strategy dynamics ($W \gg 1$). However, the result is expected to hold even when the two time scales are comparable (see below and also Refs. [Pacheco *et al.* (2006)b,a].

In general, all generic transformations are possible, as illustrated in Fig. 16.2. The transition points can be determined as follows: Strategy C is a Nash equilibrium for $R > T$. This property changes to $R' < T'$ when

$$\frac{R}{T} < \frac{\phi_{CD}}{\phi_{CC}} = \frac{\alpha_D}{\alpha_C} \frac{\alpha_C^2 + \gamma_{CC}}{\alpha_C\alpha_D + \gamma_{CD}}. \tag{16.11}$$

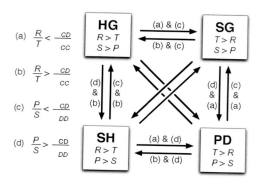

Fig. 16.2. When linking dynamics occurs much faster than strategy dynamics, the nature of the game being played changes. The arrows indicate the conditions under which a game located at the arrow start is transformed into a game located at the arrow end.

For example, ϕ_{CD} can be increased by reducing the death rate of CD links, γ_{CD}. With increasing ϕ_{CD}, the condition is fulfilled at some point. At the transition point, C is either transformed into a Nash equilibrium or loses this property. An equivalent transition for D is given by the condition

$$\frac{P}{S} < \frac{\phi_{CD}}{\phi_{DD}} = \frac{\alpha_C}{\alpha_D} \frac{\alpha_D^2 + \gamma_{DD}}{\alpha_C \alpha_D + \gamma_{CD}}. \tag{16.12}$$

However, the conditions are not entirely independent, since at least two parameters have to be varied. Usually, it is enough to vary the three link-death rates γ and fix the link-birth rates α to observe these transitions. It is also worth mentioning that, in coordination games, the transformation can change risk dominance.

16.2.4. Comparable timescales

As we have shown, active linking can lead to a wide range of scenarios that effectively change the character of the game. However, the analytical results have been obtained assuming time scale separation. Figure 16.3 shows the results of numerical simulations for a gradual change of the time scale ratio. Deviations from the analytical predictions are limited to a single order of magnitude. In other words, the time scale separation is not a very strong assumption and remains valid for a much wider range of parameters than expected. Even for moderate active linking, our analytical results are recovered, i.e. they hold when self-organising network structures and the evolutionary game dynamics on the network are intimately entangled. Having identified the relevance of time scale separation in a minimal model of linking dynamics, we now turn to more complex linking dynamics based on local rules.

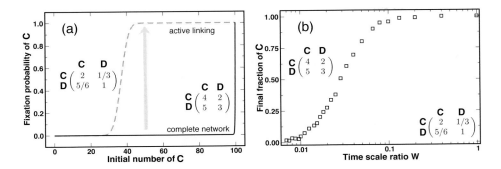

Fig. 16.3.   Active linking effectively changes the payoff matrix and the nature of the game. (a) We start from a complete network without structure dynamics ($W = 0$) and a Prisoner's Dilemma game. In this case, the fixation probability of C (full line) is essentially zero for all initial numbers of C. With active linking (dashed line), the game turns into a Stag Hunt game. In this case, C becomes risk dominant and the fixation probability of C exceeds 0.5 if the initial number of C individuals is larger than 36.   (b) Numerical simulations reveal the range of validity of our analytical approximations. We start from 50% cooperating individuals. For small W, cooperators never reach fixation. But already for $W = 0.1$, their fixation probability is close to one. Thus, moderate active linking is sufficient to make cooperation the dominant strategy here (averages over 100 realizations, population size $N = 100$, intensity of selection $\beta = 0.05$, $\alpha_C = \alpha_D = 0.4$, $\gamma_{CC} = 0.16$, $\gamma_{CD} = 0.80$ and $\gamma_{DD} = 0.32$.)

16.3. Individual Based Linking Dynamics: Local Link Formation

In the model discussed in Section 16.2, we have a fluctuating number of links and analytical results in the two limits where the time scale of linking dynamics and strategy dynamics are well separated, allowing for the mean-field treatment considered. We now introduce an alternative description in which the number of links is conserved, but in which decision to maintain or rewire a link results both from individual preference in the choice of partners and negotiation between individuals linked [Santos *et al.* (2006a); Van Segbroeck *et al.* (2008)]. Such an individual based decision making cannot be dealt with at a mean-field level and calls for a numerical implementation.

16.3.1. *Specification of the linking dynamics*

To reduce the number of parameters, let us start by restricting the space of possible games by fixing $R = 1$ and $P = 0$, while $-1 \leq S \leq 1$ and $0 \leq T \leq 2$.

$$
\begin{array}{cc}
 & \begin{array}{cc} C & D \end{array} \\
\begin{array}{c} C \\ D \end{array} & \begin{pmatrix} 1 & S \\ T & 0 \end{pmatrix} .
\end{array}
\tag{16.13}
$$

This spans the four dynamical outcomes introduced before: a) HG ($S > 0$ and $T < 1$); b) SG ($S > 0$ and $T > 1$); c) SH ($S < 0$ and $T < 1$) and d) PD ($S < 0$ and $T > 1$) (see Section 16.2.2).

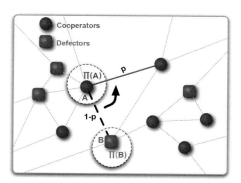

Fig. 16.4. Readjusting social ties. Cooperators and defectors interact via the links of a network. B is satisfied, since A is a cooperator ($T > 0$). On the other hand, A is unsatisfied with this situation ($S < 1$). Therefore, A wants to change the link whereas B does not. The action taken is contingent on the fitness π_A and π_B of A and B, respectively. With probability p (see Eq. (16.6)), A redirects the link to a random neighbor of B. With probability $1 - p$, A stays linked to B. Finally, if both players are dissatisfied, the same methodology is used to decide who keeps the connection.

Because $S \leq 1$ and $T \geq 0$, the payoff against a cooperator is always higher than the payoff against a defector, cf. Eq. (16.13). Thus, interacting with a cooperator is always the best possible option. Consequently, every individual will be satisfied when connected to a C and dissatisfied otherwise. Keeping the total number of links constant, all individuals are now able to decide, on an equal footing, those ties that they want to maintain and those they want to change. The co-evolution between strategy and network structure is therefore shaped by individual preferences towards interacting with one of the two strategies [Santos et al. (2006a)]. Figure 16.4 illustrates the process. If A is satisfied, she will decide to maintain the link. If dissatisfied, then she may compete with B to rewire the link (see Fig. 16.4), rewiring being attempted to a random neighbor of B. Thus, the loser in a competition for a link loses an interaction. This paves the way for the evolution of a degree-heterogeneous network. The intuition behind this reasoning relies on the fact that agents, equipped with limited knowledge and scope, look for new social ties by proxy [Kossinets and Watts (2006)]. Such a procedure can only be treated numerically and does no longer lead to a simple rescaling of a payoff matrix as the mechanism discussed in Section 16.2. On the other hand, it introduces some features characteristic of realistic social networks.

The fact that all individuals naturally seek to establish links with cooperating individuals, creates possible conflicts of interests as illustrated in Fig. 16.4. For instance, B is satisfied, because she can profit from A. Obviously, A is not satisfied and would prefer to seek for another cooperator. Decision is contingent on the payoff π_A and π_B of A and B, respectively. With probability $p = [1 + e^{-\beta[\pi_A - \pi_B]}]^{-1}$ (also used in the strategy update, cf. Eq. (16.6)), A redirects the link to a random neighbor of B. With probability $1-p$, A stays linked to B. Whenever both A and B

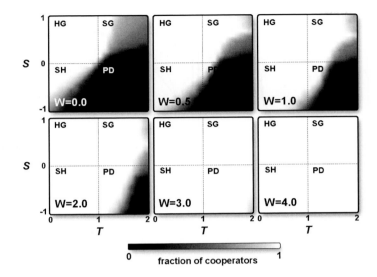

Fig. 16.5.   Final frequency of cooperators in all games for different time-scale ratios between strategy and structure dynamics. Results for the fraction of successful evolutionary runs ending in 100% of individuals with strategy C for different values of the time scale ratio W, starting from 50% of each strategy. We study the four different games in the area $2 \geq T \geq 0$ and $1 \geq S \geq -1$: HG, SG, SH and PD (see Section 16.2.2). For $W = 0$ ($N = 10^3$, $z = 30$ and $\beta = 0.005$), the results fit the predictions from well-mixed populations, although individuals only interact with a small subset of the population. With increasing W (faster structure dynamics), the rate at which individuals readjust their ties increases, and so does the viability of cooperators. Above a critical value $W_{critical} \sim 4.0$ (see also Fig. 16.6), cooperators efficiently wipe out defectors. For the strategy evolution dynamics adopted here (pairwise comparison, see Section 16.2.2), and according to [Ohtsuki *et al.* (2006)], cooperators would never be favored in static networks.

are satisfied, nothing happens. When both A and B are unsatisfied, rewiring takes place such that the new link keeps attached to A with probability p and attached to B with probability $1 - p$. Thus, the more successful individual keeps the link with higher probability.

16.3.2. *Numerical results*

As previously, this model establishes a coupling between individual strategy dynamics and population structure dynamics. This leads necessarily to a time scale associated with strategy evolution, τ_e and a second associated with structure evolution, τ_a.   When the ratio $W = \tau_e/\tau_a$ approaches 0, the network dynamics is irrelevant and we recover the fast strategy dynamics of Section 16.2.3.1. On the other hand, with increasing W, individuals become apt to adapt their ties and form a degree-heterogeneous network with increasing efficiency.

The contour plots in Fig. 16.5 illustrate the final fraction of cooperators for different values of the ratio W in networks with average connectivity $z = 30$ (this

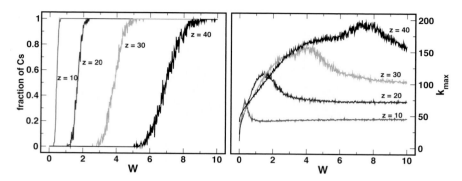

Fig. 16.6. Co-evolution of strategies and links in the game region in which defectors should dominate for different time-scales. Left panel: Final frequency of cooperators at end as a function of W for different average connectivity z. For each average connectivity z, there is a critical value of the time scale ratio $W - W_{critical}$ – above which cooperators wipe out defectors. Right panel: Connectivity k_{max} of the largest hub in the network, as a function of the time scale ratio W. With increasing z, $W_{critical}$ increases. In all cases, the heterogeneity of the associated network becomes maximal at $W_{critical}$. For higher values of W, the heterogeneity decreases again when defectors decrease in frequency. For high values of W, defectors are wiped out and only the heterogeneity generated by the rewiring mechanism in a neutral system prevails (Payoffs $R = 1$, $T = 2$, $S = -1$ and $P = 0$. Intensity of selection $\beta = 0.005$).

value reflects the mean value of the average connectivities reported in [Dorogotsev and Mendes (2003)] for socials networks). We plot the fraction of cooperators who survive evolution, averaged over 100 independent realizations for the same values of the game payoff entries (T, S) and the time scale ratio W. For $W = 0$ the results reproduce, as expected [Santos et al. (2006b)], the predictions for finite, well-mixed populations. Yet, with increasing W, cooperators gain an advantage, as they can terminate their undesirable interactions with defectors. Rewiring changes the strategy dynamics and paves the way for a radically distinct evolutionary outcome in which cooperators are now able to dominate for the entire range of games. Under structural dynamics, cooperators can cut their links to defectors, which gives them an advantage compared to the situation on a static network. The swifter the response of individuals to the nature of their ties, the easier it gets for cooperators to wipe out defectors. Note further that cooperators already dominate defectors for $W = 4$, corresponding to a situation far from the time-scale separation conditions defined in Section 16.2.3.

Additional insight is provided in Fig. 16.6 (left panel), where we show how cooperation dominates defection as a function of W when $T = 2$ and $S = -1$ (lower right corner of the panels in Fig. 16.5), which represents the most challenging case for cooperators. Different values of the average connectivity z are shown. For small W, cooperators have no chance. Their fate changes as W approaches a critical value $W_{critical}$ — which increases monotonically with connectivity z — cooperators wiping out defectors above $W_{critical}$ (the increase of $W_{critical}$ with z is expected, since there are more links to be rewired; in practice, $W_{critical}$ is determined as the value of W at which the frequency of cooperators crosses 50%). Thus, the evolutionary outcome and effective game at stake relies on the capacity of individuals to adjust to adverse ties.

Figure 16.6 also provides evidence of the detailed interplay between strategy and structure. On one hand, strategy updating promotes a local assortment of strategies, since Cs *breed* Cs and Ds *breed* Ds. On the other hand, under structural updating, one is promoting local assortative interactions between cooperators (that is, CC-links) and disassortative interactions between defectors and cooperators (that is, CD-links), which constitute *favorable steps* for cooperators, from an individual point of view. Clearly, when simultaneously active, strategy update will reinforce assortativity among Cs, but will inhibit disassortativity between Ds and Cs, which overall will promote the dominance of cooperation over defection.

16.3.3. *Graph structures under individual based linking dynamics*

For any $W > 0$, individual choices lead to heterogeneous graphs in which some individuals interact more and more often than others. The overall onset of increase of heterogeneity qualitatively follows the wave of cooperation dominance shown in Fig. 16.5 [Santos *et al.* (2006a)]. In fact, the overall heterogeneity of the graph increases as W increases reaching a maximum at $W_{critical}$, above which heterogeneity decreases again down to a stationary value determined by neutral dynamics in a system with one strategy only [Santos *et al.* (2006a)]. The results shown suggest that the adaptive dynamics of social ties introduced here coupled with social dilemmas accounts for the heterogeneities observed in realistic social networks [Amaral *et al.* (2000)].

16.4. Local Linking with Individual Linking Time Scales

In the two previous models, the linking dynamics proceeds population-wide at the same speed, determined by W. This implies that all individuals are assumed to react in the same way to adverse social ties. It is commonly observed, however, that different individuals respond differently to the same situation [Rubin (2002); Ridley (2003); Buchan *et al.* (2002)] — some have the tendency to swiftly change partner, whereas others remain connected even though they are dissatisfied with their partners' behavior. Extending the linking dynamics introduced in the previous section allows us to represent this kind of behavioral diversity [Van Segbroeck *et al.* (2008, 2009)].

16.4.1. *Specification of the linking dynamics*

We adopt the same parameterization of 2×2 games as in Section 16.3 and fix the difference between mutual cooperation and mutual defection to 1, making $R = 1$ and $P = 0$. We focus on the PD in which $S < 0$ and $T > 1$, i.e. when defection

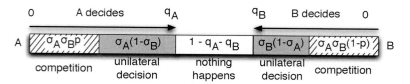

Fig. 16.7.   Decision on whose preference (either redirect the link when dissatisfied or maintain the link when satisfied) prevails in case of unilateral or mutual dissatisfaction among the interacting individuals A and B. The different colors indicate the three possible outcomes of the rewiring competition between the two individuals. First, as each individual competes for the link with probability given by her parameter σ, A and B compete both with probability $\sigma_A \sigma_B$ (indicated by the hatched zones). In this case, decision is determined by the payoff-dependent probability p (see Eq. (16.6)). Second, when only one of the individuals competes (indicated by the gray zones), this individual takes a unilateral decision. In total, A's preference prevails with probability $q_A = \sigma_A \sigma_B p + \sigma_A[1 - \sigma_B]$, B's preference with probability $q_B = \sigma_A \sigma_B(1 - p) + \sigma_B[1 - \sigma_A]$. Finally, the white zone indicates the situation in which both individuals refuse to compete, such that the link remains unchanged.

is expected to dominate, although the model could easily be applied to SH and SG games as well. Since $S \leq 1$ and $T \geq 0$, every individual prefers interacting with cooperators to interacting with defectors. Consequently, everyone attempts to maintain links with cooperators, but change links with defectors. However, unlike before, individuals are now not necessarily equally willing to engage in these conflicts. We represent their eagerness to do so by introducing an individual characteristic $\sigma \in [0, 1]$. Individuals with lower values of σ will be more resilient to change, and hence can also be viewed as more loyal towards their interaction partners. In this way, the behavior of each individual is uniquely defined by two parameters: the game strategy (C or D) and the topological strategy (σ). Note that these quantities are both transferred during strategy update. Thus, both aspects of a players strategy are subject to evolution and change over time.

Figure 16.7 illustrates how σ influences the rewiring decisions. Consider two connected individuals A and B, whose topological strategies are given by σ_A and σ_B. A potential conflict about the link arises as soon as at least one of the individuals is dissatisfied about the interaction. If this is the case, both A and B decide independently of each other whether they will compete for the link or not. Each individual competes with probability given by her topological strategy σ. As such, σ_A and σ_B define three possible outcomes for the competition over the link between A and B. First, both A and B compete for the link with probability $\sigma_A \sigma_B$. The individuals' payoffs π_A and π_B ultimately dictate the winner of this conflict. The decision of A prevails with probability $p = [1 + e^{-\beta[\pi_A - \pi_B]}]^{-1}$, the decision of B with probability $1 - p$. If decision is to redirect the link, the new partner is chosen randomly from the immediate neighbors of the former partner. Second, A competes while B does not with probability $\sigma_A[1 - \sigma_B]$. In this case, A decides the fate of the link unilaterally. Similarly, when B competes but A does not (this happens with probability $\sigma_B[1 - \sigma_A]$), B decides the fate of the link unilaterally.

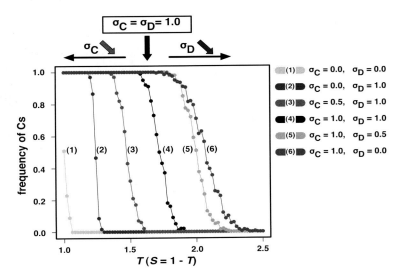

Fig. 16.8. The effect of a strategy-dependent willingness to change σ on the final frequency of cooperators. Results show the fraction of evolutionary runs ending in 100% of cooperating individuals, starting from 50% of each strategy, and this in relation to the game parameter T. The other game parameter S is chosen such that $T + S = 1$ is satisfied, bringing us into the realm of the PD. The situation in which all individuals are equally willing to react to adverse ties ($\sigma_C = \sigma_D = 1.0$) serves as a baseline. Reducing σ_D makes it easier for Cs to wipe out Ds. Reducing σ_C, on the other hand, has the opposite effect ($W = 2.5$, $N = 10^3$, $z = 30$, $\beta = 0.005$).

Hence, both individuals have the opportunity for a unilateral decision. Taken together, A's decision prevails with probability $q_A = \sigma_A \sigma_B p + \sigma_A[1 - \sigma_B]$ and B's decision with probability $q_B = \sigma_A \sigma_B(1 - p) + \sigma_B[1 - \sigma_A]$. Finally, the link remains untouched with probability $(1 - \sigma_A)(1 - \sigma_B)$, since no individual competes for the link. This last possibility encompasses the situation in which the social tie is maintained despite, e.g. mutual dissatisfaction. Overall, σ introduces a simple means to study the evolution of each individual's willingness to sever adverse ties. On the one hand, when all individuals have $\sigma = 0$, no links are rewired, reducing the population to a static society. On the other hand, when σ is maximal ($= 1$), the limits investigated in the previous section are recovered.

16.4.2. *Numerical results*

We start by associating the topological strategy σ of an individual with her strategy in the game. This means that individuals with the same game strategy will also have the same topological strategy. In the active linking model of Section 16.2, this is also included, as we assume that the propensity to form links and the lifetime of links is determined by the strategies.

When Ds are less eager to change partner ($\sigma_D = 0.5$ and $\sigma_D = 0.0$) than Cs ($\sigma_C = 1.0$), cooperators ensure the stability of favorable interactions while avoiding adverse ones more swiftly. This makes local assortment of Cs more effective, enhancing the feasibility of Cs' survival, as shown in Fig. 16.8. When Cs' willingness to change is low or absent ($\sigma_C = 0.5$ and $\sigma_C = 0.0$) compared to Ds ($\sigma_D = 1.0$), Cs' chances decrease with respect to the situation in which Cs and Ds react equally swift to adverse ties ($\sigma_C = \sigma_D = 1.0$). Comparing these results with those in which all social ties remain immutable ($\sigma_C = \sigma_D = 0.0$) does, however, show that the rewiring of DD links alone is already beneficial for cooperation.

The rewiring of links, no matter which ones, creates heterogeneous networks that are known to provide cooperators with an environment in which they may acquire an advantage over defectors. Thus, even when cooperators are resilient to change, their behavior prospers at the expense of defectors' greed.

From these results, one might expect that swift reaction to adverse social ties will evolve when σ is considered as an evolutionary trait. This intuition does, however, not always hold. We start each evolutionary run by selecting each individual's σ from a uniform distribution. We analyze the distribution of σ at the end of the evolutionary process, when the population reaches fixation (i.e., all individuals adopt the same game strategy). The lines in Fig. 16.9 correspond to the cumulative distribution $C(\sigma)$ of σ ($C(\sigma_0)$ being defined as the fraction of individuals with $\sigma > \sigma_0$) for both C (solid lines) and D individuals (dashed lines). The initial distributions of σ lead to the black diagonal line; the final distributions are shown for different values of the game parameter T. In the regime where cooperators dominate

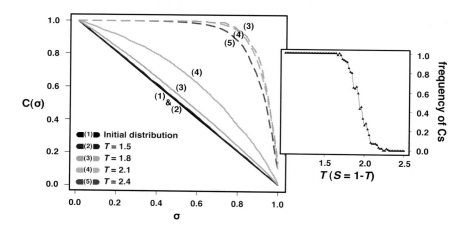

Fig. 16.9.   Evolution of σ for cooperators and defectors. The solid (dashed) lines show the fraction of cooperators (defectors) having an eagerness to change links larger than σ, and this for different values of T ($W = 5$, $\beta = 0.005$, $N = 10^3$, $z = 30$). The inset provides the fraction of runs ending in 100% of cooperators as a function of T. The values of σ are uniformly distributed in $[0, 1]$ at the start of each evolution, as indicated in black. Cooperators that react swiftly to adverse ties are only favored by natural selection when defectors start to become competitive ($1.8 \leq T \leq 2.1$). Swift defectors, on the other hand, are always selected, but the strength of this selection pressure drops as T increases. $C(\sigma_0)$ is defined as the fraction of individuals who have $\sigma \geq \sigma_0$.

($T < 1.8$), the incentive to change is low since many social ties rely on mutual satisfaction. Hence, the distribution of σ over all individuals hardly changes. For higher values of T ($1.8 \leq T \leq 2.1$) a transition occurs from cooperator dominance to defector dominance. Competition becomes fierce and it pays to respond swiftly to adverse ties, as evidenced by an increase of $C(\sigma)$ in Fig. 16.9. Ds are, however, subject to a much stronger pressure to change their links than Cs, since they can never establish social ties under mutual agreement. Thus, only Ds with high σ survive. As a consequence, defectors end up to react promptly to adverse ties, whereas cooperators will always be rather resilient to change. For even larger values of T (> 2.1), defection dominates and evolutionary competition of linking dynamics fades away. As a result, the incentive to increase swiftness reduces, a feature which is indeed reflected in the behavior of $C(\sigma)$.

16.5. Discussion

Our analysis has been limited to one-shot games. In other words, individuals interact once during the lifetime of a link as if they have never met before. But in repeated interactions, more possibilities exist. If I only take into account your behavior in the last interaction, there are already $2^2 = 4$ strategies. Since the number of strategies grows rapidly with memory [Lindgren (1991); Ebel and Bornholdt (2002b)], one often considers so called trigger strategies in which individuals keep their behavior unchanged until they are faced with an unsatisfactory partner for the first time. Such strategies can be implemented into our active linking framework, assuming that individuals act repeatedly as long as a link between them is present. This procedure leads to analytical results for evolutionary stability under active linking even in the context of repeated games [Pacheco *et al.* (2008)].

Other studies have shown numerically that network dynamics can significantly help dominated strategies. Even if only the dominant strategy can locally affect the network structure, this can help the dominated strategy under certain linking rules that put restrictions on mutual interactions of the dominant strategy [Zimmermann *et al.* (2005); Zimmermann and Eguíluz (2005); Biely *et al.* (2007)]. A recent study for growing networks has shown that the defectors in the PD have an advantage as long as a network is growing by preferential attachment. Once network growth is stopped, the cooperator strategy increases in frequency [Poncela *et al.* (2008)].

To sum up, by equipping individuals with the capacity to control the number, nature and duration of their interactions with others, we introduce an adaptive network dynamics. This leads to surprising and diverse new game dynamics and realistic social structures. We have presented approaches of how to implement network dynamics. The first one, active linking, allows to define differential equations for the numbers of links, which leads to analytical results. The second approach, individual based linking dynamics, is implemented numerically and leads to network features of empirical social networks. Both cases provide a clear and insightful message: co-evolution of population structure with individual strategy provides an efficient mechanism for the evolution of cooperation in one-shot dilemmas. Moreover, if the willingness to sever undesirable connections is also regarded as part of the individual strategy, the same principles provide an evolutionary basis for the decision of adjusting social ties. For example, in the PD cooperators evolve to maintain their interactions. But defectors are forced to seek new partners frequently, because long term relationships with defectors are undesirable.

The consideration of adaptive social networks is an important step towards more realistic models of social interactions in structured populations. Coupling the dynamics *on* networks with the dynamics *of* networks leads to emergent new phenomena outside the classical considerations of social dynamics on static networks.

References

Abramson, G. and Kuperman, M. (2001). Social games in a social network, *Phys. Rev. E* **63**, p. 030901(R).

Amaral, L. A. N., Scala, A., Barthélémy, M. and Stanley, H. E. (2000). Classes of small-world networks, *Proc. Natl. Acad. Sci. U.S.A.* **97**, 21, pp. 11149–11152.

Axelrod, R. and Hamilton, W. D. (1981). The evolution of cooperation, *Science* **211**, pp. 1390–1396.

Biely, C., Dragosits, K. and Thurner, S. (2007). The prisoner's dilemma on co-evolving networks under perfect rationality, *Phyica D* **228**, pp. 40–48.

Blume, L. E. (1993). The statistical mechanics of strategic interaction, *Games and Economic Behavior* **4**, pp. 387–424.

Branas-Garza, P., Cobo-Reyes, R., Espinosa, M. P., Jiménez, N. and Ponti, G. (2007). Altruism in the (social) network, *working paper, available at EconPapers* .

Buchan, N., Croson, R. and Dawes, R. (2002). Swift neighbors and persistent strangers: A cross-cultural investigation of trust and reciprocity in . . . , *American Journal of Sociology* **108**, 1, pp. 168–206.

Doebeli, M. and Hauert, C. (2005). Models of cooperation based on the prisoner's dilemma and the snowdrift game, *Ecology Letters* **8**, pp. 748–766.

Dorogotsev, S. and Mendes, J. (2003). *Evolution of networks: From biological nets to the Internet and WWW* (Oxford University Press).

Ebel, H. and Bornholdt, S. (2002a). Coevolutionary games on networks, *Phys. Rev. E* **66**, p. 056118.

Ebel, H. and Bornholdt, S. (2002b). Evolutionary games and the emergence of complex networks, *cond-mat/0211666* .

Gross, T. and Blasius, B. (2008). Adaptive coevolutionary networks – a review, *Interface* **5**, pp. 259–271.

Hamilton, W. D. (1996). *Narrow Roads of Gene Land Vol.1* (Freeman, New York).

Hauert, C. (2002). Effects of space in 2x2 games, *Int. J. Bifurcation and Chaos Appl. Sci. Eng.* **12**, pp. 1531–1548.

Hauert, C. and Doebeli, M. (2004). Spatial structure often inhibits the evolution of cooperation in the snowdrift game. *Nature* **428**, pp. 643–646.

Herz, A. V. M. (1994). Collective phenomena in spatially extended evolutionary games, *J. Theor. Biol.* **169**, pp. 65–87.

Hofbauer, J. and Sigmund, K. (1998). *Evolutionary Games and Population Dynamics* (Cambridge University Press, Cambridge).

Holme, P., Trusina, A., Kim, B. J. and Minnhagen, P. (2003). Prisoner's Dilemma in real-world acquaintance networks: Spikes and quasiequilibria induced by the interplay between structure and dynamics, *Phys. Rev. E* **68**, p. 030901(R).

Kossinets, G. and Watts, D. J. (2006). Empirical analysis of an evolving social network, *Science* **311**.

Lindgren, K. (1991). Evolutionary phenomena in simple dynamics, in C. G. Langton, C. Taylor, J. D. Farmer and S. Rasmussen (eds.), *Artificial Life II. SFI Studies in the Science of Complexity Vol. X* (Addison-Wesley, Redwood City), pp. 295–312.

Lindgren, K. and Nordahl, M. G. (1994). Evolutionary dynamics of spatial games, *Physica D* **75**, pp. 292–309.

Maynard Smith, J. (1982). *Evolution and the Theory of Games* (Cambridge University Press, Cambridge).

Nowak, M. A. and May, R. M. (1992). Evolutionary games and spatial chaos, *Nature* **359**, pp. 826–829.

Nowak, M. A., Sasaki, A., Taylor, C. and Fudenberg, D. (2004). Emergence of cooperation and evolutionary stability in finite populations, *Nature* **428**, pp. 646–650.

Ohtsuki, H., Hauert, C., Lieberman, E. and Nowak, M. A. (2006). A simple rule for the evolution of cooperation on graphs, *Nature* **441**, pp. 502–505.

Pacheco, J. M., Traulsen, A. and Nowak, M. A. (2006a). Active linking in evolutionary games, *Jour. Theor. Biol.* **243**, pp. 437–443.

Pacheco, J. M., Traulsen, A. and Nowak, M. A. (2006b). Co-evolution of strategy and structure in complex networks with dynamical linking, *Phys. Rev. Lett.* **97**, p. 258103.

Pacheco, J. M., Traulsen, A., Ohtsuki, H. and Nowak, M. A. (2008). Repeated games and direct reciprocity under active linking, *J. Theor. Biol.* **250**, pp. 723–731.

Poncela, J., Gómez-Gardeñes, J., Floría, L. A., Sánchez, A. and Moreno, Y. (2008). Complex cooperative networks from evolutionary preferential attachment, *PLoS One* **3(6)**, p. e2449.

Posch, M., Pichler, A. and Sigmund, K. (1999). The efficiency of adapting aspiration levels, *Proc. Roy. Soc. Lond. B* **266**, pp. 1427–1435.

Rapoport, A. and Chammah, A. M. (1965). *Prisoner's Dilemma* (Univ. of Michigan Press, Ann Arbor).

Ridley, M. (2003). *Nature Via Nurture: Genes, Experience and What Makes Us Human* (HarperCollins Publishers).

Rubin, P. (2002). *Darwinian Politics: The Evolutionary Origin of Freedom* (Rutgers University Press, New Jersey).

Santos, F. C. and Pacheco, J. M. (2005). Scale-free networks provide a unifying framework for the emergence of cooperation. *Phys. Rev. Lett.* **95**, p. 098104.

Santos, F. C. and Pacheco, J. M. (2006). A new route to the evolution of cooperation, *Jour. Evol. Biol.* **19**, pp. 726–733.

Santos, F. C., Pacheco, J. M. and Lenaerts, T. (2006a). Cooperation prevails when individuals adjust their social ties, *PLoS Comput. Biol.* **2**, pp. 1284–1291.

Santos, F. C., Pacheco, J. M. and Lenaerts, T. (2006b). Evolutionary dynamics of social dilemmas in structured heterogeneous populations. *Proc. Natl. Acad. Sci. U.S.A.* **103**, pp. 3490–3494.

Santos, F. C., Santos, M. D. and Pacheco, J. M. (2008). Social diversity promotes the emergence of cooperation in public goods games, *Nature* **454**, pp. 213–216.

Skyrms, B. (2003). *The Stag-Hunt Game and the Evolution of Social Structure* (Cambridge University Press, Cambridge).

Sugden, R. (1986). *The Economics of Rights, Co-operation and Welfare* (Blackwell, Oxford, UK).

Szabó, G. and Tőke, C. (1998). Evolutionary Prisoner's Dilemma game on a square lattice, *Phys. Rev. E* **58**, p. 69.

Szabó, G. and Vukov, J. (2004). Cooperation for volunteering and partially random partnerships, *Phys. Rev. E* **69**, p. 036107.

Taylor, C. and Nowak, M. A. (2006). Evolutionary game dynamics with non-uniform interaction rates, *Theoretical Population Biology* **69**, pp. 243–252.

Taylor, P. D. and Jonker, L. (1978). Evolutionary stable strategies and game dynamics, *Math. Biosci.* **40**, pp. 145–156.

Traulsen, A., Claussen, J. C. and Hauert, C. (2005). Coevolutionary dynamics: From finite to infinite populations, *Phys. Rev. Lett.* **95**, p. 238701.

Traulsen, A., Nowak, M. A. and Pacheco, J. M. (2006). Stochastic dynamics of invasion and fixation, *Phys. Rev. E* **74**, p. 11909.

Traulsen, A., Pacheco, J. M. and Nowak, M. A. (2007). Pairwise comparison and selection temperature in evolutionary game dynamics, *J. Theor. Biol.* **246**, pp. 522–529.

Trivers, R. (1985). *Social Evolution* (Benjamin Cummings, Menlon Park).

Vainstein, M. H. and Arenzon, J. J. (2001). Disordered environments in spatial games, *Phys. Rev. E* **64**, p. 051905.

Van Segbroeck, S., Santos, F. C., Nowé A., Pacheco, J. M. and Lenaerts, T. (2008). The evolution of prompt reaction to adverse ties, *BMC Evolutionary Biology*, **8**, p. 287.

Van Segbroeck, S., Santos, F. C., Lenaerts, T. and Pacheco, J. M. (2009). Reacting differently to adverse ties promotes cooperation in social networks, *Phys. Rev. Lett.* **102**, p. 058105.

von Neumann, J. and Morgenstern, O. (1944). *Theory of Games and Economic Behavior* (Princeton University Press, Princeton).

Wilson, E. O. (1975). *Sociobiology* (Harvard University Press, Cambridge, Massachusetts).

Zeeman, E. C. (1980). Population dynamics from game theory, *Lecture Notes in Mathematics*, p. 819.

Zimmermann, M. G. and Eguíluz, V. M. (2005). Cooperation, social networks, and the emergence of leadership in a prisoner's dilemma with adaptive local interactions, *Phys. Rev. E* **72**, p. 056118.

Zimmermann, M. G., Eguíluz, V. M. and San Miguel, M. (2005). Cooperation and emergence of role differentiation in the dynamics of social networks, *Am. J. Soc.* **110**, p. 977.

Chapter 17

From Animal Collectives and Complex Networks to Decentralized Motion Control Strategies

Arturo Buscarino[1], Luigi Fortuna[2], Mattia Frasca[2,*], Alessandro Rizzo[3,†]

[1] *Scuola Superiore di Catania, Laboratorio sui Sistemi Complessi,*
via San Nullo 5/i, 95125 Catania, Italy

[2] *Dipartimento di Ingegneria Elettrica Elettronica e dei Sistemi,*
Università degli Studi di Catania, viale A. Doria 6, 95125 Catania, Italy
* mfrasca@diees.unict.it

[3] *Politecnico di Bari, Dipartimento di Elettrotecnica ed Elettronica,*
Via E. Orabona 4, 70125 Bari, Italy
† rizzo@deemail.poliba.it

17.1. Collective Behavior and Motion Coordination in Animal Groups

In nature there is plenty of examples where many individuals join together to form a group which behaves as an entity with its own life. Spectacular examples of animal groups are: a flock of starlings flying in the evening, a V-shaped formation of geese, a fish school escaping a predator, an endless line of marching ants, swarms of locusts flying across the desert, groups of African elephants migrating in search of food or water [Sumpter (2006)]. Even more examples can be found in human groups.

Animal groups share peculiar properties. First of all, an animal group is a complex social organization which undergoes continual shape and structural changes over time and space. In a recent work [Newlands and Porcelli (2008)], employing new techniques for semi-automatic analysis of size, shape and structure of bluefin tuna schools, the authors derive some statistics on different formation types indicated as cartwheel, surface-sheet, dome, soldier, mixed, ball and oriented, and draw the conclusion that the variety of observed formations can be explained by taking into account several trade-offs arising in fish schooling. The factors that are involved in fish schooling are, in fact, competitive and cooperative, such as vision, ability of finding food, hydrodynamic benefits and energy consumption, and the risk of being eaten while being in a group compared to that taken as solitary individuals. Group formation derives from the complex interplay of such factors.

Another aspect of living in groups is discussed in [Couzin (2006)]. Animals that forage or travel in groups are able to find the location of a food source or the migration route, even if few individuals have pertinent information. Complex phenomena, such as consensus decision and effective leadership, arise in the decision-making process in animal groups so that the group as a whole can benefit of the final decision.

Group composition is also a dynamical process. For instance, in African elephant societies [Couzin (2006)] groups merge or split as the elephants move through the environment. In general, animals such as dolphins, chimpanzees and elephants take advantages from living in fission-fusion societies at different timescales; for instance, evading a predator requires continuous vigilance and rapid reactions, while locating suitable habitat requires much longer timescale.

All these examples can be explained by an unifying law, that of self-organization. The phenomena observed, in fact, arise from the local interactions between the units and not from a hierarchical organization of the system. In a certain sense, the individual is submerged by the group, but at the same time the individual variety, the information owned by the single, and the individual itself can be of fundamental importance for the group.

Besides the animal world, self-organization has been observed in many physical systems, with the emergence of complex patterns such as, for example, sand dunes in the desert or autowaves in the BZ reaction. Nevertheless, self-organization in animal world has a peculiar feature: in biological systems complexity also arises at the unit level. This is pointed out in [Sumpter (2006)], which individuates several principles underlying self-organized collective behavior in biological systems such as integrity and variability (each animal in the group is an individual different from the others), positive feedback (observed for instance when an ant finds a food source and its trail is followed by other ants), response threshold (animals change their behavior when the stimulus reaches a threshold), leadership (often assumed by individuals possessing a particular information), redundancy (animal groups, for instance, insect societies, are often formed by a vast number of replaceable units), selfishness (the cost/benefit ratio of forming a group should be advantageous for each individual) and so on.

However, many scientists have shown that the essential features of collective motion patterns of animal groups can be captured by minimal models based on the assumption that an individual can be considered a self-propelled particle. In these models local interaction rules are able to explain the coordinated behavior of the group as well as the variety of shapes and motions observed and the effective leadership by a subset of informed individuals. Early studies date to last decade. One of the most cited works on the subject is the discrete-time model discussed in [Vicsek *et al.* (1995); Czirok *et al.* (1997); Czirók *et al.* (1999)]. This was inspired by a former model previously introduced in [Reynolds (1987)] with the aim of visualizing realistic flocks and schools for the animation industry. Other early approaches are based on hydrodynamic principles [Toner and Tu (1995, 1998)], on continuous-time models with all-to-all and attractive long-range interactions [Mikhailov and Zanette (1999)], and on reaction-diffusion equations [Schenk *et al.* (1998)].

In the following, we will review with some details two of the most important models based on self-propelled particles and briefly mention some of their generalizations. Starting from the consideration that the underlying interactions between animals/particles are time-dependent and can be described by a time-variant network, we present some recent results and we focus on possible applications in autonomous robotics. We will not discuss another aspect of collective motion, i.e. the behavior of human crowds, even if a lot of interesting works have been developed in this context, e.g. the simulation of pedestrians escaping in a panic situation [Helbing and Molnar (1995); Helbing *et al.* (2000)] or the study of movement pattern through the analysis of data coming from the mobile telephone network [González *et al.* (2008)]. This is beyond the scope of this Chapter.

The field of collective motion models is in rapid evolution. In fact, only recently, theoretical considerations are beginning to be accompanied by experimental findings which were not possible in the past (for instance tracking individuals in a flock or in a school through extremely fast image processing algorithms). Technological advances are on one hand validating many aspects of these models, and on the other hand adding new valuable information which refine and add new hypotheses. In this context, the data acquisition phase is of crucial importance, and the related techniques still reveal flaws, leading to controversial conclusion, as pointed out in [Buchanan (2008)]. To mention some examples, advanced image processing techniques have been recently developed for the characterization of schools of Atlantic bluefin tuna [Newlands and Porcelli (2008)], the analysis of flocks of starlings [Ballerini *et al.* (2008)], and the evaluation of the trajectories of fish populations of giant danios (*danio aequipinnatus*) in a laboratory tank [S.V. Viscido and Grünbaum (2004); Grünbaum *et al.* (2005)].

17.2. Mathematical Models for Collective Motion

In this section we review two important models able to generate self-organized motion of autonomous agents. The first one, introduced by Vicsek and co-workers [Vicsek *et al.* (1995)], is considered by many as an ancestor of a class of collective motion models. This model has inspired many scientists for modifications and generalizations, as regards for example the second model which we will present, introduced by Couzin and co-workers [Couzin *et al.* (2002, 2005)], which is more oriented to modeling the behavior of animal groups on the move.

The Vicsek's model is a manifest example of *complexity*: a complex behavior can emerge from a population in which each element's life is governed by simple rules.

In particular, as Vicsek states in [Vicsek *et al.* (1995)], the only rule of the model is that *at each time step a given particle driven with a constant absolute velocity assumes the average direction of motion of the particles in its neighborhood of radius r with some random perturbation added.* The model is discrete in time and continuous in space, and consists of N particles which are initially distributed on a square plane of linear size L with periodic boundary conditions. Particles share the same velocity modulus ν, whereas their initial directions of motion $\theta_i(0)$ are uniformly randomly distributed. At each time step, the position $\mathbf{x}_i$ and the direction of motion θ_i of the ith particle are simultaneously updated according to

$$\mathbf{x}_i(t + \Delta t) = \mathbf{x}_i(t) + \mathbf{v}(t)\Delta t\,,$$
$$\theta_i(t + \Delta t) = \langle \theta_i(t) \rangle_r + \Delta \theta \tag{17.1}$$

where $\langle \theta_i(t) \rangle_r$ is the average direction of particles within the neighborhood of the ith particle (which is considered as a circle of radius r) and Δt is the time step width. The average direction is given by the angle $\arctan(\frac{\langle \sin \theta(t) \rangle_r}{\langle \cos \theta(t) \rangle_r})$. $\Delta \theta$ is a random number acting as a noise term and chosen with uniform probability from the interval $[-\frac{\eta}{2}, \frac{\eta}{2}]$.

The values of the noise parameter η and particle density $\rho = \frac{N}{L^2}$ affect the behavior of the system: for small values of density and noise the particles tend to form groups moving toward random directions; increasing density and noise the particles move randomly with some correlation; increasing density but maintaining small the noise it is possible to observe an ordered motion of all the particles towards the same spontaneously chosen direction. The last case is the most interesting because it shows a kinetic phase transition due to the fact that particles are driven with a constant absolute velocity. This transition was supposed to be of second order in [Vicsek *et al.* (1995)]. Recently, contradicting results in favour of a first-order transition have been obtained in [Grégoire and Chaté (2004)]. We will not enter in the debate arised, which is beyond the aim of this work.

To characterize the ordered phase it is possible to define an order parameter, the average normalized velocity, which can be calculated with the following expression:

$$v_a = \frac{1}{N\nu} \left| \sum_{i=1}^{N} \mathbf{v}_i \right| \tag{17.2}$$

where ν is the constant module of the particle velocities. Clearly, $v_a = 0$ indicates that the agents are not coordinated (there is no net transport), while $v_a = 1$ indicate that all the agents move in the same direction (the net transport is maximum). In fact, $v_a = 1$ holds when $\theta_i = \bar{\theta}$ for all i.

Vicsek's model is considered a cornerstone for the description of dynamical phenomena concerning collective motion. In his work, Vicsek himself stated that there could be many interesting variations of the model which could lead to additional nontrivial effects. With the aim of emulating a wider range of collective phenomena, relevant to the study of animal motion, Couzin and co-workers introduced a new model able to take in account typical behaviors observed in animal world [Couzin *et al.* (2002)]. In particular, Couzin's model take into account (i) the fact that an individual tends to keep a vital space free of its neighbors, (ii) the existence of two mechanisms of adjustment of the individual directions of motion, according to the distance from its neighbors.

As we did with Vicsek's model, let us summarize in few points the rules which govern the evolution of the model which is, also in this case, discrete in time and continuous in space:

(1) The highest priority task for individuals is to avoid collisions within a *repulsion radius*, r_r, thus maintaining their vital space [Krause and Ruxton (2002)].
(2) If individuals are not busy in performing the avoidance task, they tend to align with closer neighbors, contained within a given *orientation radius* r_o (this is in fact the governing rule in Vicsek's model [Vicsek *et al.* (1995)]).
(3) Together with orientation, individuals tend to approach neighbours [Partridge and Pitcher (1980); Partridge (1982a)] which are out of the orientation radius and within an *attraction radius* r_a.

For different values of its parameters, this model is able to reproduce different types of collective behaviors such as:

- swarm, i.e. a compact group with low level of parallel alignment and low angular momentum, observed in insects [Okubo and Chiang (1974); Ikawa and Okabe (1997)] and fish schools [Pitcher and Parrish (1993)];
- torus formation, in which individuals rotate around an empty core, observed in fish schools in open water [Parrish and Edelstein-Keshet (1999)];
- parallel group, in which the individuals move with highly aligned arrangement and low angular momentum, observed in bird flocks and fish schools [Major and Dill (1978); Partridge (1982b)].

Couzin *et al.* [Couzin *et al.* (2005); Couzin (2006)] formulated a further model which includes the presence of experienced individuals, who have a preferential direction of motion. This is motivated by the fact that there exist in nature examples in fish schools where a minority of individuals can influence the foraging activity of the group [Reebs (2000)]. It has also been observed that in honeybee swarms very few individuals can lead the group to a new nest site [Seeley (1985)]. The existence and the role of experienced individuals in animal collectives is discussed in more depth in [Couzin *et al.* (2005)] and in the references in there.

The presence of informed individuals is taken into account in the model with a further rule:

(4) There exist a number of *informed* individuals, who move by mixing a *social criterion* (emerging from rules 2 and 3), and an *individual* one, that is moving in the preferential direction.

Let us review the model in more detail. As for Vicsek's model, N agents move in a continuous 2D space of linear dimension L, with periodic boundary conditions.

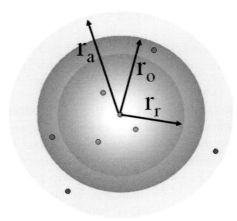

Fig. 17.1.   Repulsion, orientation, and attraction radius in the Couzin's model.

Each agent move with the same absolute velocity ν, position vector $\mathbf{c}_i$, velocity vector $\mathbf{v}_i$. The discrete-time model is run with a sampling time Δt. Figure 17.1 shows the three radii (i.e., *repulsion, orientation, attraction*) which can be individuated around a specific individual. As can be noted, the relation $r_r < r_o < r_a$ holds.

When an agent senses neighbors within his repulsion radius, it undertakes an avoidance task by turning away from its neighbors. Avoidance is the task with the highest priority and, if agents are detected in the repulsion radius, this is the only task which is performed. The avoidance task, performed by the ith agent, can be modelled with the equation

$$\mathbf{d}_{ri}(t + \Delta t) = -\sum_{j \neq i} \frac{\mathbf{c}_j(t) - \mathbf{c}_i(t)}{|(\mathbf{c}_j(t) - \mathbf{c}_i(t))|} \tag{17.3}$$

where $\mathbf{d}_{ri}$ represents the desired direction of travel under an avoidance condition.

If no agents are detected within the repulsion radius, orientation and attraction tasks are performed. Agents performing the orientation task tend to align their direction to the average direction of motion of neighbors within the orientation radius, as

$$\mathbf{d}_{oi}(t + \Delta t) = \sum_{j \neq i} \frac{\mathbf{v}_j(t)}{|\mathbf{v}_j(t)|}. \tag{17.4}$$

When an agent performs the attraction task, it tends to approach the neighbors sensed within the attraction radius (and out of the orientation radius), as

$$\mathbf{d}_{ai}(t + \Delta t) = \sum_{j \neq i} \frac{\mathbf{c}_j(t) - \mathbf{c}_i(t)}{|(\mathbf{c}_j(t) - \mathbf{c}_i(t))|}. \tag{17.5}$$

When neighbors are sensed in only one zone (i.e., either orientation or attraction), then a single task is performed. If neighbors are present in both zones, then orientation and attraction tasks are performed concurrently as

$$\mathbf{d}_{ti}(t + \Delta t) = \frac{\mathbf{d}_{oi}(t + \Delta t) + \mathbf{d}_{ai}(t + \Delta t)}{2}. \tag{17.6}$$

As stated above, informed agents have a preferential direction to pursue. Informed agents act on the basis of a combination of their individual criterion (i.e., the preferential direction), with the social criteria described above (i.e., avoidance, orientation and attraction tasks). Also for informed agents, the avoidance task has the highest priority and is performed in an exclusive fashion, and is modeled as in Eq. (17.3). The behaviour of an informed agent, not engaged in an avoidance task, is modeled as

$$\mathbf{d}'_{ti}(t + \Delta t) = \frac{\mathbf{d}_{ti}(t + \Delta t) + \omega\mathbf{g}}{|\mathbf{d}_{ti}(t + \Delta t) + \omega\mathbf{g}|} \tag{17.7}$$

where $\mathbf{g}$ is the preferential direction vector, ω regulates the balance of individual and social criteria, and $\mathbf{d}_{ti}(t + \Delta t)$ is computed as in Eq. (17.6).

The aim of Couzin's work is not limited to assessing the ability of the collective to move along a spontaneously emerging direction, as in [Vicsek *et al.* (1995)]. Besides this, the interest is also focused on the ability of the group to move along the direction of the informed individuals. Therefore, the parameter expressed in Eq. (17.2) is not sufficient to evaluate the effectiveness of the leadership of the informed elements in guiding the group towards their preferential direction. This ability can be quantified through the parameter v_d, which is defined as

$$v_d = 1 - L_{avg} \tag{17.8}$$

where L_{avg} is the angle between the average direction of all the agents, and the preferential direction, normalized between 0 and 1. L_{avg} is computed as

$$L_{avg} = \frac{\sum_{j=1}^{N} \frac{1}{N} \frac{\mathbf{v}_j(t)}{|\mathbf{v}_j(t)|} - \mathbf{g}}{\pi} \tag{17.9}$$

where $\mathbf{g}$ is the preferential direction. It is clear that v_d is closer to 1 as long as all the agents tend to move closer to the preferential direction. On the contrary, v_d approaches zero when all the agents tend to move opposite to the preferential direction.

The results obtained by Couzin's and co-workers can be summarized in the following points:

- For a given group size the accuracy of the collective direction of motion with respect to the preferential direction increases asymptotically as the proportion of informed individual increases.
- As the group size increases, a smaller proportion of informed individuals is needed to achieve a given degree of coordination (i.e. a value of parameter v_d). This means that for big groups, only a few experienced elements are needed to guide the group towards the preferential direction.

- The ω parameter, which weights the influence of the individual criterion of motion over the social one (see Eq. (17.7)) is of least importance if the proportion of informed individuals is small or large. On the contrary, at intermediate levels it is strongly correlated with the coordination capability.

Moreover, in [Couzin *et al.* (2005)], the case in which there exist two groups of informed individual with two different preferential directions is also investigated. Simulation results highlighted that if the number of components of the two groups is equal, the group behavior depends on the discrepancy between the two preferred directions: if this difference is small, the collective moves along the average direction of all informed individuals. As the difference increases, the naive individuals choose at random one of the two preferential directions. On the other hand, if the number of components of the two informed groups is different, the collective tends to assume the direction of motion of the larger informed group.

Besides being a suitable model for describing phenomena from animal world, Couzin's model has been taken as a source of inspiration to implement decentralized control schemes for pools of mobile robots [Buscarino *et al.* (2007)]. In fact, in this model many interesting issues for decentralized control are raised. In particular, the individual control law is simple and includes the avoidance task, which is fundamental for the control of a robot team. Moreover, only sensing of neighbors is needed, not identification. That is, a robot does not need to know if a sensed neighbor is an informed individual or not. On the contrary, it only has to sense its neighbor's position and velocity.

In the following Section we will highlight the networked nature of these models and present our results on the application of collective motion models and complex network theory to the problem of motion coordination.

17.3. Applying Complex Network Theory to Coordination Models

When, as in most motion coordination models, many autonomous units act on the basis of a set of rules requiring the knowledge of the state of neighboring units, a way to describe interactions is needed. The most used tool to do this is a graph, in which nodes represent agents and links represent interactions between them. Therefore, concepts and tools from graph and network theory can be exploited to study the properties of the interaction graph, and consequently to investigate the system behavior.

Some approaches tend to simplify the theoretical framework by considering a network of interactions which does not change in time (the simplest case of which is the all-to-all coupling [Sepulchre *et al.* (2007)]), even though, in general, as agents move their neighborhoods change over time and the related interaction graph is time-variant [Jadbabaie *et al.* (2003); Buscarino *et al.* (2006, 2007); Sepulchre *et al.* (2008)]. Time-variant networks are a powerful modelling and analysis tool which use is not strictly limited to motion coordination problems. For instance, such networks arise in physical, social and biological phenomena, ecosystems and telecommunications [Galstyan and Lerman (2002); Wang and Wilde (2004); Zimmermann *et al.* (2004); Frasca *et al.* (2006, 2008)].

In network theory, the effect of long-range connections (also called *weak ties*) in static networks is well known. An example is that of *small-world* networks [Watts and Strogatz (1998)], which are built by starting from a regular ring lattice network and rewiring links with a given probability. If the probability is low, this results in a network structure which is very similar to the original one, except for a few long-distance links. The presence of such links makes the characteristic path length shorter while keeping the clustering coefficient high. In other words, small-world topologies conjugate the advantages of fast information transfer typical of random networks with the robustness features typical of regular networks. Moreover, in general long-range connections enhance significantly the system performance in terms of synchronization and cooperation. One may expect that a similar result can be achieved also in time-variant networks.

In this section we report on the effects of the introduction of long-range connections in the interaction schemes of the two models introduced in Section 17.2. This is motivated by possible applications in decentralized control of autonomous mobile robots. In fact, the addition of a few long-range connections is expected to provide better performance with a moderate, if not negligible, change in the communication scheme, which is kept mostly local.

When applied to robotics, the aforementioned models can be used to control the cooperative behavior of robots through decentralized strategies. These strategies exploit mostly local connections realized through sensors able to collect information from neighboring robots, without needing a central coordination and supervision unit. Recently, technology involved in control techniques for single robots has been pushed to a high level of robustness as long as computational and communication capabilities are growing. These enhancements make possible to equip each robot with sensors, wireless transmission systems and a processor, dedicated to the processing of information collected by sensors through the communication system [Petriu *et al.* (2004)]. In this perspective, each robot is able to control its movement by exploiting information coming from other robots acting in the same environment. Thus, the development of distributed algorithms for controlling the robots is becoming very important to achieve a global coordinated and cooperative behavior without the use of central coordination [Klavins and Murray (2004)].

17.3.1. *Effect of long-range connections in the Vicsek's model*

The behavior of the Vicsek's model (17.1) in absence of noise has been investigated in several papers. In particular, in [Jadbabaie *et al.* (2003)], under a simplifying hypothesis for the heading update rule, it is proved that all particles shall move in the same heading, if the graphs describing their interactions are *periodically linked together*. A weaker condition is proved in [Li and Wang (2004)], where the sufficient

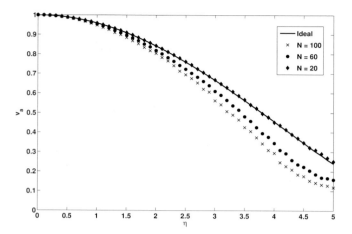

Fig. 17.2.   Order parameter v_a vs. η for different N and for the ideal case (Eq. (17.11), continuous line). The density has been fixed at $\rho = 4$ for all the cases. Other parameters are $r = 1$, $\Delta t = 1$ and $L = \sqrt{N/\rho}$. Initial conditions are random.

condition for the coordination of all agents is that the graphs are *finally linked together*, i.e. if the union of the graphs starting from any time is connected. The case of $\eta \neq 0$ has not been treated in many works. In this section we study some aspects of the Vicsek's model in presence of noise.

At first, let us consider the case in which each agent knows the heading of all the other agents, i.e. when all the agents are packed forming only one group. In this case $\langle \theta_i \rangle_r = \Theta$ for every agent and, since $v_i = \nu e^{j\theta_i}$, v_a is given by the following expression:

$$v_a = \frac{1}{N} \left| \sum_{i=1}^{N} e^{j\theta_i} \right| = \frac{1}{N} \left| \sum_{i=1}^{N} e^{j\Theta} e^{j\Delta\theta} \right| \tag{17.10}$$

and thus

$$v_a = \frac{1}{N} \left| \sum_{i=1}^{N} e^{j\Delta\theta} \right| . \tag{17.11}$$

The behavior of v_a versus the noise parameter η is shown in Fig. 17.2. It illustrates simulation results along with the curve v_a vs. η corresponding to equation (17.11). This last curve is a sort of best case in which each agent lies in the neighborhood of each other (i.e. an all-to-all communication scheme). In this case the agents move in a packed group, maximizing v_a. The average velocity v_a depends only on the noise, which degrades the overall performance of the system. This case can be achieved with an high agent density in a small environment.

In general, not all the agents will belong to each other's neighborhood, and the effects of the random perturbation on θ_i will change from agent to agent. Therefore, $\langle \theta_i \rangle_r$ will depend on i, and v_a will be smaller than the value predicted by

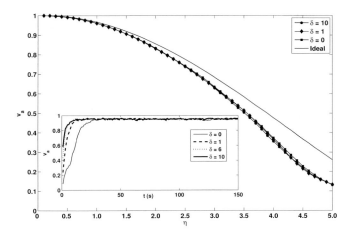

Fig. 17.3. Order parameter v_a vs. η when long-range connections are introduced. The various curves refer to different values of δ. Other parameters are $N = 100$, $L = 5$, $\rho = 4$, $r = 1$, $\nu = 0.03$, $\Delta t = 1$. Simulation time is 200 steps beyond the time necessary to reach convergence in the case $\eta = 0$. Inset: time evolution of the average velocity for the case $N = 100$, $\rho = 4$, $\eta = 1.0$, with different δ.

Eq. (17.11). With the aim of keeping the communication scheme mostly local, while approaching the performance of the ideal all-to-all scheme, we evaluate in the following the introduction of long-range connections: at each time step, some randomly selected agents are allowed to have long-range connections with other randomly chosen agents, which can lie out of the interaction radius, by including the heading of these last agents in the computation of the average heading $\langle \theta_i \rangle_r$.

The idea underlying this new model is that of obtaining a group which is much more connected, by exploiting almost the same radius of communication, plus a few long-range links. The effect expected is that of a massive connection without the communication overhead imposed by this configuration.

At first, the effects of long-range connections have been evaluated by taking into account the same set of parameters as in [Vicsek *et al.* (1995)]. In particular $N = 100$ agents are allowed to move on a plane of dimension $L = 5$, that is the density $\rho = 4$. The number of long-range connections at each time step has been indicated as δ and ranges from 0 to 10, in order to keep the number of long-range connections small. The case $\delta = 0$ corresponds to the original Vicsek's model, where interactions are only local. The performance of the system has been evaluated by taking into account the order parameter v_a with respect to increasing values of η.

Results are shown in Fig. 17.3, where it is clear that the performance is not strongly improved by increasing the value of δ. Another important effect of long-range connections is shown in the inset of Fig. 17.3, illustrating that the time required to reach the steady value of v_a decreases when long-range connections are introduced.

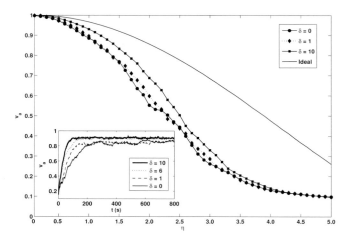

Fig. 17.4.   Order parameter v_a vs. η. Curves refer to different values of δ. Other parameters are $N = 100$, $L = 10$, $\rho = 1$, $r = 1$, $\nu = 0.03$, $\Delta t = 1$. Simulation time is 200 step beyond the time necessary to reach convergence in the case $\eta = 0$. Inset: time evolution of the average velocity for the case $N = 100$, $\rho = 1$, $\eta = 1.0$ for different values of δ.

These considerations allow us to conclude that in the case of $\rho = 4$ the system shows similar values of v_a independently of the value of δ, but there is a small improvement of the time necessary to reach the steady speed. The behavior of the system has been further investigated with respect to different values of the density, leading to the conclusion that long-range connections introduce clear advantages for low density values. This is a very important point since, as shown in [Vicsek et al. (1995)], v_a decreases with ρ. At low densities, coordination performance is poorer than at high densities. Therefore, it is at low density levels that the introduction of long-range connections leads to the greatest benefits.

The case of low density ($\rho = 1$) is dealt with in Fig. 17.4, where it is clearly shown that the improvement is more significant. Also in the case of low density, long-range connections decrease the time needed to reach an ordered behavior. The inset of Fig. 17.4 shows the trend of the average velocity v_a for $N = 100$ and $\eta = 1$. Three curves corresponding to three values of δ are shown.

In [Vicsek et al. (1995)] the emergence of a kinetic transition between ordered and disordered phase is described in terms of the power law which expresses the relationship between v_a and the critical noise $\eta_c L$ (i.e. the value of noise beyond which the phase transition occurs) as follows:

$$v_a \sim ((\eta_c(L) - \eta)/\eta_c)^{\beta} . \tag{17.12}$$

The improvement due to long-range connections does not modify this relationship, but leads to a decreased β exponent in the power law. We have numerically verified this result as shown in Fig. 17.5.

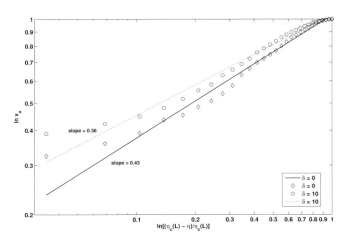

Fig. 17.5.   Power-law approximation of v_a vs. $(\eta_c(\mathrm{L})\text{-}\eta)/\eta_c$ in logarithmic scale. It can be seen that the slope of the fitting curve decreases from 0.43 to 0.36 introducing 10 long-range connections. $\eta_c = L^{-0.5}$ as in [Czirok *et al.* (1997)].

17.3.2.   *Effect of long-range connections in the Couzin's model*

The behavior of the Couzin's model is here investigated by carrying out numerical simulations for different values of the model parameters, and by evaluating the order parameters v_a and v_d.

At first, model behavior is characterized with respect to different values of the density ρ. When long-range connections are not considered, v_a increases as a function of ρ, as shown in Fig. 17.6(a). In this case, in fact, the coordination between individuals is favored by an increased density, because agents may exchange more information.

The comparison between the cases with $\delta = 0$ and $\delta \neq 0$ shows that the presence of long-range connections improves system performance in terms of v_a, i.e. the degree of group coordination. It should be noticed that v_a for $\delta \neq 0$ decreases as ρ increases. This is due to the fact that at high densities avoidance tasks occur more frequently. Moreover, Fig. 17.6(a) shows that the improvements due to the introduction of long-range connections are more important for low density values, as observed also in the Vicsek's model. It should be noticed that a few long-range connections improve system performance, but a further increase in the number of long-range connections leads to poor advantages. This behavior also occurs in small-world networks, when the changes in the characteristic path length are considered with respect to the number of long-range connections. In fact, in small-world networks, the introduction of a few long-range links decreases the characteristic path length, but further increases of the number of long-range links have small effects on this parameter.

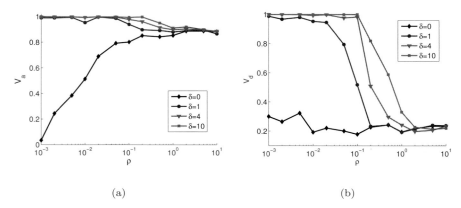

(a)   (b)

Fig. 17.6.   (a) Order parameter v_a vs. ρ for different δ. Other parameters are $N = 100$, $N_i = 10$, $r_r = 0.3$, $r_o = 0.8$, $r_a = 1$, $\omega = 0.3$, $\nu = 0.03$. Each simulation is for $T = 250s$. (b) Order parameter v_d vs. ρ with respect to different values of δ. Other parameters are as in Fig. 17.6(a).

For high density, v_a tends to become independent of the number of long-range connections. In fact, when the density is high, the agents are close to each other and are mainly engaged in performing avoidance. Therefore, the advantages introduced by long-range connections are more important at low density as observed in the Vicsek's model.

The effects of long-range connections on the parameter v_d are similar. In fact, the presence of long-range connections causes an increase of the value of v_d, which means that the mean motion direction of the group tends to be similar to the preferential motion direction adopted by informed individuals. Fig. 17.6(b) shows v_d as a function of ρ with respect to different values of δ: long-range connections favor the tendency to form a group with a homogeneous direction, thus enhancing leadership of informed individuals. As in Vicsek's model, this is effective only at low density values.

Figures 17.6(a) and 17.6(b) refer to the case of ten informed individuals ($N_i = 10$). In the following, we address the case of increasing the number of informed individuals. Figures 17.7(a) and 17.8(a) show that v_a and v_d increase with N_i in absence of long-range connections. The advantages of having a high percentage of informed individuals are clear at low density values, where groups mainly constituted by informed individuals form. For higher density, increasing the number of informed individuals has small effects on v_a and v_d. Figures 17.7(b) and 17.8(b) show v_a and v_d in the presence of long-range connections. The introduction of long-range connections has greater effects when N_i is small. On the other hand, increasing the number of N_i in presence of long-range connections does not show great improvements, unlike the case with $\delta = 0$. Furthermore, when $\delta = 0$ the transition between high values of v_d and low values of v_d is sharper than in the case

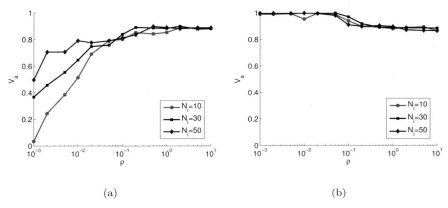

(a)   (b)

Fig. 17.7.   Order parameter v_a vs. ρ for different values of N_i and δ: $\delta = 0$ (a) and $\delta = 1$ (b). Other parameters (except for N_i) are as in Fig. 17.6(a).

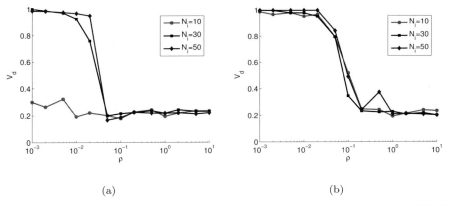

(a)   (b)

Fig. 17.8.   Order parameter v_d vs. ρ for different values of N_i and δ: $\delta = 0$ (a); $\delta = 1$ (b). Other parameters (except for N_i) are as in Fig. 17.6(a).

$\delta \neq 0$. Thus in the latter case the group is able to follow the direction of informed individuals for higher values of the density.

Similarly to Vicsek's model, we analyze the system behavior in presence of a noise term, characterized by a Gaussian distribution with zero mean and standard deviation σ. Figure 17.9(a) shows that, when the noise level is increased, v_a decreases, as in Vicsek's model. Moreover, also in this case the introduction of long-range connections improves the system performance (in terms of v_a) in the presence of noise.

In Fig. 17.9(b) v_d as a function of σ is shown for different values of δ. When the noise level is increased, v_d also decreases. It is worth noticing that in the case of

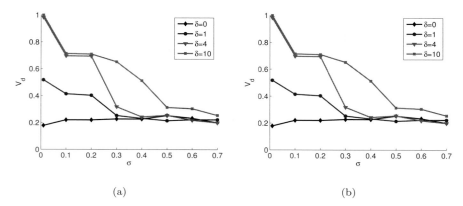

(a)   (b)

Fig. 17.9.   Order parameters v_a (a) and v_d (b) vs. the noise level σ for different values of δ. Other parameters (except for σ) are as in Fig. 17.6(a).

$\delta = 0$, even low values of σ lead to poor leadership effectiveness, while when $\delta \neq 0$ leadership is effective at least for low values of the noise level.

17.3.3. *Proximity graphs and topological interactions*

Although most of the models share a networked structure to model the interactions between agents, the definition of such network is not univocal. In [Bullo *et al.* (2008)] many examples of construction of the interaction network (also called *proximity graph*) are given. Among them, we cite:

- the r-disk graph in which two agents are neighbors if their locations are within a distance r;
- the Delaunay graph in which two agents are neighbors if their corresponding Voronoi cells [Aurenhammer (1991)] intersect;
- the r-limited Delaunay graph in which two agents are neighbors if their corresponding r-limited Voronoi cells intersect;
- the visibility graph in which two agents are neighbors if their positions are visible to each other.

Beyond the definitions stated above, another technique of construction of the proximity graph, emerging from experimental observation, should be considered. In fact, in a recent paper, Ballerini *et al.* (2008) showed that the interaction between birds in airborne flocks is governed by a topological distance rather than a metric one. In particular, through an accurate study based on image processing performed on bird flocks they showed that each bird interacts on average with a fixed number of neighbors (six-seven), rather than with all neighbors within a fixed metric distance.

In the simulations and in the results discussed so far, the r-disk graph has been considered. In this Section, we compare some features of the Vicsek's model in

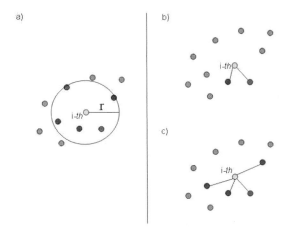

Fig. 17.10.   Examples of r-disk neighboourhood (a), topological neighbourhood with two (b) and four (c) neighbours.

which the interaction network is built by two different techniques: the r-disk and the topological graph. In this last case, only the nearest N_n agents are considered neighbors and consequently part of the interaction network. In Fig. 17.10 an example of r-disk neighborhood (a), topological neighborhood with two (b) and four (c) neighbors, respectively, is illustrated.

To compare the two models, at first the order parameter v_a is evaluated with respect to the agent density ρ in the r-disk case, and to the number of neighbors N_n in the topological case, for different values of the noise parameter η. The parameter N_n seems to be a reasonable counterpart of ρ in the topological case, since an increase of both parameters leads in both cases to the inclusion of more agents in the computation of the average heading. In Fig. 17.11 a similar behavior can be observed for both r-disk (Fig. 17.11(a)) and topological (Fig. 17.11(b)) cases: v_a tends to increase for increasing values of ρ and N_n, and to decrease for increasing values of the noise parameter η. The choice of the range for ρ in the r-disk case is made to achieve an average number of neighbors in the r-disk case $(\rho \pi r^2)$ equal to the fixed number of neighbors N_n in the topological case (i.e., $\rho = N_n/\pi r^2$). With this in mind, it can be observed from Fig. 17.11 that the topological graph performs slightly better.

An interesting feature which distinguishes the behavior of the two models is the reaction of the group to an attack by a predator. In [Ballerini *et al.* (2008)] the authors claim that, although both ways of defining the interaction network provide cohesion for suitable values of the parameters, topological interactions lead to a more robust cohesion. This is shown by simulating the attack of a predator (the approach of which causes a repulsive reaction in the other agents) in both models, and evaluating the probability that the group does not break up or, on the opposite, breaks into two or more connected components. Ballerini *et al.* (2008) conclude that

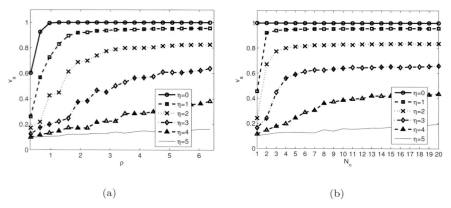

Fig. 17.11.   Order parameters v_a. The parameters have been chosen as: $N = 100$, $v = 0.03$, $\rho = 4$.

groups based on r-disk interactions very often break into more than one component, while topological groups exhibit a greater cohesion.

The smashing observation of topological interactions between the elements of an animal group opens the way to new investigations. On the other hand, in the perspective of developing distributed control algorithms inspired to animal grouping, exploiting the advantages of this topological construction of the proximity graph requires the adoption of sensory and communication equipments able to locate and communicate with arbitrarily distant units.

17.4. From Biological Networks to Engineering Problems

As discussed above, the development of theory and applications on motion coordination encompasses a twofold interest. On one hand, the interest is focused on the development of suitable models able to emulate and, to some extent, predict, the behavior of animal collectives, such as flocks, herds, swarms, schools, etc. On the other hand, engineers are interested in taking inspiration from the animal world to develop strategies for coordinating the motion of embedded systems, like teams of mobile robots, mobile sensor networks, unmanned vehicles, and so on. Generally speaking, the interest is focused on those applications which require the adoption of a team of autonomous agents with computing, communication, and mobility capabilities. The field of application is wide, covering distributed sensing, search and rescue, environmental modeling, surveillance, exploration [Martínez et al. (2007)]. In this framework, the trend is to design and implement *decentralized* coordination and control strategies. That is, the control algorithm implemented on a given agent must rely on information about the agent itself and a few agents laying on a given *neighborood.* In this way, at least four advantageous features can be achieved:

- **Simplicity.** The decentralized control strategy to be implemented on each agent is often simple and require a low computational effort. Moreover, very often many control schemes do not require the *identification* of the neighbors and rather rely on a simpler *sensing* of some relevant variables (*e.g.* position, velocity, etc.).
- **Robustness.** The absence of a central coordination unit and the exploitation of a reduced communication scheme allow the system to keep working even in case of failure of one or more agents. Moreover, the problem of the existence of a single vulnerable point (i.e., the coordination unit) is not present.
- **Scalability.** The control effort is independent of the number of agents.
- **Limited communication.** As the communication between agents is performed mostly locally (i.e. within a neighborhood), many real-world applications with communication constraints can be dealt with. Moreover, the autonomous character of the control laws often allows to cope with limited and/or uncertain communication, or with finite delays [Paley *et al.* (2008); Fiorelli *et al.* (2006); Moreau (2005)].

Although very often the control law aboard the single agent seems "simple", the approaches towards the construction of a system and control theory of coordinated motion involves several advanced mathematical tools. A rigorous formalization of the framework is necessary, at least to:

- *model* the phenomena involved;
- *code* the motion coordination tasks;
- *design* the distributed control strategy;
- *prove* the correctness and convergence of the strategy;
- *assess* the overall performance of the control strategy.

With this in view, many approaches have been considered. Justh and Krishnaprasad (2004) exploit the Lie group theory to set up a framework for the analysis of the motion of a planar formation, emphasizing the fact that the configuration space of the vehicles consists of N copies of the group $SE(2)$ (where N is the number of vehicles). The Frenet-Serret equations of motion are used to describe the evolution of both position and orientation of a vehicle moving at unit speed under a steering control. Based on this model, an effective framework for the analysis and control of the formation motion has been developed in which each moving agent is modeled with an oscillator. The control problem is in fact paralleled with a problem of synchronization of a network of oscillators, which is another fascinating topic of complex system science [Strogatz (2000, 2003)]. For a review of this approach we address the reader to [Paley *et al.* (2007)] and to the references in there.

Another interesting approach has been introduced in [Bullo *et al.* (2008); Martínez *et al.* (2007)], which essentially exploits tools from computational geometry, such as the notion of proximity graph. Several coordination tasks, such as deployment, consensus, rendezvous, cohesiveness, etc., are encoded by using aggregate objective functions, then different techniques to design a suitable control law are provided, by exploiting the design of gradient flows, the analysis of emergent behaviors, the identification of meaningful local objective functions, the composition of basic behaviors (heuristics).

Finally, different strategies have been introduced to assess the correctness and performance of coordination algorithms. This can result in a complicated task, because of the time-variant nature of the connections, the presence of uncertainties, the heterogeneity in the agent population, or even the nondeterministic nature of some control algorithms. Several strategies have been introduced, such as stochastic linear techniques [Jadbabaie *et al.* (2003)], circulant matrices [Marshall *et al.* (2004)], algebraic graph theory [Olfati-Saber and Murray (2004)], differential equations [Justh and Krishnaprasad (2004)], invariance principles [Cortés and Bullo (2005)], graph grammars [Klavins (2007)], Monte Carlo methods [Pallottino *et al.* (2007)], partial difference equations [Bliman and Ferrari-Trecate (2008)], and time-scale separation [Nabet *et al.* (2007)].

The references mentioned in this work depict a variegate range of techniques to analyze and solve coordination motion problems. Nevertheless, it seems that the underlying (either implicitly or explicitly) network structure constitutes a common feature shared by most, if not all, of them. It is our opinion that network science can be a unifying and cross-fertilizing element between several approaches, towards the development of both effective analyses of biological collectives and powerful motion coordination algorithms.

References

Aurenhammer, F. (1991). Voronoi diagrams: A survey of a fundamental geometric data structure, *ACM Computing Survey* **23**, 3, pp. 345–405.

Ballerini, M., Cabibbo, N., Candelier, R., Cavagna, A., Cisbani, E., Giardina, I., Lecomte, V., Orlandi, A., Parisi, G., Procaccini, A., Viale, M. and Zdravkovic, V. (2008). Interaction ruling animal collective behavior depends on topological rather than metric distance: Evidence from a field study, *Proceedings of the National Academy of Sciences* **105**, pp. 1232–1237.

Bliman, P.-A. and Ferrari-Trecate, G. (2008). Average consensus problems in networks of agents with delayed communications, *Automatica* **44**, pp. 1985–1995.

Buchanan, M. (2008). The mathematical mirror to animal nature, *Nature* **453**, pp. 714–716.

Bullo, F., Cortés, J. and Martínez, S. (2008). *Distributed Control of Robotic Networks - A mathematical approach to motion coordination algorithms* (Electronic version, http://coordinationbook.info).

Buscarino, A., Fortuna, L., Frasca, M. and Rizzo, A. (2006). Dynamical network interactions in distributed control of robots, *Chaos* **16**, p. 015116.

Buscarino, A., Fortuna, L., Frasca, M. and Rizzo, A. (2007). Effects of long-range connections in distributed control of collective motion, *International Journal of Bifurcation and Chaos* **17**, 7, pp. 2411–2417.

Cortés, J. and Bullo, F. (2005). Coordination and geometric optimization via distributed dynamical systems, *SIAM Journal on Control and Optimization* **44**, 5, pp. 1543–1574.

Couzin, I. D. (2006). Behavioral ecology: social organization in fission-fusion societies, *Current Biology* **16**, pp. 169–171.

Couzin, I. D., Krause, J., Franks, N. R. and Levin, S. A. (2005). Effective leadership and decision-making in animal groups on the move, *Nature* **433**, pp. 513–516.

Couzin, I. D., Krause, J., James, R., Ruxton, G. D. and Franks, N. R. (2002). Collective memory and spatial sorting in animal groups, *Journal of theoretical biology* **218**, pp. 1–11.

Czirók, A., Barabasi, A. L. and Vicsek, T. (1999). Collective motion of self-propelled particles: Kinetic phase transition in one dimension, *Physical Review Letters* **82**, pp. 209–212.

Czirok, A., Stanley, H. E. and Vicsek, T. (1997). Spontaneously ordered motion of self-propelled particles, *Journal of Physics A: Mathematical and General* **30**, 5, pp. 1375–1385.

Fiorelli, E., Leonard, N. E., Bhatta, P., Paley, D. A., Bachmayer, R. and Fratantoni, D. M. (2006). Multi-AUV control and adaptive sampling in Monterey Bay, *IEEE Journal of Oceanic Engineering* **31**, 4, pp. 935–948.

Frasca, M., Buscarino, A., Rizzo, A., Fortuna, L. and Boccaletti, S. (2006). Dynamical network model of infective mobile agents, *Physical Review E* **74**, p. 036110.

Frasca, M., Buscarino, A., Rizzo, A., Fortuna, L. and Boccaletti, S. (2008). Synchronization of moving chaotic agents, *Physical Review Letters* **100**, p. 044102.

Galstyan, A. and Lerman, K. (2002). Adaptive boolean networks and minority games with time-dependent capacities, *Physical Review E* **66**, p. 015103.

González, M. C., Hidalgo, C. A. and Barabási, A.-L. (2008). Understanding individual human mobility patterns, *Nature* **453**, pp. 779–782.

Grégoire, G. and Chaté, H. (2004). Onset of collective and cohesive motion, *Physical Review Letters* **92**, p. 025702.

Grünbaum, D., Viscido, S. and Parrish, J. (2005). Extracting interactive control algorithms from group dynamics of schooling fish, in V. Kumar, N. Leonard and A. Morse (eds.), *Cooperative Control: A Post-Workshop Volume, 2003 Block Island Workshop on Cooperative Control* (Springer-Verlag), pp. 104–117.

Helbing, D., Farkas, I. and Vicsek, T. (2000). Simulating dynamical features of escape panic, *Nature* **407**, pp. 487–490.

Helbing, D. and Molnar, P. (1995). Social force model for pedestrian dynamics, *Physical Review E* **51**, pp. 4282–4286.

Ikawa, T. and Okabe, H. (1997). Three-dimensional measurements of swarming mosquitoes: a probabilistic model, measuring system, and example results, in J. K. Parrish and W. M. Hamner (eds.), *Animal Groups in Three Dimensions* (Cambridge University Press, Cambridge).

Jadbabaie, A., Lin, J. and Morse, A. S. (2003). Coordination of groups of mobile autonomous agents using nearest neighbor rules, *IEEE Transactions on Automatic Control* **48**, 6, pp. 988–1001.

Justh, E. W. and Krishnaprasad, P. S. (2004). Equilibria and steering laws for planar formations, *Systems and Control Letters* **52**, pp. 25–38.

Klavins, E. (2007). Programmable self-assembly, *IEEE Control Systems Magazine* **27**, 4, pp. 43–56.

Klavins, E. and Murray, R. M. (2004). Distributed algorithms for cooperative control, *IEEE Pervasive Computing* **3**, 1, pp. 56–65.

Krause, J. and Ruxton, G. D. (2002). *Living in Groups* (Oxford University Press, Oxford, UK).

Li, S. and Wang, H. (2004). Multi-agent coordination using nearest neighbor rules: revisiting the vicsek model, *CoRR* **cs.MA/0407021**.

Major, P. F. and Dill, L. M. (1978). The three-dimensional structure of airborne bird flocks, *Behavioral Ecology and Sociobiology* **4**, pp. 111–122.

Marshall, J. A., Broucke, M. E. and Francis, B. A. (2004). Formations of vehicles in cyclic pursuit, *IEEE Transactions on Automatic Control* **49**, 11, pp. 1963–1974.

Martínez, S., Cortés, J. and Bullo, F. (2007). Motion coordination with distributed information, *IEEE Control Systems Magazine* , pp. 75–88.

Mikhailov, A. S. and Zanette, D. H. (1999). Noise-induced breakdown of coherent collective motion in swarms, *Physical Review E* **60**, 4, pp. 4571–4575.

Moreau, L. (2005). Stability of multiagent systems with time-dependent communication links, *IEEE Transactions on Automatic Control* **50**, 2, pp. 169–182.

Nabet, B., Leonard, N. E., Couzin, I. D. and Levin, A. (2007). Dynamics of decision making in animal group motion, Submitted, available online at http://www.princeton.edu/ naomi/publications/2007/JNS07.html.

Newlands, N. K. and Porcelli, T. A. (2008). Measurement of the size, shape and structure of atlantic bluefin tuna schools in the open ocean, *Fisheries Research* **91**, pp. 42–55.

Okubo, A. and Chiang, H. C. (1974). An analysis of the kinematics of swarming behavior of Anarete pritchardi Kim (Diptera: Cecidomyiidae), *Researches on Population Ecology* **16**, pp. 1–42.

Olfati-Saber, R. and Murray, R. M. (2004). Consensus problems in networks of agents with switching topology and time-delays, *IEEE Transactions on Automatic Control* **49**, 9, pp. 1520–1533.

Paley, D. A., Leonard, N. E., Sepulcre, R., Grünbaum, D. and Parrish, J. K. (2007). Oscillator models and collective motion, *IEEE Control Systems Magazine* , pp. 89–105.

Paley, D. A., Zhang, F. and Leonard, N. E. (2008). Cooperative control for ocean sampling: The glider coordinated control system, *IEEE Transactions on Control Systems Technology* **16**, 4, pp. 735–744.

Pallottino, L., Scordio, V. G., Frazzoli, E. and Bicchi, A. (2007). Decentralized cooperative policy for conflict resolution in multi-vehicle systems, *IEEE Transactions on Robotics* **23**, 6, pp. 1170–1183.

Parrish, J. K. and Edelstein-Keshet, L. (1999). Complexity, pattern, and evolutionary trade-offs in animal aggregation, *Science* **284**, pp. 99–101.

Partridge, B. L. (1982a). The structure and function of fish schools, *Scientific American* **245**, pp. 114–183.

Partridge, B. L. (1982b). The structure and function of fish schools, *Scientific American* **245**, pp. 90–99.

Partridge, B. L. and Pitcher, T. J. (1980). The sensory basis of fish schools: Relative role of lateral line and vision, *Journal of Comparative Physiology* **135**, pp. 315–325.

Petriu, E. M., Whalen, T. E., Abielmona, R. and Stewart, A. (2004). Robotic sensor agents: a new generation of intelligent agents for complex environment monitoring, *IEEE Instrumentation and Measurement Magazine* **7**, 3, pp. 46–51.

Pitcher, T. J. and Parrish, J. K. (1993). The functions of shoaling behavior, in T. J. Pitcher (ed.), *The Behavior of Teleost Fishes* (Chapman and Hall, London), pp. 363–439.

Reebs, S. G. (2000). Can a minority of informed leaders determine the foraging movements of a fish shoal? *Animal Behavior* **59**, p. 403–409.

Reynolds, C. (1987). Flocks, bird and schools: A distributed behavioral model, *Computer Graphics* **21**, pp. 25–34.

Schenk, C. P., Schutz, P., Bode, M. and Purwins, H. G. (1998). Interaction of self organized quasi particles in a two dimensional reaction-diffusion system: The formation of molecules, *Physical Review E* **58**, pp. 6480–6486.

Seeley, T. D. (1985). *Honeybee Ecology: a Study of Adaptation in Social Life* (Princeton University Press, Princeton).

Sepulchre, R., Paley, D. and Leonard, N. E. (2007). Stabilization of planar collective motion: All-to-all communication, *IEEE Transactions on Automatic Control* **52**, 5, pp. 811–824.

Sepulchre, R., Paley, D. and Leonard, N. E. (2008). Stabilization of planar collective motion with limited communication, *IEEE Transactions on Automatic Control* **53**, 3, pp. 706–719.

Strogatz, S. H. (2000). From Kuramoto to Crawford: exploring the onset of synchronization in populations of coupled oscillators, *Physica D* **143**, pp. 1–20.

Strogatz, S. H. (2003). *SYNC: the emerging science of spontaneous order* (Hyperion, New York).

Sumpter, D. J. T. (2006). The principles of collective animal behaviour, *Philosophical Transactions of the Royal Society B* **361**, pp. 5–22.

S.V. Viscido, J. P. and Grünbaum, D. (2004). Individual behavior and emergent properties of fish schools: A comparison of observation and theory, *Marine Ecology Progress Series* **273**, pp. 239–249.

Toner, J. and Tu, Y. (1995). Long-range order in a two-dimensional dynamical xy model: How birds fly together, *Physical Review Letters* **75**, 23, pp. 4326–4329.

Toner, J. and Tu, Y. (1998). Flocks, herds and schools: A quantitative theory of flocking, *Physical Review E* **58**, 4, pp. 4828–4858.

Vicsek, T., Czirók, A., Ben-Jacob, E., cohen, I. and Shochet, O. (1995). Novel type of phase transitions in a system of self-driven particles, *Physical Review Letters* **75**, 6, pp. 1226–1229.

Wang, J. and Wilde, P. D. (2004). Properties of evolving e-mail networks, *Physical Review E* **70**, p. 066121.

Watts, D. J. and Strogatz, S. H. (1998). Collective dynamics of 'small-world' networks, *Nature* **393**, pp. 440–442.

Zimmermann, M., Eguìluz, V. and Miguel, M. S. (2004). Coevolution of dynamical states and interactions in dynamic networks, *Physical Review E* **69**, p. 065102.

Chapter 18

Interplay of Network State and Topology in Epidemic Dynamics

Thilo Gross

Max-Planck Institute for the Physics of Complex Systems,
Nöthnitzer Straße 38, 01187 Dresden, Germany

18.1. Introduction

Throughout history epidemic diseases have been a constant threat to humans (Cartwright, 1972; Oldstone, 1998). There is evidence that even our earliest ancestors suffered from disease-related mortality (Hart, 1983). Many of the old diseases have by now become harmless childhood maladies or have disappeared entirely. However, as humankind prospered and spread out new epidemic diseases continued to arrive in the human population (Karlen, 1995). The extend to which epidemic diseases have shaped our culture and politics can be guessed from religious texts, which contain numerous references to epidemics and rules for their avoidance.

In the 20th century we have witnessed a brief episode in which it seemed that mankind was about to win the struggle and free itself from epidemics. The widespread use of antibiotics drove many infectious diseases back to hiding places in remote locations and animal populations (so-called *disease vectors*). Some sources hold that the second world war was the first major war in history in which far more lives were lost due to fighting than epidemics (Karlen, 1995). But, even when the belief that biomedical progress would conquer epidemics was still widespread, there was evidence to the contrary: In the 1950 new hemorrhagic fevers appeared in several parts of the world (Daubney *et al.*, 1993). The subsequent decades brought Legionnaires disease (Breiman *et al.*, 1990), Ebola fever (Bowen *et al.*, 1977) and a return of syphilis (Sell and Norris, 1983; Grassly *et al.*, 2005). Nevertheless, the awareness to epidemic diseases remained low until AIDS appeared in the 1980s (Grmek, 1990). Since then new diseases, such as SARS (Omi, 2006) and ESME (Haglund and Günther, 2003), have emerged and old killers, such as Malaria (Pampana and Russel, 1955), Tuberculosis (Bloom and Christopher, 1992) and Influenza (Kilbourne, 1987), have returned; some of them in new forms.

At present old and new diseases seem to arrive at a increasing rate. Apart from increased attention in the media, there are several fundamental reasons for

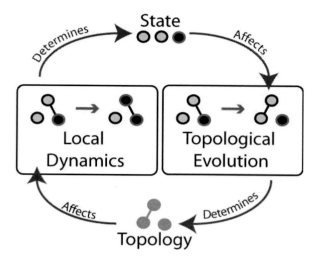

Fig. 18.1.   In an adaptive network the evolution of the topology depends on the dynamics of the nodes. Thus a feedback loop is created in which a dynamical exchange of information is possible. Figure reprinted from Gross and Blasius (2008).

this "epidemic of epidemics" (Karlen, 1995). Today the world's population density is higher than ever, which means that epidemics can spread at a higher rate. We also change our environment faster than ever, which promotes epidemics on many scales: Global change drives host populations of pathogens into more densely populated areas; New farming methods and transport of livestock entail an increased probability of the transmission of animal diseases to humans (*zoonoses*); Finally, artificial environments, such as air-conditioning systems and cooling towers provide new ecological niches in which pathogens (in this example Legionella) can survive.

As we will see below another reason for the emergence and reemergence of diseases lies in the nature of human contact networks. Since neither ecological change, population growth, or the contact networks are likely to change significantly over the next decades, more efficient methods to combat epidemics have to be found. In this context networks appear on two levels. On the molecular level, the study of gene regulatory, metabolic or signaling networks can lead to the development of new medicines and pesticides that target pathogens and disease vectors. On the population level the study of contact networks can lead to new insights into the spreading of epidemics across social networks and to the improvement of vaccination schemes. In the present chapter I focus on this latter level.

One idea that has recently attracted considerable attention is to use insights from network research to optimize vaccination campaigns. Previous research has shown that the amount of vaccine that is necessary to stop an epidemic can be significantly reduced by targeted vaccination (Albert *et al.*, 2000; Holme, 2004). However, the success of targeted vaccination schemes depends strongly on certain network properties. Currently, attempts are made to measure these properties in real world social networks (Caldarelli, 2007). But, is it safe to assume that these properties will remain unchanged in the face of a major epidemic? The example of SARS has shown that epidemics can induce behavioral changes that feed back to the contact network (Omi, 2006). Including this feedback in a model gives rise to an adaptive network: a system on which the dynamics of the nodes depends on the network topology, while the evolution of the network topology depends on the state of the nodes (Gross and Blasius, 2008).

While the dynamics OF networks and the dynamics ON networks have been studied for a long time in physics (Albert and Barabasi, 2002; Dorogovtsev and Mendes, 2003; Newman, 2003; Boccaletti *et al.*, 2006; Newman *et al.*, 2006), adaptive networks have only very recently come into focus (Gross and Blasius, 2008). The dynamical interplay between the state and topology has been shown to give rise to several phenomena: In particular adaptive networks have been shown to self-organize robustly to critical states (Bornholdt and Rohlf, 2000) and exhibit the spontaneous emergence of distinct classes of nodes (Ito and Kaneko, 2002) and complex topologies (Holme and Ghoshal, 2006; Rosvall and Sneppen, 2006) based on simple local rules. Moreover, in adaptive networks bifurcations (Gross *et al.*, 2006) and phase transitions (Holme and Newman, 2007) appear that involve local as well as topological degrees of freedom.

In this chapter I discuss a simple conceptual adaptive network model of epidemic spreading. It is shown that, even in this very simple model, adaptive feedback leads to qualitative changes in the dynamics. The core of the chapter is formed by results that have been previously published in Gross *et al.* (2006). In contrast to the original publication the results are discussed in the context of subsequent works (Zanette, 2007; Shaw and Schwartz, 2008; Gross and Kevrekidis, 2008; Risau-Gusman and Zanette, 2008; Zanette and Risau-Gusman, 2008) which provide a broader perspective.

The chapter starts in Sec. 18.2 with a, necessarily brief, introduction to epidemics on networks. The subsequent section, Sec. 18.3, introduces a simple model of epidemics on adaptive networks. Furthermore, numerical results are discussed that show the adaptive response of the contact network to the emergence of the disease. Section 18.4 contains a detailed introduction to the moment closure approximation which is then used to study the dynamics of the adaptive network analytically on an emergent level. In the subsequent section, Sec. 18.5, the focus shifts back to the detailed level as we discuss the mechanisms behind the observed dynamics in the context of subsequent investigations. The final section, Sec. 18.6, summarizes the results.

18.2. Epidemics on Networks

Traditionally epidemics are modeled in the mean field limit of a well-mixed population or by systems of partial differential equations that account for physical space

(Anderson and May, 2005; von Festenberg *et al.*, 2007). However, over the recent years the modeling of epidemics on artificial, i.e. computer generated, contact networks has attracted increasing attention (Pastor-Satorras and Vespignani, 2001; May and Lloyd, 2001).

Previous research has shown that the structure of social networks can have a significant impact on the dynamics of epidemics. A quantity that is of particular interest is the epidemic threshold, which is roughly speaking the the minimum infectiousness that a disease has to have in order to invade and persist on a network (Anderson and May, 2005). This threshold is generally lower on networks that have a broader degree distribution and vanishes in scale-free networks with a scaling exponent between two and three (Pastor-Satorras and Vespignani, 2001; May and Lloyd, 2001). Model epidemics can thus survive on a scale-free contact network even if their infectiousness is arbitrarily low.

It has been frequently pointed out that real contact networks are not perfectly scale-free. In particular the maximum degree that can be found on in a real populations is finite and hence non-zero epidemic thresholds do exist (May and Lloyd, 2001). In former times also technological constraints on the size of cities and the speed of travel imposed strong limits on disease transmission due to geographic embedding. However, in the present day these constraints have largely been removed by technological progress. Today scale-free-like behavior is observed in many social networks (Caldarelli, 2007). It can be expected that in the mega-cities of the future the social networks will exhibit scale-free behavior over many orders of magnitude.

Scale-free networks, and more generally networks with a broad degree distribution, are very robust against random attacks, but relatively susceptible to targeted attacks (Albert *et al.*, 2000; Holme, 2004). In the context of diseases this means that targeted vaccination which focuses on the most important, i.e., most central, nodes of the network can be much more effective than vaccination of random individuals. Vaccinating the most highly connected nodes can split the network into small disconnected components and thus prevent large outbreaks. A caveat of this approach is that social networks generally exhibit a assortative degree correlations (Newman, 2002); on these networks nodes of high degree are preferentially linked to other nodes of high degree. Thus the highly connected nodes form a densely connected community. Attacking these nodes by vaccination is therefore unlikely to disconnect large parts of the network unless a large fraction of the highly connected nodes is removed.

In summary the degree distribution and degree correlations are important network properties that can strongly affect both the dynamics of the epidemic and the success of counter-measures. In the subsequent section I argue that these properties can change due to a simple, intuitive mechanism.

18.3. An Adaptive SIS Model

In order to study the interplay of epidemic spreading and topological evolution we consider a network of N nodes and L links. Each of the nodes represents an individual while each of the links represents a contact over which the disease can potentially spread. In epidemiological models discrete states are generally assigned to the individuals that signify the stage of the disease. In the simplest case there are only two of these states called S, for susceptible, and I, for infected. Thus the network effectively consists of boolean nodes, and the links can be denoted as SS-links, SI-links, and II-links according to the states of the nodes that they connect.

Susceptible individuals can become infected if they are in contact with an infected individual. The transmission of the disease along a given SI-link is assumed to occur at a rate p. Once an individual has been infected she has a chance to recover, which happens at a rate r and immediately returns the individual to the susceptible state. Together these processes of infection and recovery constitute the dynamical rules for a so-called *SIS-model*, a standard model of epidemiology. In the model proposed in Gross *et al.* (2006) another process is included: If a susceptible individual is connected to an infected individual she may want to break the link and instead establish a new link to another susceptible individual. On a given SI-link this *rewiring* occurs at a rate rate w.

Rewiring, rather than cutting links, captures that the number of social contacts cannot be reduced arbitrarily; If you knew that the owner of your neighborhood bakery has some infectious disease, you might decide to buy your bread somewhere else while stopping to eat entirely is usually not the preferred option.

In the adaptive SIS-model the rewiring process has been introduced 'optimistically': Only susceptible nodes rewire, and they manage unerringly to rewire to a node that is also susceptible. Under these conditions rewiring always reduces the number of links that are accessible for epidemic spreading and therefore the *prevalence* of the disease, i.e., the density of infected, is always reduced by this form of rewiring behavior.

Despite the positive primary effect of rewiring, it has also some adverse consequences, that may limit the benefit from rewiring if the epidemic is already established in the population. Figure 18.2 shows results from simulations that have previously reported in Gross *et al.* (2006). The results were obtained by explicit individual-based simulation of a network with $N = 10^5$ nodes and $L = 10^6$ links. In this case the average degree of a node is $\langle k \rangle = 2L/N = 20$, as every link connects to two nodes. The figure shows in the bottom row the degree distributions of susceptible and infected nodes and, in the top row, the mean degree of the neighbors $\langle k_{nn} \rangle$ of a node with given degree. Let us first consider the left and center column which correspond to (1) frozen network topology $w = 0$ and (2) frozen epidemic dynamics $p = r = 0$.

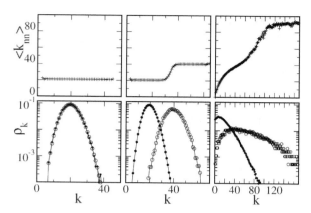

Fig. 18.2.   Structure of adaptive networks. Plotted is the mean nearest-neighbor degree $\langle k_{nn} \rangle$ (top) and the degree distribution ρ_k for susceptible nodes (bottom, circles) and infected nodes (bottom, dots) depending on the degree k. (Left) Indiscriminate rewiring: the network is a random graph with Poissonian degree distributions and vanishing degree correlation. (Center) No local dynamics ($p = r = 0$): the infected and the susceptible nodes separate into two unconnected random subgraphs. (Right) Adaptive network with rewiring and local dynamics ($w = 0.3$, $r = 0.002$, $p = 0.008$): the degree distributions are broadened considerably and a strong assortative degree correlation appears. Figure reprinted from Gross et al. (2006).

In the first case (left column) rewiring is done independently of the epidemic state, in this case there is no adaptive topological change and the network becomes a random graph. Although nodes of high degree have a slightly higher probability of being infected, the effect of this asymmetry is hardly visible in the degree distribution; both the infected and the susceptible nodes independently follow almost exactly the Poisson distribution, that would be expected in a random graph. The neighbor-degree is $\langle k_{nn} \rangle = 21$, independently of the degree of the node under consideration, which implies that the degree correlation vanishes.

In the second case (center column) all nodes remain in their initial state, however all SI-links are converted into SS-links by rewiring. To investigate this in more detail let us denote the *prevalence* of the disease, i.e. the density of infected nodes by $I = 1 - S$, where S is the density of of susceptible nodes. Likewise, the average density of the links per node is denoted by $[SS]$, $[SI]$, and $[II]$, respectively. So, effectively all quantities are normalized to the total number of nodes N. In the initial random configuration the link densities are approximately $[SS] = \langle k \rangle S^2/2$, $[II] = \langle k \rangle I^2/2$, and $[SI] = \langle k \rangle/2 - [SS] - [II] = \langle k \rangle SI$. Since all links that are initially SI-links end up as SS-links we obtain the final link densities $[SI] = 0$, $[SS] = \langle k \rangle (1 - I^2)/2$, and $[II] = \langle k \rangle (I^2)/2$. Note that, although the susceptible nodes are isolated from the infected, the links within each of the two subpopulations are still placed randomly, so that the infected nodes as well as the susceptible nodes both follow a Poissonian degree distribution, albeit with different mean. Consequently the degree correlation is still zero within each subpopulation. However, since susceptible individuals have a higher degree and are connected to other susceptible individuals a positive degree correlation appears if the whole population is considered.

An even stronger effect of state-sensitive rewiring can be observed if rewiring and epidemic spreading take place on the same time-scale. The right column of Fig. 18.2 shows that the degree distribution becomes very wide (note the different scale) and a strong positive (assortative) degree correlation appears. To understand this consider the following: In the case of fast rewiring, the degree of susceptible nodes increased because of links that were rewired into the susceptible population. However, this rewiring had to stop once all SI-links had been converted into SS-links effectively limiting the width of the degree distribution. If both epidemic and topological dynamics take place on the same timescale either the disease will go extinct or rewiring will continue indefinitely. In the latter case the degree of a susceptible node growth linearly until it eventually becomes infected, giving rise to a wide degree distribution.

It is interesting to note that there are several parallels between the adaptive SIS model studied here and models of opinion formation on adaptive networks (Do and Gross, 2009). Also in opinion formation the strongest impact of the dynamics on the topology is observed if topological evolution and local dynamics take place on approximately the same time scale (Gil and Zanette, 2006; Holme and Newman, 2007). In this case the appearance of a broad degree distribution can be linked to a single continuous phase transition. In case of the adaptive SIS model one could suspect the corresponding phase transition is actually the epidemic threshold. However, as I will show in the next section the situation is slightly more complex.

Finally, note that the shape of the degree distribution is reminiscent of the one observed by Ito and Kaneko in an adaptive network of coupled oscillators (Ito and Kaneko, 2002). However, while Ito and Kaneko start with a homogeneous population that spontaneously divides into two distinct topological classes, it clear that in the adaptive SIS model these classes correspond to susceptible and infected nodes. Hence the emergence of the bimodal degree distribution is less surprising in the present case.

18.4. Extracting Dynamics by Moment Closure

Let us now study the dynamics of the adaptive SIS model with the tools of nonlinear dynamics. For this purpose we need to derive a low-dimensional emergent-level description of the system. A convenient tool to achieve this is the so-called *moment-closure approximation* which is frequently used in epidemiology (Keeling *et al.*, 1997; Parham *et al.*, 2008; Peyrard *et al.*, 2008).

18.4.1. *Basic moment-expansion of the model*

Based on previous section it is reasonable to assume that the state of the system can be characterized by the density of infected nodes, I; the density of SS-Links per node, $[SS]$; and the density of II-Links per Node, $[II]$. The density of susceptible nodes, S, and the density of SI-Links, $[SI]$, are then given by the conservation relations $S + I = 1$ and $[SS] + [SI] + [II] = \langle k \rangle$. An advantage of this normalization is that we can write all subsequent equations as if we were dealing with a number of individual nodes and links instead of densities.

Let us start by writing a balance equation for the density of infected nodes. Infection events occur at the rate $p[SI]$ increasing the number of infected nodes by one; Recovery events occur at a rate rI and reduce the number of infected nodes by one. This leads to

$$\frac{\mathrm{d}}{\mathrm{dt}} I = p[SI] - rI . \tag{18.1}$$

The equation contains the (presently unknown) variable $[SI]$ and therefore does not yet constitute a closed model. One way to close the model were a mean field approximation, in which the density of SI-Links is approximated by $[SI] \approx \langle k \rangle SI$. However, in the present case this procedure is not feasible: Rewiring does not alter the number of infected and hence does not show up in Eq. (18.1). Thus the mean-field approximation is not able to capture the effect of rewiring. Instead, we will treat $[SI]$, $[SS]$, and $[II]$ as dynamical variables and capture their dynamics by additional balance equations. This approach is often called *moment expansion* as the link densities can be thought of as the first moments of the network.

For the sake of conciseness it is advantageous to write balance equations only for the densities of SS- and II-links and obtain the density of SI-links by the conservation relation stated above. First the II-links: A recovery event can destroys II-links if the recovering node was part of such links. The expected number of II-links in which a given infected node is involved is $2[II]/I$. (Here, the two appears since a single II-link connects to two infected nodes.) Taking the rate of recovery events into account, the total rate at which II-links are destroyed is simply $2r[II]$.

To derive the rate at which II-links are created is only slightly more involved. In an infection event the infection spreads across a link, converting the respective link into an II-link. Therefore every infection event will create at least one II-link. However, additional II-links may be created if the newly infected node has other infected neighbors in addition to the infecting node. In this case the newly infected node was previously the susceptible node in one or more ISI-triplets. In the following we denote the density of triplets with a given sequence of states A, B, C as $[ABC]$. Using this notation we can write the number of II-links that are created in an infection event as $1 + [ISI]/[SI]$. In this expression the '1' represents the link over which the infection spreads while the second term counts the number of ISI-triplets that run through this link. Given this relation we can write the total rate at which II-links are created as $p[SI](1 + [ISI]/[SI]) = p([SI] + [ISI])$.

Now the SS-links: Following a similar reasoning as above we find that infection destroys SS-links at the rate $p[SSI]$. Likewise SS-links are created by recovery at the rate $r[SI]$. In addition SS-links can also be created by rewiring of SI-links. Since rewiring events occur at a rate $w[SI]$ and every rewiring event gives rise to exactly one SS-link, the total rate at which rewiring creates SS-links is simply $w[SI]$.

Summing all the terms, the dynamics of the first moments can be described by the balance equations

$$\frac{\mathrm{d}}{\mathrm{d}t}[SS] = (r + w)[SI] - p[SSI] \,, \tag{18.2}$$

$$\frac{\mathrm{d}}{\mathrm{d}t}[II] = p([SI] + [ISI]) - 2r[II] \,. \tag{18.3}$$

Again, these equations do not yet constitute a closed model, but depend on the unknown third moments $[SSI]$ and $[ISI]$. However, the first order-moment expansion captures the effect of rewiring. While we will return to the equation above later, a feasible way of closing the system is to approximate the second moments by a mean-field-like approximation: the *moment-closure approximation*.

Let us start by approximating $[ISI]$. One half of the ISI-triplet is actually an SI-link, which we know occurs at the density $[SI]$. In order to approximate the number ISI-triplets running through a given link we have to calculate the expectation value of the number of *additional* infected nodes that are connected to the susceptible node. For this purpose let us assume that the susceptible node of the given SI-link has an expected number of $\langle q \rangle$ links in addition to the one that is already occupied in the SI-link. Every one of these links is an SI-link with probability $[SI]/(\langle k \rangle S)$. (Here, we have neglected the fact that we have already used up one of the total number of SI-links. This assumption is good if the number of SI-links is reasonably large.) Taking the density of SI-links and the probability that they connect to additional SI-link into account we obtain

$$[ISI] = \kappa \frac{[SI]^2}{S} \tag{18.4}$$

where $\kappa = \langle q \rangle / \langle k \rangle$ remains to be determined. The quantity $\langle q \rangle$ that appears in κ is the so-called *mean excess degree*. Precisely speaking it denotes the expected number of additional links that are found by following a random link.

The mean excess degree of a network is governed by two opposing effects (Newman, 2003): One the one hand we are only counting the additional links, so that for a node of given degree k the excess degree is $q = k - 1$. On the other hand, we have reached the node by following a link and therefore have a higher probability to arrive at a node of high degree. Depending on the network topology the $\langle q \rangle$ can therefore be larger or smaller than $\langle k \rangle$. It is a special property of Erdös-Renyi random graphs that both effects cancel, so that $\langle q \rangle = \langle k \rangle$ and $\kappa = 1$. Smaller values of κ are found in homogeneous networks such as regular lattices while networks with a wider degree distribution generally correspond to larger (and in scale-free networks even diverging) values of κ.

In the present example the network topology changes dynamically in time. For the sake of simplicity I will assume that $\kappa \approx 1$. Note, that this could be called a random-graph-like approximation. It is *not* a high-degree approximation as is sometimes held in the literature and therefore holds also in sparse networks.

Setting $\kappa = 1$ we can approximate the density of triplets by $[ISI] = [SI]^2/S$, and following a similar reasoning $[SSI] = 2[SS][SI]/S$. Substituting these relations into the balance equations we obtain a closed system of differential equations

$$\frac{\mathrm{d}}{\mathrm{d}t}I = p[SI] - rI\,,\tag{18.5}$$

$$\frac{\mathrm{d}}{\mathrm{d}t}[SS] = (r + w)[SI] - 2p[SI]\frac{[SS]}{S}\,,\tag{18.6}$$

$$\frac{\mathrm{d}}{\mathrm{d}t}[II] = p[SI](1 + \frac{[SI]}{S}) - 2r[II]\,.\tag{18.7}$$

18.4.2. *Dynamics of the adaptive SIS model*

Systems of ordinary differential equations (ODEs) can be studied by the standard tools of dynamical systems theory. In Gross *et al.* (2006) the stationary solutions of these equations and their stability are computed analytically. This analysis reveals several transitions in which the dynamics changes qualitatively. In the context of the low-dimensional emergent-level ODEs these transitions appear as bifurcations, while in the context of the individual-based detailed-level simulations they correspond to phase transitions. Note, that even the equilibrium solutions of the ODE system correspond in general to highly dynamic states on the detailed level, in which individual nodes undergo infection and recovery and links are continuously rewired.

In Fig. 18.3 results of the analytical investigation of the system of ODEs are shown in comparison to results from individual-based simulation of the full network. Without rewiring, there is only a single, continuous dynamical transition, which occurs at the epidemic threshold and corresponds to a transcritical bifurcation. As the rewiring is switched on, this threshold increases. The epidemic threshold still marks the critical value of infectiousness, p, for the invasion of the disease. However, another lower threshold, marked by a fold bifurcation point, appears. Above this threshold an epidemic that is already established in the network can persist (endemic state). In the following we distinguish between the invasion threshold and the persistence threshold for epidemics. In contrast to the case without rewiring the two thresholds are discontinuous (first order) transitions. Between them a region of bistability is located, in which both the healthy and the endemic state are stable. Thus, a hysteresis loop is formed. First order transitions, bistability, and hysteresis are generic features of the model that can be observed at all finite rewiring rates.

Increasing the rewiring rate further hardly reduces the size of the epidemic in the endemic state, however both thresholds are shifted toward higher infection rates. At higher rewiring rates the the nature of the persistence threshold changes. First, a subcritical Hopf bifurcation emerges. In this bifurcation the stability of a stationary solution is lost by the interaction with an unstable limit cycle. Although the fold

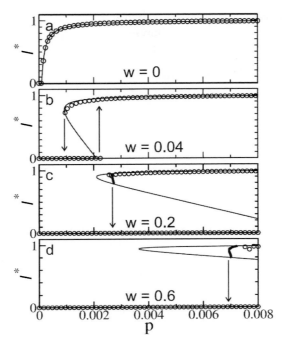

Fig. 18.3. Bifurcation diagram of the stationary density of the infected I^* as a function of the infection probability p for different values of the rewiring rate w. In each diagram I^* has been computed analytically from the ODE system (thin lines). Along the stable branches these results have been confirmed by the explicit simulation of the full network (circles). Without rewiring only a single continuous transition occurs at p=0.0001 (a). By contrast, with rewiring a number of discontinuous transitions, bistability, and hysteresis loops (indicated by arrows) are observed (b), (c), (d). Fast rewiring (c), (d) leads to the emergence of limit cycles (thick lines indicate the lower turning point of the cycles), which have been computed numerically with the bifurcation software AUTO. Parameter: $r = 0.002$. Figure reprinted from Gross *et al.* (2006).

bifurcation still occurs it no longer marks the persistence threshold as the stability of the endemic state is lost in the Hopf bifurcation, before the fold bifurcation is reached.

At higher rewiring the Hopf bifurcation becomes supercritical. While this bifurcation still marks the threshold for stationary persistence of the disease, it gives rise to a stable limit cycle on which non-stationary persistence is possible at lower infectiousness. However, the oscillatory parameter region is very narrow. Toward lower infection rates it is bounded by a fold bifurcation of cycles, which leads to the extinction of the disease.

The findings from the bifurcation analysis are summarized in a two-parameter bifurcation diagram shown in Fig. 18.4. A more detailed understanding can be gained if one considers what happens on the network level. Indeed I will come back to this point in the subsequent section. But, before let us discuss some possible extensions of the moment closure approximation.

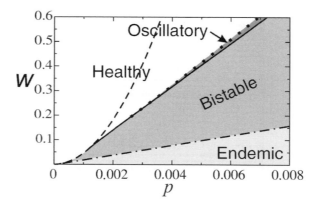

Fig. 18.4. Two parameter bifurcation diagram showing the dependence on the rewiring rate w and the infection probability p. In the white and light gray regions there is only a single attractor, which is a healthy state in the white region and an endemic state in the light gray region. In the medium gray region both of these states are stable. Another smaller region of bistability is shown in dark gray. Here, a stable healthy state coexists with a stable epidemic cycle. The transition lines between these regions correspond to transcritical (dash-dotted), fold (dashed), Hopf (continuous), and cycle fold (dotted) bifurcations. Note that the fold and transcritical bifurcation lines emerge from a cusp bifurcation at $p = 0.0001$, $w = 0$. The rewiring rate is $r = 0.0002$. Figure reprinted from Gross et al. (2006).

18.4.3. Accuracy and extensions of the moment closure approximation

The derivation of the Eqs. (18.5)–(18.7) has involved approximations at various stages. Nevertheless the predicted bifurcation diagrams are in very good agreement with the numerical results. In the literature much attention has been paid to inaccuracies in the factor κ. However, the results presented here show that the approximation $\kappa = 1$ yields good results even for networks with a relatively wide degree distribution.

One reason for the good agreement between theory and simulation is that adaptive networks are particularly well suited to be treated by the moment-closure approximation; The ongoing topological evolution of an adaptive network means that there is a constant mixing of the topology. One could say that, over time, the adaptive network is an ensemble of itself. It is therefore more accessible to mean-field-like approximations. In addition the mixing reduces spatial correlations. This may explain why moment closure seems to yield better results on adaptive networks than on static ones.

In my experience ODEs found by moment closure are in good agreement with the explicit numerical simulation of adaptive networks as long as the network under consideration is sufficiently large ($N \approx 10^5$). In small networks deviations can appear because of stochastic fluctuations, which for instance in the SIS model can lead to premature extinction of the disease. However, as real world networks are mostly large this deviation can be seen as a shortcoming of the simulation rather than the moment-closure approximation. Nevertheless, in large networks the closure approximation (rather than the moment expansion itself) is the main source of inaccuracy in the derivation of the ODE system. Let us therefore discuss three approaches which avoid or amend this approximation. The starting point for this discussion are the Eqs. (18.2), (18.3) and Eq. (18.1).

One way to avoid the approximation of the second moments is to treat them also as dynamical variables. In this case their dynamics have to be captured by balance equations which will in turn depend on the on the third moments. While this only shifts the problem of closure up by one level. It is reasonable to assume that closure at the level of third moments will yield a more precise approximation than closure at the level of second moments. Apart from the gained accuracy, closure at the third level would allows include processes that act on triplets or triangles rather than on nodes or links. However, apart from a discussion in Peyrard *et al.* (2008) moment closure at higher moments has so far received little attention because of the difficulties involved. In particular the number of equations in the model grows combinatorially with the order of the expansion and the number of states in the model.

Instead of extending the moment expansion to higher orders one can also try to increase the accuracy of the approximation at the second order. In this case the challenge is to find a way to accurately compute the expected density of certain triplets (e.g. $[SSI]$ and $[ISI]$) from a given density of nodes and links(e.g. I, $[SS]$, $[II]$). This may be done for instance by a hybrid analytical/numerical approach proposed in Gross and Kevrekidis (2008). In this case the bifurcations are computed numerically on the level of ODEs. But, whenever the bifurcation software needs to evaluate the system-level equations, it starts a series of a few, very short, individual-based detailed-level simulations. From appropriately initialized simulations good approximations of the true triplet densities are then extracted. While this procedure provides more accurate results than analytical closure it is faster than full simulation of the system and yields information that would be difficult to extract by simulation alone.

Yet another approach would be to treat the second moments as unknown functions of the link and node densities. Even in this general case the methods of dynamical system theory could be applied to extract many dynamical properties of the system. For instance the approach of generalized models can be used to investigate the stability of stationary solutions, find transitions to oscillatory behavior, and provide evidence for chaotic dynamics without the need to restrict the unknown functions in the model to specific functional forms (Gross and Feudel, 2006).

18.5. Interpretation and Extensions of the Model

The adaptive SIS model which we have considered so far is clearly conceptual in nature. In order to facilitate the mathematical analysis we have focused on the simplest framework at hand in which epidemic dynamics and topological evolution can be combined. The purpose of these investigations was to demonstrate that the adaptive interplay between local dynamics and topological evolution can give rise to certain phenomena that can not be observed in static networks. The analysis has revealed three such phenomena: (1) a significant shift of the invasion threshold toward higher infection rates; (2) the appearance of a persistence threshold below the invasion threshold, giving rise to bistability and hysteresis; and (3) the emergence of an oscillatory phase.

The next logical step is to search for these phenomena in more realistic models. Indeed, since the original publication of the adaptive SIS-model in Gross *et al.* (2006) subsequent works have appeared that add realism by extending the model in several ways (Zanette, 2007; Shaw and Schwartz, 2008; Gross and Kevrekidis, 2008; Risau-Gusman and Zanette, 2008; Zanette and Risau-Gusman, 2008). In order to understand the common themes and differences between these works and to extrapolate to real world situations, it is conductive to revisit the phenomena listed above and consider them from an individual-based perspective.

As we have seen in Sec. 18.3 rewiring affects epidemic spreading in two opposing ways. The primary effect of rewiring is two reduce the number of links through which the epidemic can spread. But, in time rewiring can lead to the formation of a large densely-linked cluster of susceptible nodes in which the infection can spread rapidly once it has been invaded.

For the phenomenon (1), the increase of the invasion threshold, only the primary effect of rewiring is of importance. By definition, only dynamics that take place at a low density of infected nodes are relevant for invasion. However, in this limit the number of links that are rewired is also low and hence cannot lead to a significant build-up of connectivity in the susceptible subpopulation. At least under our optimistic assumptions the epidemic threshold is therefore always increased by rewiring.

In more realistic scenarios the epidemic threshold could in principle be decreased by rewiring if rewiring led to an increased density of SI-links. In reality this can happen in several ways. Susceptible nodes may erroneously rewire SS-links to infected nodes. Also, infected nodes may want to avoid contact with other infected nodes and therefore rewire II-links to susceptible nodes. More relevant is probably, the existence of additional disease states. While there are many infectious diseases that follow SIS-dynamics, these are mostly too harmless or too easily controlled to trigger a strong rewiring response in the population. Real world diseases that trigger a stronger behavioral response often have pronounced exposed (E) or asymptomatic (A) phases. Nodes in these phases appear to be susceptible, but have already been in contact with the infectious agent, and will eventually progress to the infected phase. While they are still in the E or A phase they act like susceptible nodes, i.e., they rewire their links to other susceptible nodes, whom they can subsequently (E-phase) or immediately (A-phase) infect.

Another phase that often appears in epidemic models is the R-phase, which denotes either recovered individuals that have acquired (possibly temporary) immunity or removed nodes that have died from the disease. However, in either case R-nodes should not affect the invasion threshold, as their density vanishes in the relevant limit.

The impact of additional epidemic states and rewiring rules was investigated in Shaw and Schwartz (2008); Risau-Gusman and Zanette (2008); Zanette and Risau-Gusman (2008). While it was found that rewiring can have detrimental effects under certain conditions these were relatively mild.

Phenomenon (two), the emergence of the persistence threshold is clearly related to the appearance of highly connected susceptible nodes; Below the invasion threshold, an infectious disease can persist on the networks if a large fraction of the population is already infected. In this case SI-links are rewired into a relatively small population of susceptible nodes. The number of SS-links per susceptible node increases correspondingly. Therefore many SI-links are created in every infection event, which almost compensates the SI-links lost due to rewiring. One could say that, if the susceptible subpopulation is too small, rewiring becomes ineffective as it is unlikely that the rewired link will remain out of reach of the disease for a significant time.

The mechanism described above is robust as long as links are rewired and not just cut. One could argue that few real world diseases reach sufficiently high prevalence for the effect to be of relevance. However, in particular the new emerging diseases on which our main focus lies are known to reach locally high densities of infected (Karlen, 1995).

The third phenomenon, the existence of an oscillatory phase, is caused by the interplay of the two opposing effects of rewiring; First, the number of SI-links are reduced by rewiring. However, even as the density of infected nodes decreases a densely linked cluster of susceptible nodes is build up. Eventually the epidemic invades this cluster which causes a sharp increase in the density of infected nodes, thus completing the cycle. What appears as a smooth oscillation on the level of the ODE is therefore really a series of outbreaks and avalanches if considered on the detailed level.

At first glance it seems surprising that in the oscillatory phase the primary epidemic-suppressing effect of rewiring governs the dynamics when the density of infected is high, while the secondary epidemic-promoting effect rules the increase that sets in at the low point. However, as it is often the case in autonomous oscillations, the cyclic behavior becomes possible due to the differences in the time scales on which the two effects act; The primary reduction of SI-links is immediately effective, but the build-up of links in the susceptible subpopulation requires some time before it can affect the dynamics. Nevertheless the mechanism that causes the oscillatory behavior is relatively fragile. If the density of infected at the low point of the cycle becomes too low the epidemic-promoting effect of rewiring is too weak

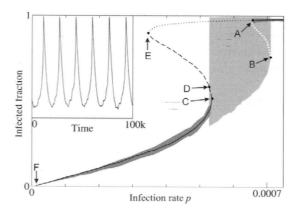

Fig. 18.5.   Bifurcation diagram showing that awareness-dependent rewiring gives rise to a large oscillatory region. The rewiring rate is assumed to change proportional to the prevalence so that $w = w_0 I$. Lines mark branches of stable steady states (solid) and saddle points (dashed/dotted). The white line tentatively marks a branch of saddle limit cycles. Shaded regions indicate ranges of I observed during long individual-based simulations in the neighborhood of the attractive limit cycle (light gray) and of stable stationary solutions (dark gray). Computation of the Jacobian eigenvalues reveal a subcritical Hopf bifurcation (A), two fold bifurcations (C,E) and a transcritical bifurcation (F). One can suspect the presence of a fold bifurcation of cycles (B) and a homoclinic bifurcation (D). Inset: time series on the limit cycle attractor at $p = 0.0006$. Other parameters: $w_0 = 0.06$, $r = 0.0002$. Figure reprinted from Gross and Kevrekidis (2008).

to launch the epidemic into the growth phase and the disease goes extinct. Hence only cycles of relatively small amplitude at high disease prevalence were observed in the present model.

In many subsequent works no cyclic regime was found for two reasons: First, the basin of attraction of the limit cycle is relatively small. Therefore in simulations stochastic fluctuations can cause the system to depart from the cycle unless the number of simulated nodes is relatively high (e.g., $N \approx 10^6$). Moreover, in some model variants the SI-links are rewired to random targets and not just to susceptible nodes. As the density of infected nodes increases, so does the quota of unsuccessful rewiring, in which a susceptible node rewires an SI-link link from one infected node to another. Therefore, the epidemic-suppressing effect of rewiring becomes weaker with increasing density of infected. Rewiring can in this case no longer control the epidemic at the high point of the cycle and hence oscillations are avoided at the cost of higher prevalence. By contrast if there is an effect that increases the efficiency of rewiring with increasing density of infected then oscillatory behavior is promoted. For instance in Gross and Kevrekidis (2008) it was assumed that the awareness of the population is increased if the density of infected is high and thus rewiring activity is increased. In this case the authors find robust large-amplitude oscillations over a considerable parameter range (see Fig. 18.5). Some evidence suggests that such awareness-driven cycles occur in nature (Grassly *et al.*, 2005).

It is well known that cycles can also appear in models that involve an R-stage of the epidemic, such as SIRS models. It appears likely that in an adaptive SIRS model the interaction of rewiring-induced oscillations and recovery-induced oscillations can give rise to more complex, i.e., chaotic or quasiperiodic, dynamics.

18.6. Summary and Conclusions

In this chapter we have investigated a simple conceptual model of the topological response of contact networks to the emergence of an infectious disease and the feedback of topological change on the epidemic. The analysis of this model has revealed that state-dependent rewiring is likely to increase the invasion threshold for diseases but at the same time introduces a persistence threshold below the invasion threshold. The consequence are discontinuous transitions, bistability, hysteresis and the existence of an oscillatory phase.

From an abstract point of view rewiring is certainly beneficial as it generally reduces the number of infected and can potentially drive the disease to extinction. However, from another perspective a word of caution is in order. If we observe small sub-threshold outbreaks of an epidemic in the real world we should ask whether the low prevalence is maybe a result of rewiring and not of low infectiousness of the disease as such. If rewiring plays a significant role in keeping the epidemic below threshold then a discontinuous transition can be expected if the threshold is eventually crossed. This may result in a large outbreak which can be difficult to combat because of hysteresis.

Also, if network properties are taken into account in the planning of vaccination campaigns, one should consider that these properties can change in response to emergence of a major epidemic. In particular the widening of the degree distribution and the appearance of strong degree correlations are detrimental for targeted vaccination.

Let me emphasize that the concerns stated above are based on the analysis of very simple models and may therefore prove to be unfounded. To assess the role of state-topology interplay in epidemic dynamics in the real world, more research is certainly necessary. The structure of real contact networks is shaped by many social processes and constraints not related to epidemics. For minor epidemics, disease-induced changes in the contact network are therefore probably negligible. However, if a disease is perceived as a major threat, it may induce risk-avoidance behavior that has a strong impact on the contact network.

The moment closure approximation that has been discussed here in detail, appears to be a powerful tool for future investigations. It is interesting to note that this tool is also applicable to certain simple models of social processes which could therefore be included in models of epidemics. A discussion of the differences and similarities between epidemic and social contact processes on adaptive networks is presented in Do and Gross (2009). Constraints that cannot be captured easily by the moment closure approach arise from the physical space in which personal contact networks are necessarily embedded. To clarify the impact of this will probably require explicit spatial simulations. Another factor that is not captured by moment closure is the role of stochastic fluctuations. An informative investigation of fluctuations in epidemics on adaptive networks is presented in Shaw and Schwartz (2008, 2009).

The detrimental effects of rewiring discussed here is absent if links are cut instead of rewired. Whether cutting or rewiring links is the main response to a real world disease depends strongly on its epidemiology as well as on choices made by the individuals. Two prominent responses to SARS, wearing face masks and leaving the affected region, correspond to cutting and rewiring links respectively. Also in the case of AIDS links can be rewired, say, by finding a new partner, or cut, say, by the use of condoms.

In summary epidemics on adaptive networks still pose many open questions. To answer these questions will require further work in the physics of adaptive networks, but also in field epidemiology and the sociology of contact networks. The results that have been obtained so far suggest that future interdisciplinary work could shed light on the physics of the adaptive interplay between state and topology, on the evolution of social networks, and on the dynamics of real world diseases.

References

Albert, R. and Barabasi, A. (2002). Statistical mechanics of complex networks, *Rev. Mod. Phys.* **74**, 1, pp. 1–54.

Albert, R., Jeong, H. and Barabasi, A. (2000). Error and attack tolerance of complex networks, *Nature* **406**, pp. 378–382.

Anderson, R. M. and May, R. M. (2005). *Infectious diseases of humans. Dynamics and Control* (Oxford University Press, Oxford).

Bloom, B. and Christopher, J. M. (1992). Tuberculosis: Commentary on a reemergent killer, *Science* **257**, pp. 1055–1064.

Boccaletti, S., Latora, V., Moreno, Y., Chavez, M. and Hwang, D. (2006). Complex networks: Structure and dynamics, *Physics Reports* **424**, p. 175.

Bornholdt, S. and Rohlf, T. (2000). Topological evolution of dynamical networks: Global criticality from local dynamics, *Phys. Rev. Lett.* **84**, 26, pp. 6114–6117.

Bowen, E. T. W., Lloyd, G., Harris, W. J., Platt, G. S., Baskerville, A. and Vella, E. E. (1977). Viral haemorrhagic fever in southern sudan and northern zaire, *The Lancet* **1**, pp. 571–573.

Breiman, R., Fields, B. S., Sanden, G. N., Volmer, L., Meier, A. and Spika, J. S. (1990). Association of shower use with legionaires' disease, *JAMA* **263**, pp. 2924–2926.

Caldarelli, G. (2007). *Scale-free networks* (Oxford University Press, Oxford).

Cartwright, F. F. (1972). *Disease and History* (Dorset, New York).

Daubney, R., Hudson, J. R. and Garnham, P. C. (1993). Enzootic hepatitis of rift valley fever: an undescribed disease of sheep, cattle and man from east africa, *J. Pathol. Bacteriol.* **34**, pp. 545–549.

Do, L. and Gross, T. (2009). Contact processes and moment closure on adaptive networks, To appear in T. Gross and H. Sayama: Adaptive Networks, Springer Verlag, Heidelberg.

Dorogovtsev, S. N. and Mendes, J. F. F. (2003). *Evolution of Networks* (Oxford University Press, Oxford).

Gil, S. and Zanette, D. H. (2006). Coevolution of agents and networks: Opinion spreading and community disconnection, *Phys. Lett. A* **356**, pp. 89–95, (DOI: 10.1016/j.physleta.2006.03.037).

Grassly, N. C., Fraser, C. and Garnett, G. P. (2005). Host immunity and synchronized epidemics of syphilis across the united states, *Nature* **433**, pp. 417–421.

Grmek, M. (1990). *History of AIDS – Emergence and origin of a modern pandemic* (Princeton University Press, Princeton, NJ).

Gross, T. and Blasius, B. (2008). Adaptive coevolutionary networks: A review, *Journal of the Royal Society Interface* **5**, pp. 259–271.

Gross, T., Dommar D'Lima, C. and Blasius, B. (2006). Epidemic dynamics on an adaptive network, *Physical Review Letters* **96**, pp. 208701–4.

Gross, T. and Feudel, U. (2006). Generalized models as a universal approach to the analysis of nonlinear dynamical systems, *Physical Review E* **73**, pp. 016205–14.

Gross, T. and Kevrekidis, I. G. (2008). Robust oscillations in sis epidemics on adaptive networks: Coarse graining by automated moment closure, *European Physics Letters* **82**, pp. 38004–6.

Haglund, M. and Günther, G. (2003). Tick-borne encephalitis: Virus, disease, and prevention, *Vaccine* **21**, pp. S11–S18.

Hart, G. (1983). *Disease in ancient Man* (Clarke & Irwin, Toronto).

Holme, P. (2004). Efficient local strategies for vaccination and network attack, *Europhys. Lett.* **68**, pp. 908–914.

Holme, P. and Ghoshal, G. (2006). Dynamics of networking agents competing for high centrality and low degree, *Phys. Rev. Lett.* **96**, pp. 908701–4, (DOI: 10.1103/PhysRevLett.96.098701).

Holme, P. and Newman, M. E. J. (2007). Nonequilibrium phase transition in the coevolution of networks and opinions, *Phys. Rev. E* **74**, pp. 056108–5, (DOI: 10.1103/PhysRevE.74.056108).

Ito, J. and Kaneko, K. (2002). Spontaneous structure formation in a network of chaotic units with variable connection strengths, *Phys. Rev. Lett.* **88**, 2, pp. 028701–4.

Karlen, A. (1995). *Man and microbes: Diseases and plagues in history and modern times* (Touchstone, New York).

Keeling, M. J., Rand, D. A. and Morris, A. (1997). Dyad models for childhood epidemics, *Proc. R. Soc. B* **264**, p. 11491156.

Kilbourne, E. (1987). *Influenza* (Plenum, New York).

May, R. M. and Lloyd, A. L. (2001). Infection dynamics on scale-free networks, *Phys. Rev. E* **64**, pp. 066112–4, (DOI: 10.1103/PhysRevE.64.066112).

Newman, M. E. J. (2002). Assortative mixing in networks, *Phys. Rev. Lett.* **89**, pp. 208701–4, doi:10.1103/PhysRevLett.89.208701.

Newman, M. E. J. (2003). The structure and function of complex networks, *SIAM Review* **45**, 2, pp. 167–256.

Newman, M. E. J., Barabasi, A. and Watts, D. J. (2006). *The structure and dynamics of networks* (Princeton University Press, Princeton).

Oldstone, M. B. A. (1998). *Viruses, Plagues, and History* (Oxford University Press, Oxford).

Omi, S. (2006). *SARS – How a global epidemic was stopped* (WHO Press, Geneva).

Pampana, E. J. and Russel, P. F. (1955). *Malaria:A world problem* (WHO Press, Geneva).

Parham, P. E., Singh, B. K. and Ferguson, N. M. (2008). Analytical approximation of spatial epidemic models of foot and mouth disease, *Theor. Popul. Bio.* **73**, p. 349–368.

Pastor-Satorras, R. and Vespignani, A. (2001). Epidemic spreading in scale-free networks, *Phys. Rev. Lett.* **86**, 14, pp. 3200–3203, (DOI: 10.1103/PhysRevLett.86.3200).

Peyrard, N., Dieckmann, U. and Franc, A. (2008). Long-range correlatons improve understanding of the influence of network structure on contact dynamics, *Theor. Popul. Bio.* **73**, p. 383–394.

Risau-Gusman, S. and Zanette, D. H. (2008). Contact switching as a control strategy for epidemic outbreaks, ArXiv:0806.1872.

Rosvall, M. and Sneppen, K. (2006). Modeling self-organization of communication and topology in social networks, *Physical Review E* **74**, pp. 16108–4, arXiv: physics/0512105.

Sell, S. and Norris, S. J. (1983). The biology, pathology, and immunology of syphilis. *Int. Rev. Exp. Pathol.* **24**, pp. 204–276.

Shaw, L. B. and Schwartz, I. B. (2008). Fluctuating epidemics on adaptive networks, *Phys. Rev. E* **77**, pp. 0661011–10.

Shaw, L. B. and Schwartz, I. B. (2009). Fluctuating epidemics on adaptive networks, To appear in T. Gross and H. Savama: Adaptive Networks, Springer Verlag, Heidelberg.

von Festenberg, N., Gross, T. and Blasius, B. (2007). Seasonal forcing drives spatio-temporal pattern formation in rabies epidemics, *Mathematical Modelling of Natural Phenomena* **2**, 4, pp. 63–73.

Zanette, D. H. (2007). Coevolution of agents and networks in an epidemiological model, ArXiv:0707.1249.

Zanette, D. H. and Risau-Gusman, S. (2008). Infection spreading in a population with evolving contacts, To appear in *J. Biol. Phys.*

Index